Applications of Crystal Plasticity in Forming Technologies

Applications of Crystal Plasticity in Forming Technologies

Editors

Ulrich Prahl
Sergey Guk
Faisal Qayyum

MDPI • Basel • Beijing • Wuhan • Barcelona • Belgrade • Manchester • Tokyo • Cluj • Tianjin

Editors
Ulrich Prahl
Department of metal forming,
TU Bergakademie Freiberg,
Freiberg, Germany

Sergey Guk
Institut für Metallformung,
TU Bergakademie Freiberg,
Freiberg, Germany

Faisal Qayyum
Institut für Metallformung,
Technische Universität
Bergakademie Freiberg,
Freiberg, Germany

Editorial Office
MDPI
St. Alban-Anlage 66
4052 Basel, Switzerland

This is a reprint of articles from the Special Issue published online in the open access journal *Crystals* (ISSN 2073-4352) (available at: https://www.mdpi.com/journal/crystals/special_issues/plasticity_technologies).

For citation purposes, cite each article independently as indicated on the article page online and as indicated below:

LastName, A.A.; LastName, B.B.; LastName, C.C. Article Title. *Journal Name* **Year**, *Volume Number*, Page Range.

ISBN 978-3-0365-5733-5 (Hbk)
ISBN 978-3-0365-5734-2 (PDF)

Contents

About the Editors

Ulrich Prahl

Ulrich Prahl is a full Professor at Institute of Metal Forming, Techniche Universität Bergakademie Freiberg. He is an established scientist in ICME—and has developed numerical material and process modeling for metallic materials, production processes (forming and heat treatment) for modern steels, development of alloy concepts and process technologies for the nanostructuring of structures as well as for the adjustment of metastable micro-structural phases. He has extensive experience in a combination of experimental laboratory techniques with numerical simulations to model, evaluate and optimize industrial forming and heat treatment processes.

Sergey Guk

Sergey Guk is part of the High Strength Material Group at Institute of Metal Forming, Techniche Universität Bergakademie Freiberg. He is an established scientist working on the engineering and characterization of multiphase high strength metallic materials. His expertise includes thermomechanical processes, metallic materials, materials processing, mechanical behavior, microstructure, crystallography, mechanical testing, and mechanical properties at different length scales.

Faisal Qayyum

Faisal Qayyum is a Research Associate at Institute of Metal Forming, Techniche Universität Bergakademie Freiberg. He is an emerging scientist with expertise in crystal plasticity modeling of polycrystalline multiphase materials. His expertise also includes multi-physics modeling, fatigue and fracture, crack propagation, crystal plasticity, metal matrix composite, material characterization and finite element analysis.

Preface to "Applications of Crystal Plasticity in Forming Technologies"

Materials science advancements have led to the development of complex materials for targeted applications and have pushed manufacturing boundaries. However, because the microstructural characteristics are responsible for its bulk deformation behavior and life after failure, it is critical to acquire the appropriate material properties required for safe performance during service life when engineering the microstructural attributes. Recently, the development of microstructurally informed detailed models to investigate the global and local deformation behavior of single- and multi-phase materials has been facilitated by crystal plasticity-based numerical simulation models. They have helped to study the effect of the microstructural features on deformation and damage behavior under multiaxial loading conditions. Furthermore, these models can be used with machine learning techniques to optimize microstructural features for materials application or in a process route.

In this Special Issue, we have gathered work on simulations of polycrystalline metals and alloys at various length scales to model multiscale localization phenomena such as slip bands, cracks, and twins. This collection of articles discusses cutting-edge methods that integrate simulation and experiments to capture the creation of materials and use materials informatics to analyze sizable datasets and direct the development of continuum or microstructural theories.

Ulrich Prahl, Sergey Guk, and Faisal Qayyum
Editors

crystals

Editorial

Applications of Crystal Plasticity in Forming Technologies

Faisal Qayyum *, Sergey Guk * and Ulrich Prahl *

Institut für Metallformung, Techniche Universität Bergakademie Freiberg, Bernhard-von-Cotta-Str. 4, 09599 Freiberg, Germany
* Correspondence: faisal.qayyum@imf.tu-freiberg.de (F.Q.); sergey.guk@imf.tu-freiberg.de (S.G.); ulrich.prahl@imf.tu-freiberg.de (U.P.); Tel.: +49-3731-393698 (F.Q.); Fax: +49-3731-393656 (F.Q.)

Citation: Qayyum, F.; Guk, S.; Prahl, U. Applications of Crystal Plasticity in Forming Technologies. *Crystals* **2022**, *12*, 1466. https://doi.org/10.3390/cryst12101466

Received: 11 October 2022
Accepted: 13 October 2022
Published: 17 October 2022

The Special Issue on 'Crystal Plasticity in Forming Technologies' is a collection of 11 original articles dedicated to theoretical and experimental research that provides new insights and practical findings in topics related to crystal plasticity.

Advancements in materials science have led to the development of complex materials for targeted applications and have pushed manufacturing boundaries. As the microstructural attributes of any material are responsible for the bulk deformation behavior and life after failure, it is important to engineer them to obtain the desired material properties necessary for their safe functionality during their service life. Furthermore, the formability limits of such materials play a huge role in dictating bulk deformation process limits during manufacturing and hence can significantly affect the cost of production. In the recent past, crystal plasticity-based numerical simulation models have paved the way for developing microstructurally informed detailed models to analyze the global and local deformation behavior of a wide variety of single- and multi-phase metallic and non-metallic materials. Such models can be used to study the effect of microstructural artifacts on the deformation and damage behavior of materials under multiaxial loading conditions. In conjunction with advanced computer algorithms, these models can be applied to optimize microstructural attributes for the desired material application or a process route.

With the advent of electromobility and the requirement for lightweight structures, magnesium alloys are becoming an attractive choice for industries. However, the manufacturing process and loading conditions can drastically affect the deformation behavior of these alloys during service. Crystal-plasticity-based phenomenological or physical numerical simulation models can help in this case. Yaghoobi et al. [1] reviewed recent advances in the crystal plasticity modeling of magnesium and its alloys. They highlighted the benefits and limitations of different models that capture detwinning in such simulations and the advances of several experimental techniques that complement the development and validation of such detailed models. Continuing with experimental techniques, Sun et al. [2] investigated the effect of grain-boundary misorientation on slip transfer in Mg–Gd–Y magnesium alloy. They proposed that the ductility of such materials can be significantly improved by the initial yielding of the components to increase the local misorientation index in medium strain regimes. In high-strain regimes of the same material class, Liu et al. [3] investigated the local strain heterogeneity using electron backscattered diffraction (EBSD) measurement and digital image correlation (DIC) analysis of miniature tensile test samples. They showed that damage initiation occurs at triple points of the grains with highly contrasting Schmidt factors. Such experimental trials on the mesoscale are necessary to identify simulation model parameters and validate the results.

Multiphase steels are extensively used because of their high strength and improved stability at higher temperatures. Qayyum et al. [4] worked on analyzing the influence of non-metallic inclusions on the local deformation and damage behavior of Modified 16MnCrS5 Steel. They carried out full-phase crystal plasticity simulations on plain strain 2D periodic RVEs constructed using EBSD data from differently processed samples. These simulations investigated the effect of different processing routes on the local microstructure and the distribution of inclusions, resulting in local heterogeneity during deformation. Qayyum et al. [4] argued that the local results might be affected by the assumption of 2D RVEs. To fill this gap, Tseng et al. [5] worked on understanding the difference in the local difference between simulations of crystal plasticity of 2D and 3D RVE. They then provided a methodology to transform local stress and strain distributions from 2D to 3D observations in full-phase crystal plasticity simulations of dual-phase steels.

A comparison of structural, microstructural, elastic, and microplastic properties of all-aluminum alloy conductor (A50) and aluminum conductor steel-reinforced (AC50) cables after various operation periods in power transmission lines was presented in [6]. They used outer clippings of actual transmission wires that were in operation for 0–20 years in the Volgograd region of Russia. Appropriate samples were manufactured and analyzed for microstructural, crystallographic, or density changes. They found that the AC50-reinforced wires performed slightly better than the A50 wires in microstructural integrity over the years. Furthermore, they pointed out that recrystallization in the wire strains exposed to long aging times under external load is a very important consideration for structures with long expected service lives. Continuing with this estimation, Trusov et al. [7] tried to describe dynamic recrystallization by employing an advanced statistical multilevel model. In this article, they propose an original method of statistical modeling to form and reconstruct the grain structure by applying the Laguerre polyhedral. The proposed model has high computational efficiency and can describe the peculiarities of physical mechanisms, where the decisive factor is the interaction between contacting grains.

Zhang et al. [8] developed a multiscale model that connects discrete dislocation dynamics to the finite element method to study the plastic behavior of materials on small scales. This is a user-subroutine-based model with a complicated boundary problem for discrete dislocation dynamics (DDD) simulation. The model was solved using the finite element method (FEM). The plastic strain was calculated in discrete dislocation dynamics (DDD) and transferred to the FEM to participate in the constitutive law computation. Uniaxial compression tests on single-crystal micropillars were carried out to validate the generated model. The yield stress was shown to depend on sample size and underlying deformation mechanisms.

In addition to the works mentioned above, some cross-discipline work has been published in this Special Issue related to scale bridging simulations for other cases and materials. These articles provide a comprehensive understanding of how crystal-plasticity-based models can be used in other branches of materials science to improve understanding of material deformation and damage behavior.

Solder joints in electronic systems comprise only a few crystals and have significantly anisotropic thermo-mechanical behavior. Therefore, it is important to understand their reliability under varying loading conditions to ensure the robust functionality of the electronic systems that now power our lives. The reliability of SAC305 Individual Solder Joints during stress–fatigue conditions at room temperature was investigated by Abueed et al. [9]. They used a specially designed fixture with an Instron 5948 micromechanical tester and analyzed the stress–strain loops from different loading conditions to distinguish damage from fatigue from damage from creep. Their findings show that using higher stress levels or longer dwell times greatly reduces fatigue while significantly improving plastic work and strain.

Skakunova and Rychkov [10], in their article, used Raman spectroscopy and optical microscopy to analyze L-leucinium hydrogen maleate (LLHM). This first molecular crystal retains its unusual plasticity at freezing temperatures. LLHM was cooled to 11 K without undergoing a phase transition, while high-pressure impact caused noticeable changes in crystal structure between 0.0 and 1.35 GPa. Surprisingly, the pressure transmission medium (PTM) significantly impacted how the LLHM system behaved under difficult circumstances. High pressures have been associated with LLHM phase transitions into an amorphous form or solid–solid phase transitions that cause crystal fracture. The researchers showed that the low-temperature stability of LLHM raises the intriguing hypothesis that LLHM maintains flexibility below 77 K.

Lastly, Jiandong and Lianhe [11] investigated the Griffith crack problem and the relationship between a screw dislocation and semi-infinite fracture in cubic quasicrystal piezoelectric materials using a complex variable function approach. They provided an in-depth discussion regarding the linear force and coupling elastic coefficient that affect the stress intensity factor of phonon and phason fields. They pointed out that numerical examples that the linear force and the coupling elastic constant significantly influence the stress intensity factor.

We hope this collection of papers will meet the expectations of readers looking for new advances in applications of crystal plasticity in forming technologies and bring inspiration for further research work.

Author Contributions: Data curation, F.Q.; writing—original draft preparation, F.Q.; writing—review and editing, F.Q., S.G., U.P.; visualization, S.G.; supervision, S.G., U.P.; project administration, U.P. All authors have read and agreed to the published version of the manuscript.

Funding: We acknowledge the DAAD Faculty Development for Candidates (Balochistan), 2016 (57245990)-HRDI-UESTP's/UET's funding scheme in cooperation with the Higher Education Commission of Pakistan (HEC) for sponsoring the stay of Faisal Qayyum at IMF TU Freiberg. Part of the work was conducted within the DFG-funded collaborative research group TRIP Matrix Composites (SFB 799). The authors gratefully acknowledge the German Research Foundation (DFG) for the financial support.

Acknowledgments: The contributions of all authors are gratefully acknowledged. The authors thank the *Crystals* Editorial Office and Autumn Duo for their excellent communication, support, and friendly and fully professional attitude. The authors also acknowledge the support of Martin Diehl and Franz Roters (MPIE, Düsseldorf) for their help regarding the functionality of DAMASK.

Conflicts of Interest: The authors declare no conflict of interest.

References

1. Yaghoobi, M.; Voyiadjis, G.Z.; Sundararaghavan, V. Crystal Plasticity Simulation of Magnesium and Its Alloys: A Review of Recent Advances. *Crystals* **2021**, *11*, 435. [CrossRef]
2. Sun, J.; Liu, J.; Chen, Q.; Lu, L.; Zhao, Y. Study of the Effect of Grain-Boundary Misorientation on Slip Transfer in Magnesium Alloy Using a Misorientation Distribution Map. *Crystals* **2022**, *12*, 388. [CrossRef]
3. Liu, J.; Sun, J.; Chen, Q.; Lu, L. Intergranular Cracking in Mg-Gd-Y Alloy during Tension Test. *Crystals* **2022**, *12*, 1040. [CrossRef]
4. Qayyum, F.; Umar, M.; Elagin, V.; Kirschner, M.; Hoffmann, F.; Guk, S.; Prahl, U. Influence of Non-Metallic Inclusions on Local Deformation and Damage Behavior of Modified 16MnCrS5 Steel. *Crystals* **2022**, *12*, 281. [CrossRef]
5. Tseng, S.; Qayyum, F.; Guk, S.; Chao, C.; Prahl, U. Transformation of 2D RVE Local Stress and Strain Distributions to 3D Observations in Full Phase Crystal Plasticity Simulations of Dual-Phase Steels. *Crystals* **2022**, *12*, 955. [CrossRef]
6. Levin, A.A.; Narykova, M.V.; Lihachev, A.I.; Kardashev, B.K.; Kadomtsev, A.G.; Prasolov, N.D.; Panfilov, A.G.; Sokolov, R.V.; Brunkov, P.N.; Sultanov, M.M.; et al. Comparison of Structural, Microstructural, Elastic, and Microplastic Properties of the AAAC (A50) and ACSR (AC50/8) Cables after Various Operation Periods in Power Transmission Lines. *Crystals* **2022**, *12*, 1267. [CrossRef]
7. Trusov, P.; Kondratev, N.; Podsedertsev, A. Description of Dynamic Recrystallization by Means of an Advanced Statistical Multilevel Model: Grain Structure Evolution Analysis. *Crystals* **2022**, *12*, 653. [CrossRef]
8. Zhang, Z.; Tong, Z.; Jiang, X. Development of the Concurrent Multiscale Discrete-Continuum Model and Its Application in Plasticity Size Effect. *Crystals* **2022**, *12*, 329. [CrossRef]
9. Abueed, M.; Al Athamneh, R.; Tanash, M.; Hamasha, S.D. The Reliability of SAC305 Individual Solder Joints during Creep–Fatigue Conditions at Room Temperature. *Crystals* **2022**, *12*, 1306. [CrossRef]
10. Skakunova, K.D.; Rychkov, D.A. Low Temperature and High-Pressure Study of Bending L-Leucinium Hydrogen Maleate Crystals. *Crystals* **2021**, *11*, 1575. [CrossRef]
11. Pi, J.; Li, L. The Griffith Crack and the Interaction between Screw Dislocation and Semi-Infinite Crack in Cubic Quasicrystal Piezoelectric Materials. *Crystals* **2022**, *12*, 1250. [CrossRef]

crystals

MDPI

Review

Crystal Plasticity Simulation of Magnesium and Its Alloys: A Review of Recent Advances

Mohammadreza Yaghoobi [1], George Z. Voyiadjis [2,*] and Veera Sundararaghavan [3]

1 Materials Science and Engineering, University of Michigan, Ann Arbor, MI 48109, USA; yaghoobi@umich.edu
2 Computational Solid Mechanics Laboratory, Department of Civil and Environmental Engineering, Louisiana State University, Baton Rouge, LA 70803, USA
3 Aerospace Engineering, University of Michigan, Ann Arbor, MI 48109, USA; veeras@umich.edu
* Correspondence: voyiadjis@eng.lsu.edu;

Abstract: Slip and extension twinning are the dominant deformation mechanisms in Magnesium (Mg) and its alloys. Crystal plasticity is a powerful tool to study these deformation mechanisms. Different schemes have incorporated crystal plasticity models to capture different properties, which vary from the simple homogenization Taylor model to the full-scale crystal plasticity finite element model. In the current study, a review of works available in the literature that addresses different properties of Mg and its alloys using crystal plasticity modes is presented. In addition to slip and twinning, detwinning is another deformation mechanism that is activated in Mg and its alloys. The different models that capture detwinning will also be addressed here. Finally, the recent experimental frameworks, such as in-situ neutron diffraction, 3D high energy synchrotron X-ray techniques, and digital image correlation under scanning electron microscopy (SEM-DIC), which are incorporated along crystal plasticity models to investigate the properties of Mg and its alloys, are addressed. Future research directions towards improving the deformation response of Mg and its alloys are identified, which can lead to increased deployment of the lightest structural metal in engineering applications.

Keywords: crystal plasticity; twinning; detwinning; dislocation; X-ray diffraction; SEM-DIC; Magnesium

Citation: Yaghoobi, M.; Voyiadjis, G.Z.; Sundararaghavan, V. Crystal Plasticity Simulation of Magnesium and Its Alloys: A Review of Recent Advances. *Crystals* **2021**, *11*, 435. https://doi.org/10.3390/cryst11040435

Academic Editor: Sergio Brutti

Received: 17 March 2021
Accepted: 15 April 2021
Published: 17 April 2021

1. Introduction

Mg alloys can significantly impact many industrial applications such as automobile, aerospace, and transportation, not to mention the energy sector, by effectively reducing the weight of the designed system. This can lead to a significant improvement in fuel economy and reducing emissions [1]. However, the current understanding of commercial Mg alloys and their physical metallurgy is less developed compared to other established structural metals and alloys. To overcome many of the challenges in using Mg alloys in these applications, different aspects of these alloys, including strength, formability, and fatigue resistance, should be enhanced. To fulfil any of these objectives, the underlying deformation mechanisms of Mg alloys should be thoroughly understood.

Two important deformation mechanisms of Mg alloys are plastic slip and extension twinning. Mg alloys have Hexagonal Closest Packed (HCP) crystal structure. In the case of slip, different modes can be activated, including basal $\langle a \rangle$ $(\{0001\}\langle 11\bar{2}0\rangle)$, prismatic $\langle a \rangle$ $(\{10\bar{1}0\}\langle 11\bar{2}0\rangle)$, pyramidal $\langle a \rangle$ $(\{10\bar{1}1\}\langle 11\bar{2}0\rangle)$, and pyramidal $\langle c+a \rangle$ $(\{\bar{1}\bar{1}22\}\langle\bar{1}\bar{1}23\rangle)$. In the case of unalloyed Mg polycrystals, the basal mode $\langle a \rangle$ has a very low critical resolved shear stress (CRSS) and becomes the dominant plasticity deformation mechanism in most of the loading conditions. In the case of Mg alloys, although the basal mode CRSS increases, it still becomes the lowest among the slip modes. Although the basal mode CRSS is very low, its corresponding resolved shear stress disappears during the tensile deformation along the c-axis. The extension twinning $\{10\bar{1}2\}\langle\bar{1}011\rangle$ becomes an important deformation mechanism of Mg alloys in this loading condition [2–9]. Similarly, the extension twinning is the controlling deformation mechanism during the compression loading parallel to the basal plane. The highly asymmetric response of Mg alloys for some special orientations can be attributed to the activation of extension twinning, which has a complex morphology and micromechanics [3,10–16]. Although consideration of slip and twinning mechanisms is enough to describe the response of Mg alloys in most of the loading types, which are monotonic, this is not the case for the loading types, which involve unloading, such as cyclic loading. These loadings are especially important in many applications, such as fatigue. In the case of cyclic loading parallel to the basal plane, twinning occurs during compression loading, which results in the 86.3° reorientation of the basal pole. During subsequent reversed cyclic loading, i.e., tensile loading parallel to the basal plane, the size of the twinned regions shrinks, and in some cases, the twin fully disappears, a phenomenon that is commonly known as detwinning [7,17–22]. This twinning and detwinning occur alternately during the cyclic loading of Mg alloys.

In-situ experiments have a key role in unraveling the underlying deformation mechanisms of Mg alloys. Three in-situ experimental schemes, that were used to investigate the Mg alloy response were Digital Image Correlation (DIC), Synchrotron X-ray techniques, and neutron diffraction. In-situ DIC is another experimental technique that is used to address the deformation mechanisms of Mg alloys. Different aspects of Mg alloys have been studied by in-situ DIC, including the effect of twin nucleation on strain field [23], twin morphology [24], and the effect of aging on the accumulation of microscale plasticity [25]. Two common in-situ Synchrotron X-ray techniques of three-dimensional X-ray diffraction (3DXRD) and high-energy diffraction microscopy (HEDM) were incorporated in the investigation of Mg alloy behavior. Aydıner et al. [26] investigated the response of the AZ31 Mg alloy using the 3DXRD technique and analyzed the evolution of stress during deformation. It was observed that the stress state inside the twinned region was different from the untwinned grain. Other studies of Mg alloys have incorporated the in-situ 3DXRD and HEDM to analyze the micromechanics of twinning [27], the effect of age hardening on the deformation behavior [28], evaluate the CRSS for deformation modes [29,30], twin growth [31], and detwinning [7,22,32]. Using in-situ neutron diffraction, it was shown that the twinning in extruded Mg-7.7 at.% Al alloy was manifested as the change in the intensity of diffraction peaks during loading [33]. The in-situ neutron diffraction experiment on the Mg alloy showed that the stress relaxation occurred in the twinned region compared to the untwinned region of grain [34]. In-situ neutron experiments were also used to address the deformation detwinning in Mg alloys [35–39]. Although these in-situ experiments can provide invaluable information regarding the underlying deformation mechanisms of Mg alloys, they are very expensive, time-consuming, and complex. Potential simulation frameworks benefit the research community to complement these experimental observations.

Namakian and Voyiadjis [40] incorporated atomistic simulations and proposed a new mechanism for $\{10\bar{1}2\}\langle\bar{1}011\rangle$ twinning in which it was postulated that partial stacking faults (PSFs) play a key role in the formation of $\{10\bar{1}2\}$ twins. A detailed crystallographic study was conducted for different HCP metals. Accordingly, PSFs were created by 13 $\langle 10\bar{1}0\rangle$ displacement vectors on every other basal plane and created a faulting plane on $\{10\bar{1}2\}$ plane. Figure 1 shows the geometrical properties associated with the transformation of the untwinned parent into the twinned child unit cell. The model defines the properties of HCP crystals in a way that atoms can undergo spontaneous cooperative movements within these crystals. Accordingly, considerable stress relaxation occurs due to twin formation. This can be described using a deformation gradient, which shows that the macroscopic effect of the current mechanism is a simple shear process. Namakian et al. [41] incorporated atomistic simulations to model both deformation mechanisms of slip and twinning in single crystal Mg along with the interrelationships on the loose π_1L and dense π_1D first-order crystallographic planes. The generalized stacking fault energy (GSFE) analysis was used to investigate the slip mechanisms to study the $\langle c + a\rangle$ dislocation's core structure, dissociation mechanism, and mobility. Namakian et al. [41] stated that the screw component of a dissociated pyramidal-I $\langle a\rangle$ dislocation had a key role in compression twinning (CTW) nucleation. They also showed that CTW can grow by activating pyramidal-I ⟨a⟩ slip on the preexisting twin boundaries. Figure 2 shows the minimum energy path (MEP) of four deformation mechanisms associated with π_1 plane, including homogeneous CTW nucleation, heterogeneous nucleation mechanism of pyramidal-I ⟨c + a⟩ slip, i.e., $1/3\left[\bar{1}\bar{1}23\right]\left(10\bar{1}1\right)$ slip to CTW $\left[\bar{1}012\right]\left(10\bar{1}1\right)$ (or zonal twinning mechanism), heterogeneous nucleation mechanism of pyramidal- I ⟨a⟩ slip $1/3\left[\bar{1}2\bar{1}0\right]\left(10\bar{1}1\right)$ to CTW $\left[10\bar{1}\bar{2}\right]\left(10\bar{1}1\right)_{\pi 1\_D}$, and pyramidal-I ⟨c + a⟩ slip $1/3\left[\bar{1}\bar{1}23\right]\left(10\bar{1}1\right)$ using Liu et al. [42] embedded-atom method (EAM) and Wu et al. [43] modified embedded-atom method (MEAM) interatomic potentials. Namakian et al. [41] showed the importance of the interrelationships of non-basal slips and CTW on π_1 plane. They concluded that these interrelationships should be incorporated in simulation frameworks with larger length scales, including crystal plasticity, to accurately capture the mechanical response of Mg.

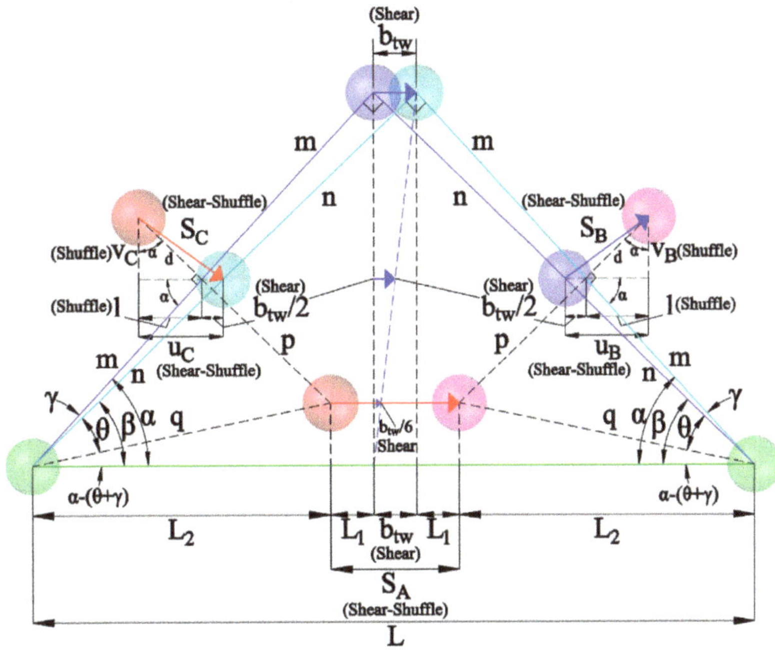

Figure 1. The schematics of the transformation of the untwinned parent unit cell (blue segments and red and blue atoms) into the twinned children unit cell (cyan segments and magenta and cyan atoms for the first two layers close to the twin boundary) during $\left(10\bar{1}2\right)\left[\bar{1}011\right]$ twinning (After Namakian and Voyiadjis [40]).

The deformation mechanisms and response of metallic systems and alloys have been studied using frameworks at different length scales spanning from density functional theory (DFT) [44,45], atomistic simulations [11,40,41,46–56], discrete dislocation dynamics [57,58], and continuum mechanics [59]. However, all these schemes except continuum mechanics suffer from length and time scale gaps to model a real-size sample or experiment. Crystal plasticity (CP) is a very powerful continuum mechanics framework to model the response of Mg alloys with both slip and twinning deformations. Various CP models have been developed to simulate slip and deformation in HCP polycrystals [60–69]. A combination of CP and finite elements, which is commonly known as CP finite element (CPFE), benefits the advantages of both frameworks. Two main categories of CPFE frameworks were developed to capture both slip and twinning deformations, which were developed by Staroselsky and Anand [63] using rate-independent formulation and Kalidindi [62] using rate-dependent formulation. In the case of Mg and its alloys, the rate-independent framework developed by Staroselsky and Anand [63] was used to model the slip along with twinning [66,68,69]. Kalidindi [62] introduced a modified plastic velocity gradient tensor by incorporating twinning. Various researchers have included twinning into their CPFE framework following the formulations introduced by Kalidindi [62], such as Abdolvand and his coworkers [65,70–72] and Zhang and Joshi [67]. Qiao et al. [73] included the stress relaxation due to twinning into the CPFE formulation, a CPFE framework that can capture the stress relaxation effect for twinning. Hama et al. [74] incorporated twinning into the CPFE framework to model the anisotropy of Mg alloy sheets subjected to a two-step loading. Prasad et al. [75] simulated the ductile fracture in pure Mg by adding the twinning to plane strain CPFE.

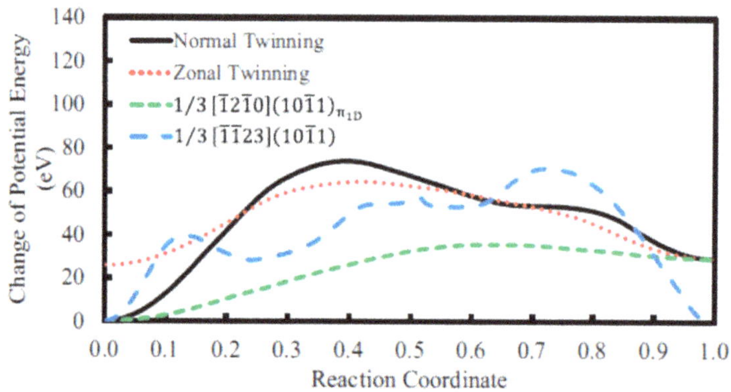

(a)

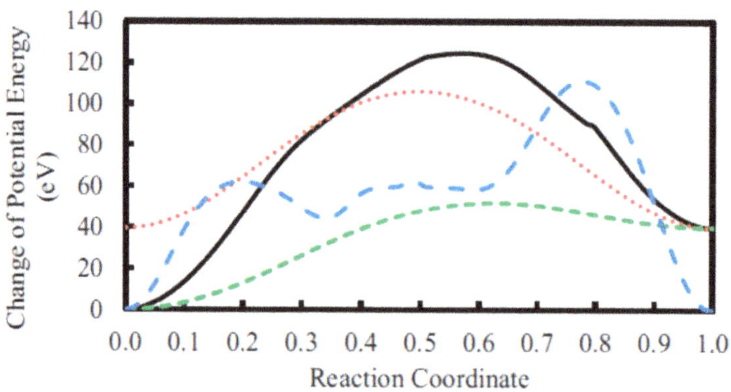

(b)

Figure 2. Minimum energy paths of four deformation mechanisms associated with π_1 plane, i.e., homogeneous CTW nucleation, heterogeneous nucleation mechanism of pyramidal-I ⟨c + a⟩ slip, i.e., $1/3\left[1\bar{1}\bar{2}3\right]\left(101\bar{1}\right)$ slip to CTW $\left[1\bar{0}12\right]\left(101\bar{1}\right)$, heterogeneous nucleation mechanism of pyramidal-I ⟨a⟩ slip $1/3\left[1\bar{2}1\bar{0}\right]\left(101\bar{1}\right)$ to CTW $\left[101\bar{2}\text{–}\right]$_(π1_D), and pyramidal-I ⟨c + a⟩ slip $1/3\left[1\bar{1}\bar{2}3\right]\left(101\bar{1}\right)$ (**a**) Liu et al. [42] EAM potential (**b**)Wu et al. [43] MEAM potential (After Namakian et al. [41]).

The CP models were used in combination with other schemes to model twinning. One of these candidates was a viscoplastic self-consistent (VPSC), which was first used by Lebensohn and Tomé [76] to model the twinning that occurred in zirconium alloys. Agnew et al. [77] used the VPSC to model the slip and twinning in Mg alloys. Beyerlein and Tomé [12] simulated the twinning in pure zirconium using VPSC by introducing a probabilistic model for the twin nucleation. Kumar et al. [78] used a full-field elasto-viscoplastic model based on fast Fourier transformation (FFT), which included twinning to model the response of Mg. Lévesque et al. [79] introduced twinning into a Taylor-type CP model to simulate the response of AM30 and AZ31B Mg alloys. The multiscale framework of VPSC-FE was incorporated to simulate the slip and twinning deformations [80–82]. Accordingly, the response of each FE material point was defined using an RVE, which was modeled by VPSC. A similar multiscale scheme has been used by Ardeljan et al. [83] and Feather et al. [84] to simulate the response of Mg alloys in which an RVE modeled by a Taylor-type CP defines the constitutive model of an FE material point. A CP model has also been coupled with a phase field method to better describe the twin morphology in HCP metals [85,86].

Most of the studies have focused on the response of the Mg alloys during monotonic loading, which only requires the deformation slip and twinning. In the case of more complex strain paths such as cyclic loading, however, detwinning should also be defined in the CP model. A meso-scale composite grain (CG) model was the first detwinning model, which was included in the VPSC framework [87,88]. An empirical model defines the twin nucleation and propagation in the CG model. Guillemer et al. [89] used a simple phenomenological detwinning model along a self-consistent model to capture the cyclic behavior of extruded Mg. The first physically-based model, which includes both mechanisms, was the Twinning-Detwinning (TDT) model developed by Wang and his coworkers [90–92], which was used along the EVPSC framework. Qiao et al. [93] modeled the cyclic response of the ZK60A Mg alloy using the TDT/EVPSC scheme. The twinning and detwinning can be effectively handled by EVPSC framework by introducing new grains as twin nucleation and removing that twinned grain for complete detwinning, while the change in the volume of the twinned grain manifests the twin growth or shrinkage. However, in the case of studying the local variable information in a polycrystal, one cannot use the EVPSC model. One way to obtain the local information is using the CPFE framework to capture twinning and detwinning mechanisms. However, most of the CPFE models introduced to capture the twinning and detwinning mechanisms are phenomenological and neglect some of the main elements of the deformation twinning and detwinning. The biggest challenge in CPFE is that the introduction of twin nucleation, growth, shrinkage, and detwinning is not as straightforward as EVPSC. In the case of Mg and its alloys, the CPFE method has been used recently to model the cyclic behavior, including twinning and detwinning [22,94–97]. A material point is treated in these models as two states of twinned or not twinned, and the reorientation criteria are commonly based on the most active twinning system, which is called the predominant twinning reorientation (PTR) scheme [61]. Before the reorientation, the stress inside the twinned region does not contribute to the stress state of the material point, while after reorientation, the stress state inside the parent grain is neglected. However, the stress state of the twinned region and untwinned region are not similar, which was shown using in-situ experiments [26,34]. Yaghoobi et al. [9] introduced a multiscale CPFE framework in which the deformation twinning and detwinning were captured using a physically-based TDT model proposed by Wang and his coworkers [90–92]. They used a Taylor-type homogenization scheme to model both twinned and untwinned regions, coexisting at a material point.

Traditional crystal plasticity models were developed largely without a connection to grain size and shape effects. In Magnesium alloys, it has been observed that the grain size strengthening follows the Hall–Petch effect [98]. The incorporation of grain size effect into constitutive models for single slip began in 1962 with Armstrong et al. [99], who modified the Hall–Petch equation to correspond to the flow stress on a slip system (the "micro-Hall–Petch relation"). The interrelationship between grain size and texture was not considered until 1983, when Weng [100] employed the mean grain size in the equation for slip system resistance through the micro-Hall–Petch relation. Sun and Sundararaghavan [101] presented a crystal plasticity model for grain size and shape effects where the slip length of each slip system at a material point was considered in a micro-Hall–Petch strengthening term for each slip system. Recent HREBSD experiments [102] in Mg-4Al alloy have revealed the interaction of slip systems and grain boundaries is dependent on not only the slip length but also strongly on the angle between the incoming and emerging slip planes at a grain boundary. In the future, there is a need to better incorporate such insights in crystal plasticity.

In the current work, a review of works available in the literature, which addresses different properties of Mg and its alloys using crystal plasticity models, is presented. In addition to slip and twinning, detwinning is another deformation mechanism, which is activated in Mg and its alloys. The different models that capture detwinning will also be addressed here. Finally, the recent experimental frameworks, such as in-situ neutron diffraction, 3D high energy synchrotron X-ray techniques, and digital image correlation under scanning electron microscopy (SEM-DIC), are incorporated along crystal plasticity models to investigate the properties of Mg and its alloys are addressed. Future research directions towards improving the deformation response of Mg and its alloys are identified, which can lead to increased deployment of the lightest structural metal in engineering applications.

2. Crystal Plasticity Models

2.1. Overview

The CP theory is developed according to the assumption that the plasticity in metals and alloys occurs as a result of slip-on prescribed slip systems. The finite deformation continuum mechanics is usually incorporated to describe the formulation. The deformation gradient tensor **F** is decomposed as below:

$$\mathbf{F} = \mathbf{F}^e\mathbf{F}^p \quad (1)$$

where $\mathbf{F}^e$ and $\mathbf{F}^p$ are the elastic and plastic deformation gradients, respectively. Figure 3 shows the deformation described in Equation (1). Elastic distortion of the crystal lattice and plastic slip are two deformation mechanisms, which sustain the applied deformation in this formulation. Accordingly, the macroscopic velocity gradient **L** can be decomposed as below:

$$\mathbf{L} = \mathbf{L}^e + \mathbf{L}^p \quad (2)$$

where $\mathbf{L}^e$ and $\mathbf{L}^p$ are the elastic and plastic velocity gradients, respectively. The novel relation of CP is to link the macroscopic velocity gradient $\mathbf{L}^p$ to micro deformation $\dot{\gamma}^\alpha$ as below:

$$\mathbf{L}^p = \sum_{\alpha=1}^{N_s} \dot{\gamma}^\alpha \mathbf{S}^\alpha \quad (3)$$

where $\dot{\gamma}^\alpha$ is the shearing rate on slip system α, N_s is the number of slip systems, and $\mathbf{S}^\alpha$ is the Schmid tensor for the slip system α, which can be defined as follows:

$$\mathbf{S}^\alpha = \mathbf{m}^\alpha \bigotimes \mathbf{n}^\alpha \quad (4)$$

where unit vectors $\mathbf{m}^\alpha$ and $\mathbf{n}^\alpha$ denote the slip direction and slip plane normal, respectively, in the deformed configuration.

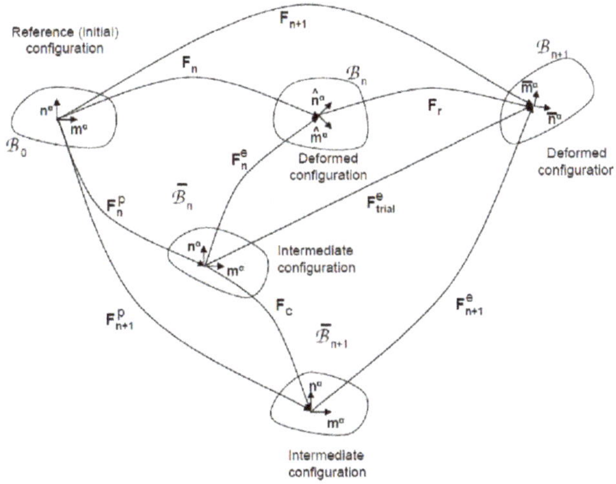

Figure 3. Different configurations of the body in finite strain framework (After Yaghoobi et al. [103]).

The resolved shear stress on the slip system α can be obtained as follows:

$$\tau^{\alpha} = \boldsymbol{\sigma} \cdot \mathbf{S}^{\alpha} \tag{5}$$

where $\boldsymbol{\sigma}$ is the Cauchy stress tensor and $\cdot$ operator denotes the standard inner product of tensors. The shearing rate on the αth slip system γ^{α} can be calculated depending on if the CP formulation is rate-independent or rate-dependent. In the case of a rate-independent model, the yield surface is defined for each system, i.e., f^{α}, as follows:

$$f^{\alpha} = |\tau^{\alpha}| - s^{\alpha} \tag{6}$$

where s^{α} is the slip resistance for slip system α. The values of $\dot{\gamma}^{\alpha}$ is then obtained using the yield surface and considering the hardening model, which describes the evolution of slip resistance.

In the case of rate-dependent formulation, the evolution of shearing rate on the αth slip system γ^{α} can be defined as follows:

$$\dot{\gamma}^{\alpha} = \dot{\gamma}_0 \left| \frac{\tau^{\alpha}}{s^{\alpha}} \right|^{1/m} \text{sign}[\tau^{\alpha}] \tag{7}$$

where $\dot{\gamma}_0$ and m are the material parameters denoting the reference shearing rate and rate sensitivity of the material. One should note that the kinematic hardening is neglected in both Equations (6) and (7).

2.2. Twinning

In the case of Mg and its alloys, extension twinning $\{10\bar{1}2\}\langle\bar{1}011\rangle$ is very important, along with the plastic slip (Figure 4). The simplest modification to include twinning into the CP formulation is to consider twinning as pseudo-slip systems (See, e.g., Staroselsky and Anand [64]), and their contribution to the plastic velocity gradient tensor can be described as follows:

$$\mathbf{L}^{\text{p}} = \sum_{\alpha=1}^{N_s+N_t} \dot{\gamma}^{\alpha} \mathbf{S}^{\alpha} \tag{8}$$

where N_t is the number of twinning systems. Figure 5 shows the kinetics of slip and twinning defined in Equation (8). The relation between the plastic slip of twinning systems and their corresponding twin volume fraction can be described as follows:

$$\dot{f}^{\beta} = \frac{\dot{\gamma}^{\beta}}{S} \tag{9}$$

where $\dot{f}^{\beta}$ is the rate of change in twin volume fraction of twin pseudo-slip system β and S is the characteristic twin shear strain, which defines the amount of shear associated with twinning. The value of S depends on the c/a ratio, which can be defined as follows [104]:

$$S = \frac{\sqrt{3}}{c/a} - \frac{c/a}{\sqrt{3}} \tag{10}$$

where a and c are depicted in Figure 4. In the case of Mg and its alloys, the $S = 0.129$.

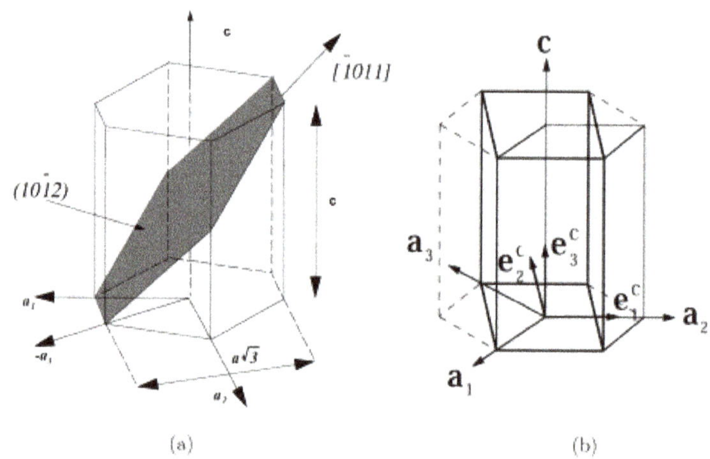

Figure 4. (**a**) The crystallography of the extension twinning $\{10\bar{1}2\}\langle\bar{1}011\rangle$ in Mg and its alloys. (**b**) Description of deformation systems in the orthonormal system $\{e_i^c | i = 1,2,3\}$ (After Staroselsky and Anand [64]).

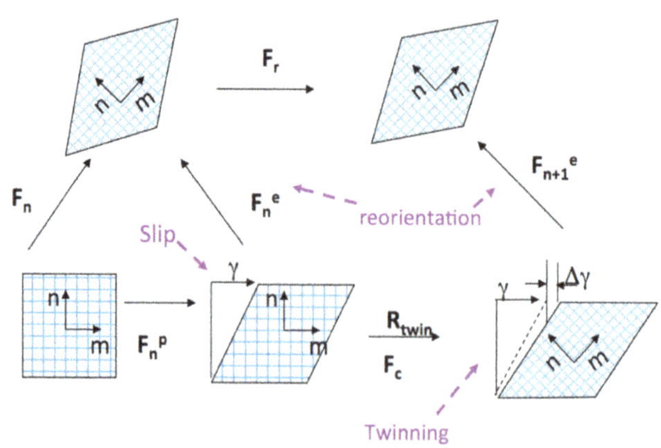

Figure 5. The deformation modes kinematics in finite strain framework (After Yaghoobi et al. [103]).

Although the twin pseudo-slip system's contribution to plastic velocity gradient tensor is similar to that of the slip systems, they are not precisely treated similarly due to the polar nature of twinning. In other words, only the tensile component of stress along the c-axis can trigger the extension twinning. This is reflected in the CP formulation for the twinning pseudo-slip systems as follows:

$$\begin{cases} if\ \boldsymbol{\sigma} : \mathbf{S}^{\beta} \leq 0 \rightarrow \tau^{\beta} = 0 \\ if\ \boldsymbol{\sigma} : \mathbf{S}^{\beta} > 0 \rightarrow \tau^{\beta} = \boldsymbol{\sigma} : \mathbf{S}^{\beta} \end{cases} \tag{11}$$

Kalidindi [62] further enhanced the effect of twinning on crystal plasticity formulation. The work considers the contribution of stress in both untwinned and twinned region as follows:

$$\boldsymbol{\sigma} = \left(1 - \sum_{\beta=1}^{N_t} f^{\beta}\right)\boldsymbol{\sigma}^{\mathrm{mt}} + \sum_{\beta=1}^{N_t} f^{\beta}\boldsymbol{\sigma}^{\mathrm{tw},\beta} \tag{12}$$

where f^{β} is the reoriented volume fraction of grain according to the twin system β. Kalidindi [62] then presented the modified macroscopic plastic velocity gradient tensor $\mathbf{L}^{\mathrm{P}}$, which includes the contribution of multiple twinned systems as below:

$$\mathbf{L}^{\mathrm{P}} = \left(1 - \sum_{\beta=1}^{N_t} f^{\beta}\right)\sum_{\alpha=1}^{N_s} \dot{\gamma}^{\alpha}\mathbf{S}^{\alpha}_{\mathrm{sl}} + \sum_{\alpha=1}^{N_t} s\dot{f}^{\beta}\mathbf{S}^{\beta}_{\mathrm{tw}} + \sum_{\beta=1}^{N_t} f^{\beta}\left(\sum_{\alpha=1}^{N_{s-tw}} \dot{\gamma}^{\alpha}\mathbf{S}^{\alpha}_{\mathrm{sl-tw}}\right) \tag{13}$$

where $\mathbf{S}^{\alpha}_{\mathrm{sl}}$, $\mathbf{S}^{\beta}_{\mathrm{tw}}$, and $\mathbf{S}^{\alpha}_{\mathrm{sl-tw}}$ denote the Schmid tensors for the slip systems in parent grain, twin systems in parent grain, and slip systems in twinned children, respectively. N_{s-tw} denotes the number of slip systems in the twinned region. Equation (10) describes three mechanisms contributing to the plastic velocity gradient $\mathbf{L}^{\mathrm{P}}$. The first term defines the contribution of slip inside the parent grain, the second term defines the contribution of twin systems in parent grain, and the third term is the summation of the contributions of slip systems in all twinned children.

In the twinning models, there is a key part that should be defined as the reorientation. Figure 6 shows the crystallography of the deformation twinning, which includes the untwinned region (parent grain) and twinned region (child). While Equation (9) defines the amount of twin volume required for a certain amount of shear strain, it does not define the reorientation in the model. Different criteria have been introduced for the reorientation. Van Houtte [60] was the first researcher to propose a method for reorientation. In this method, reorientation of the grain is decided depending on the volume fractions calculated at every incremental step along with using a random criterion. Tomé et al. [61] modified the model presented by Van Houtte [60] using the predominant twin reorientation scheme (PTR) in which the grain is reoriented according to its predominant twin variant as the integration of twin volume throughout the time satisfy specific criteria.

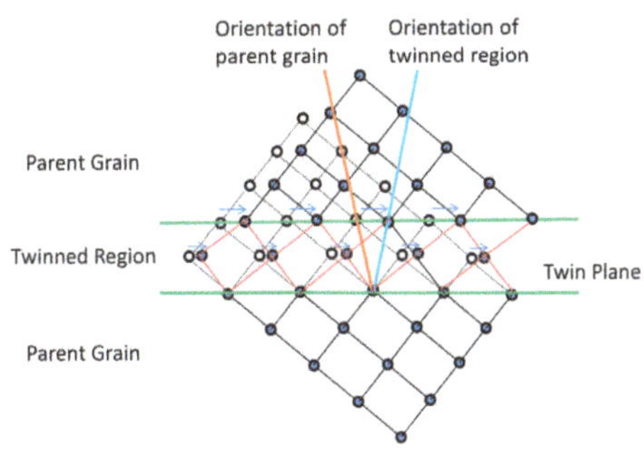

Figure 6. The crystallographic description of extension twinning (After Yaghoobi et al. [103]).

2.3. Stress Relaxation

The in-situ neutron diffraction experiment on the Mg alloy showed that the stress relaxation occurs in the twinned region compared to the untwinned region of grain [34]. Different crystal plasticity models were incorporated to capture the stress relaxation during twinning. The simplest model is to consider larger CRSSs for twin nucleation than growth [12,93]. Wu et al. [105] presented a comprehensive physically-based twin model implemented in the EVPSC framework. They presented this twin model, which was called 'TNPG,' to capture twin nucleation, propagation, and growth for magnesium alloys (Figure 7). In Figure 7a, the twinning process initiates with twin nucleation. It is assumed that the grain boundary is the preferred nucleation site. As the twin propagates according to Figure 7b, stress relaxation occurs. Different notation is used throughout the current review for twinning resistance as s^α compared to the one used by Wu et al. [105], which used $\hat{\tau}^\alpha$ (Figure 7). The twinning resistance s^α decreases during the propagation phase until it reaches its minimum value of s_g^α when the twin volume is $f^\alpha = f_g^\alpha$. Afterward, the twinning resistance increases due to the hardening occurring in the twin thickening process. Wu et al. [105] captured the stress relaxation by modifying the twin systems resistances s^α as follows:

$$s^\alpha = \begin{cases} s_0^\alpha - \frac{s_0^\alpha - s_g^\alpha}{f_g^\alpha} f^\alpha & \left(\text{if } f^\alpha \leq f_g^\alpha\right) \\ s_g^\alpha + \left(s_1^\alpha + h_1^\alpha \Gamma^\alpha\right)\left(1 - \exp\left(-\frac{h_0^\alpha}{s_1^\alpha}\Gamma^\alpha\right)\right) & \left(\text{if } f^\alpha > f_g^\alpha\right) \end{cases} \tag{14}$$

where Γ^α is the accumulated shear of αth twinning system, which is calculated after its corresponding twin volume f^α surpass f_g^α (see Figure 7). They used this model to capture the stress-strain response of AZ31B Mg alloy obtained by Lou et al. [106]. They also compared the predicted twin volume versus the experimental results. Figure 8 shows that the model developed by Wu et al. [105] can successfully capture both the stress-strain and twin volume during uniaxial compression and tension. Qiao et al. [73] reviewed the available scheme to model stress relaxation in the CPFE framework. They used the model developed by Wu et al. [105] for EVPSC and incorporated it in the CPFE model to capture the stress relaxation in Mg single crystal during twinning.

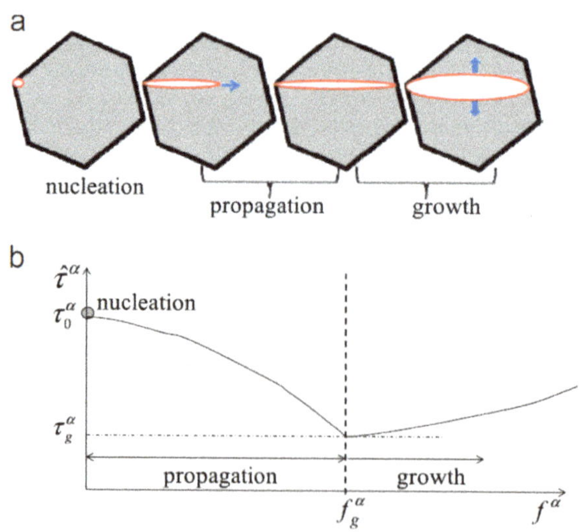

Figure 7. Extension twinning in Mg and its alloys. (**a**) The schematics of extension twin nucleation, propagation, and growth, (**b**) variation of slip resistance for extension twinning system α at different stages of deformation twinning (After Wu et al. [105]).

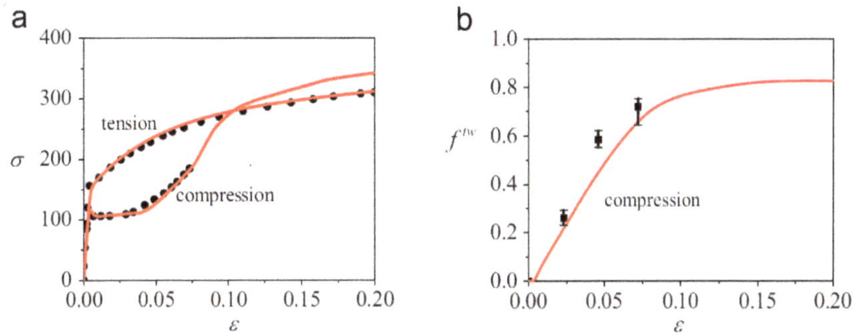

Figure 8. Comparison of the experimental and simulated response of the AZ31B Mg alloy during uniaxial loadings along the rolling direction. (**a**) Stress-strain curves of uniaxial tension and compression, (**b**) twin volume fraction of uniaxial compression (After Wu et al. [105]).

2.4. Detwinning

In the case of more complex strain paths, the detwinning mechanism should also be included in addition to twinning and slip modes. Various models address the detwinning along twinning and slip modes using crystal plasticity [22,87–89,94–97]. Here, a physically-based Twinning-Detwinning (TDT) model is elaborated, which was developed by Wang and his coworkers [90–92] for the EVPSC framework and extended later on for the CPFE framework [9]. In this model, the twinning and detwinning mechanisms can be divided into 4 major operations as follows (Figure 9):

- *Operations A*: Twin nucleation and growth due to parent grain reduction.
- *Operations B*: Twin growth due to the twinned child propagation.
- *Operations C*: Twin shrinkage due to the parent propagation.
- *Operations D*: Detwinning in the twinned child.

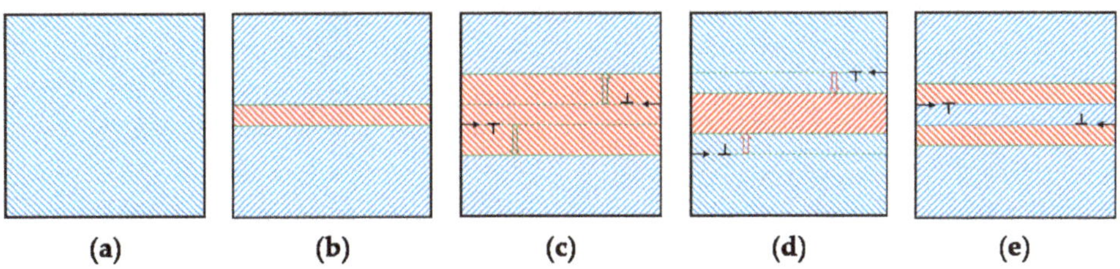

Figure 9. Deformation twinning and detwinning operations: (**a**) Initial twin-free parent grain (**b**) *Operation A*: As soon as the twin volume reaches a threshold of f_0, twin nucleation occurs. This nucleated twin can grow according to the reduction in the volume of parent grain. (**c**) *Operation B*: The growth of the twin region can occur according to the child propagation. (**d**) *Operation C*: As the parent region grows, the twin volume reduces. (**e**) *Operation D*: Detwinning inside the twinned child leads to the twin volume reduction (After Yaghoobi et al. [9]).

The polarity of simple twinning model should be modified for these operations. In the parent grain, the crystallographic systems include N_s slip systems, N_t twin systems for *Operation A*, and N_t twin systems for *Operation C*. In the case of the twin child, the crystallographic systems include N_{s-tw} slip systems, one twin system for *Operation D*, and one twin system for *Operation B*. *Operation A* can get activated in parent grain for when the resolved shear in the twin system k is larger than the corresponding slip resistance, i.e., $\tau^{N_s+k} > s^{N_s+k}$. *Operation B* is activated in the twin child, which is nucleated by the activation of βth twin variant, when $\tau_\beta^{N_s-tw+2} < 0$ and $|\tau_\beta^{N_{s-tw}+2}| > s_\beta^{N_{s-tw}+2}$. *Operation C* initiates in the parent grain for the kth twin variant when $\tau^{N_s+N_t+k} < 0$ and $\left|\tau^{N_s+N_t+k}\right| > s^{N_s+N_t+k}$.

Operation D initiates inside the twin child, which is nucleated by the activation of βth twin variant, when $\tau_{\beta}^{N_{s-tw}+1} > s_{\beta}^{N_{s-tw}+1}$.

In the TDT/EVPSC model developed by Wang and his coworkers [90–92], the twinned child is manifested as a new grain, and the *Operations A-D* is reflected by the change in the volume of the parent and twin grains. Wang et al. [90] used the developed TDT model to capture the cyclic response of AZ31B Mg alloy for complex loading paths. They compared the model prediction versus the experimental data provided by Lou et al. [106] (Figure 10). They also showed the variation of twin volume and how the deformation twinning and detwinning governs the twin volume and stress-strain curve shape. Furthermore, they compared the prediction of TDT/EVPSC model versus the CG results reported by Proust et al. [88] for the experimental results of AZ31B Mg alloy subjected to the case of in-plane compression followed by through-thickness compression (Figure 11). Although the CG model was able to capture some features of twinning-detwinning mechanisms, the TDT/EVPSC model considerably enhances the accuracy of the simulation results.

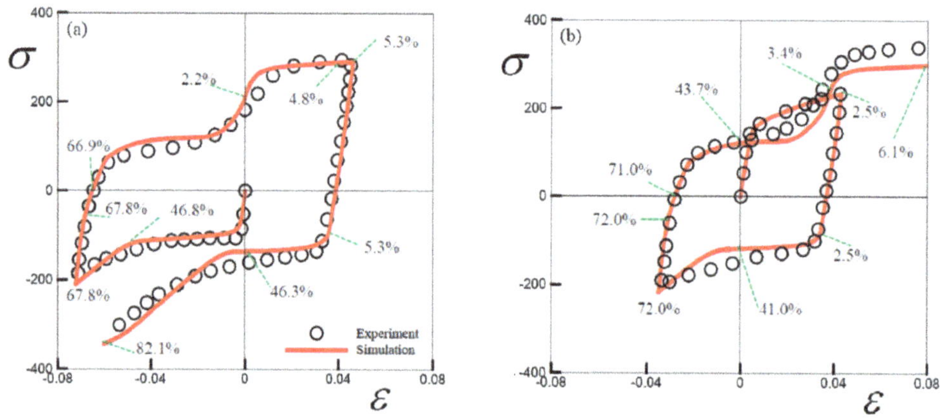

Figure 10. Comparison of predicted cyclic response of AZ31B Mg alloy versus the experimental data subjected to different loading paths of: (**a**) Uniaxial compression-tension-compression, (**b**) uniaxial tension-compression- tension. The twin volume is predicted by simulation and demonstrated at different strains (After H. Wang et al. [90]).

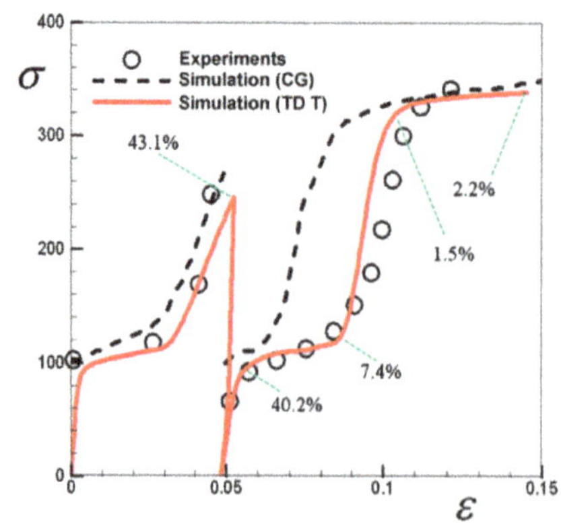

Figure 11. Comparison of predicted stress versus the absolute value of the accumulated strain in AZ31B Mg alloy versus the experimental data subjected to in-plane compression followed by compression loading through the thickness along the RD. The twin volume is predicted by simulation and demonstrated at different strains (After H. Wang et al. [90]).

Unlike the EVPSC framework, the TDT operations are not trivial to apply to a material point in the case of CPFE framework. Yaghoobi et al. [9] used a multiscale framework in which the untwinned and twinned regions were considered within a sub-scale model at a material point and the stresses homogenized using a Taylor-type scheme (Figure 12). This allows multiple variants to coexist at a material point, which avoids the need for the selection of predominant variants as in the PTR models. They implemented the model in an open-source CPFE software PRISMS-Plasticity [103]. The twin volume evolution of a material point for the βth twin variant is obtained as below:

$$\dot{f}^{\beta} = \left[1 - \sum_{\beta=1}^{N_t} f^{\beta}\right] \frac{\left(\dot{\gamma}^{N_s+\beta} - \dot{\gamma}^{N_s+N_t+\beta}\right)}{S} - f^{\beta} \frac{\left(\dot{\gamma}_{\beta}^{N_{s-tw}+1} - \dot{\gamma}_{\beta}^{N_{s-tw}+2}\right)}{S} \tag{15}$$

where $\dot{\gamma}^{N_s+\beta}$ and $\dot{\gamma}^{N_s+N_t+\beta}$ are the shear rate inside the parent corresponds to operations A and C, and $\dot{\gamma}_{\beta}^{N_{s-tw}+1}$ and $\dot{\gamma}_{\beta}^{N_{s-tw}+2}$ are the shear rates within the child nucleated due to the activation of the βth twin system.

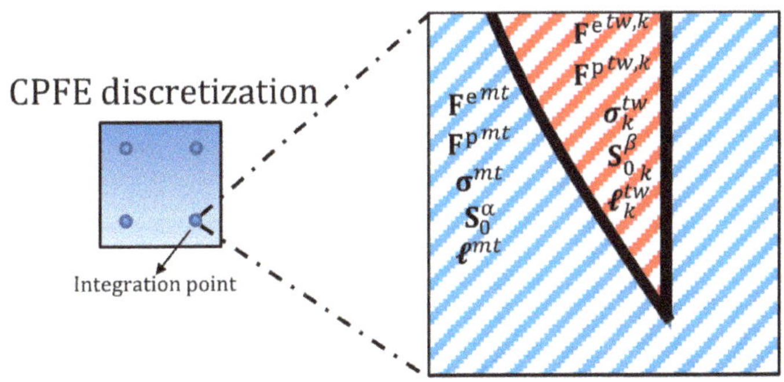

Figure 12. A partially twinned material point in the CPFE framework (After Yaghoobi et al. [9]).

Yaghoobi et al. [9] calibrated the model using the uniaxial experimental data of extruded ZK60A Mg alloy presented by Wu [107] (Figure 13). The predictions of the modeled cyclic response of the ZK60A alloy along the extrusion direction were compared with the experimental data by Wu et al. [17] and Wu [107]. Yaghoobi et al. [9] further studied the cyclic response of ZK60A by comparing the predicted twin variation versus the normalized intensity of the {0002} diffraction peak along the longitudinal direction measured by Wu et al. [35] and Wu [107]. The result showed that not only can the multiscale TDT CPFE framework successfully capture the cyclic stress-strain response of the ZK60A (Figure 14), it can also capture the deformation twinning and detwinning during the cyclic loading (Figure 15). Furthermore, Yaghoobi et al. [9] showed the evolution of predicted basal (0001) pole figures at strains of $\varepsilon = \pm 1.2\%$ (Figure 16). It was shown that the deformation twinning resulted in the reorientation of the basal pole, which intensifies in the basal {0002} peak in the loading direction at max compressive strains of $\varepsilon = -1.2\%$ (Figure 16b,d). On the other hand, the detwinning mechanism removed the increased basal {0002} peak intensity at a max tensile strain of $\varepsilon = 1.2\%$ (Figure 16c,e).

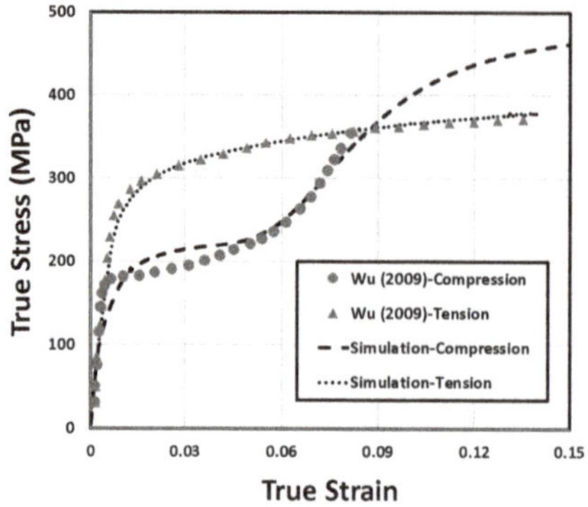

Figure 13. The comparison of the simulated stress-strain response responses of ZK60A Mg alloy subjected to the uniaxial loading along the extrusion direction versus the experimental data of Wu [107] (Yaghoobi et al. [9]).

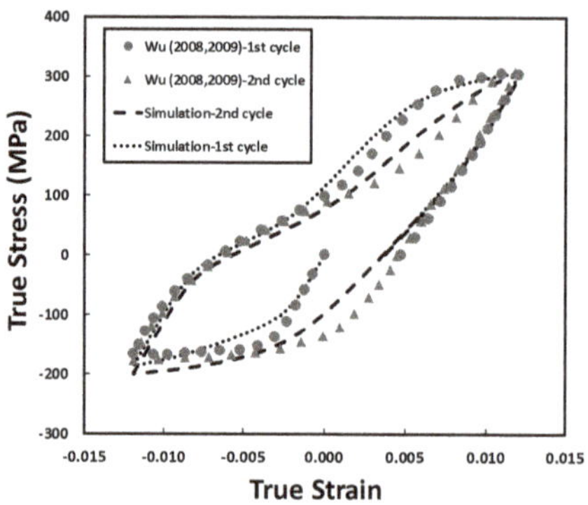

Figure 14. Comparison of the predicted stress-strain curve versus the experimental results of Wu et al. [17] and Wu [107] during the cyclic loading along the extrusion direction in ZK60A Mg alloy (After Yaghoobi et al. [9]).

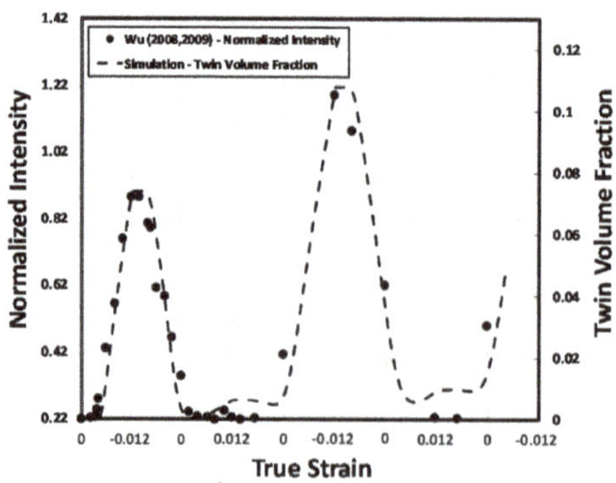

Figure 15. The variation of predicted twin volume versus the strain compared to the experimentally measured change in the normalized intensity of the {0002} diffraction peak along the longitudinal direction obtained by Wu et al. [35] and Wu [107] in ZK60A Mg alloy during the cyclic loading along the extrusion direction (After Yaghoobi et al. [9]).

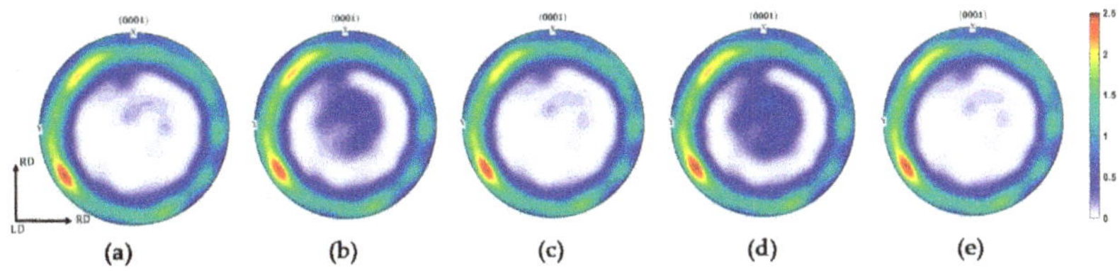

Figure 16. The evolution of basal (0001) pole figures predicted by CPFE simulation at different strains: (**a**) Initial texture, (**b**) first maximum compression ($\varepsilon = -1.2\%$) in cycle 1, (**c**) first maximum tension ($\varepsilon = 1.2\%$) in cycle 1, (**d**) second maximum compression ($\varepsilon = -1.2\%$) in cycle 2, (**e**) second maximum tension ($\varepsilon = 1.2\%$) in cycle 2 (After Yaghoobi et al. [9]).

3. In-situ Experiments

3.1. In-Situ DIC Experiments

3.1.1. Microscale Deformation Mechanisms

Githens et al. [25] used the in-situ DIC under scanning electron microscopy (SEM-DIC) to study the deformation mechanisms of WE43-T5 Mg alloy with weak basal texture subjected to uniaxial tension and compression along the rolling direction. The SEM-DIC technique provided the full-field strain maps. They first investigated the response of the WE43 alloy during uniaxial tension (Figure 17). They showed that the heterogeneous pattern of strain does not vary after the yield point. They also studied the strain probability distribution, which confirms that the strain map does not vary after yielding ($\varepsilon = 2.91\%$ and 4.86%). However, this pattern is different from the one before yielding ($\varepsilon = 1.23\%$). They simulated the sample using CPFE simulation open-source software of PRISMS-Plasticity [103] and compared the predicted strain map with the experimental observations, which agrees well (Figure 18).

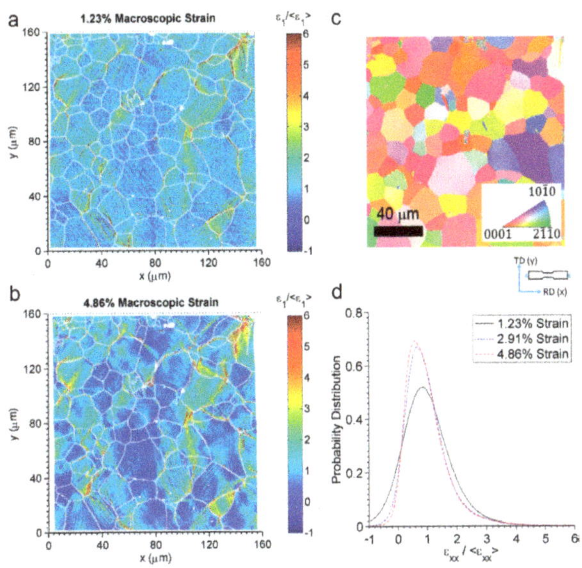

Figure 17. Using SEM-DIC technique to investigate the underlying deformation mechanisms in WE43-T5 Mg alloy during tensile loading along the rolling direction: (**a**) The map of normalized maximum principal strain maps at $\varepsilon = 1.23\%$, (**b**) the map of normalized maximum principal strain maps at $\varepsilon = 4.86\%$, (**c**) the inverse pole figure map, (**d**) a probability distribution of the strain at different applied tensile strains of $\varepsilon = 1.23\%$, 2.91%, and 4.86% (After Githens et al. [25]).

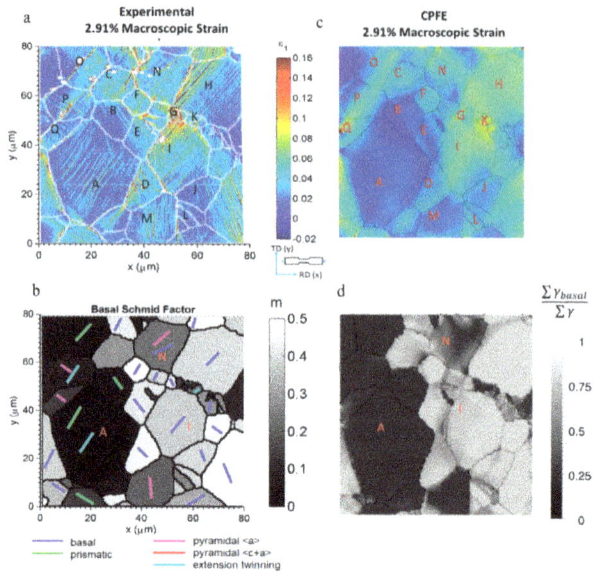

Figure 18. Identification of active deformation mode using SEM-DIC experiment and CPFE simulation at a tensile strain of $\varepsilon = 2.91\%$ (**a**) SEM-DIC strain map, (**b**) slip traces for different deformation mode along with the basal Schmid factor map, (**c**) CPFE simulation strain map, (**d**) the activity of basal slip system obtained from CPFE simulation (After Githens et al. [25]).

They not only verified the CPFE framework using the SEM-DIC results, but they also resolved and categorized the strain from individual slip traces for the first time in any Mg alloy, as shown in Figure 18a. They used this information to categorize the active slip/twinning modes with respect to their nominal Schmid factors in both SEM-DIC and CPFE results (Figure 19). The results show that the basal modes sustain the most plastic deformation during uniaxial tension. The non-basal slip modes are less active. The least active is the extension twinning, which is because of the tested sample texture and loading direction. Githens et al. [25] also studied the distribution of active slip/twin modes during the compression loading (Figure 20). The activity of extension twinning is considerably increased during compression compared to the tensile loading (Figure 19). In addition, the non-basal slip modes contribution to the deformation is very limited during compression (Figure 20).

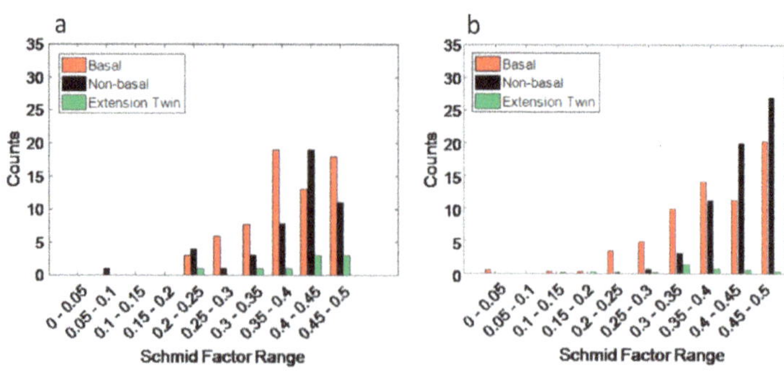

Figure 19. Comparison of active deformation modes versus corresponding Schmid factor at a uniaxial tensile strain of $\varepsilon = 4.86\%$ (**a**) SEM-DIC experimental data, (**b**) CPFE simulation results (After Githens et al. [25]).

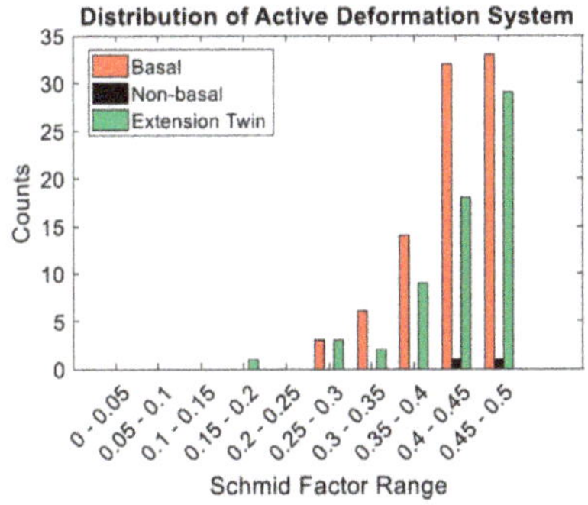

Figure 20. Active deformation modes versus corresponding Schmid factor at uniaxial compression strain of $\varepsilon = -4.2\%$ obtained using the SEM-DIC experiment (After Githens et al. [25]).

3.1.2. Effects of Heat Treatment

Ganesan et al. [8] incorporated the in-situ DIC along the SEM technique to investigate the effect of heat treatment on the response of WE43 Mg alloy with the weak basal texture. They started with the WE43-T5 sample and applied different heat treatment conditions of the solution treated (ST) and aging times of 15 min, 2 h, 4 h, and 16 h (T6 condition). They investigated the effect of heat treatment on the macro responses of WE43 Mg (Figure 21). The results showed that the T5 condition had the highest yield stress, which can be attributed to its finer grain size. Within the heat-treated samples, the T6 condition had the maximum yield strength, which was due to the role of precipitates. Each of the samples was then subjected to the in-situ uniaxial tension loading along the rolling direction, where the SEM-DIC technique was used to gather the deformation maps. Figure 22 shows the normalized maximum principal strain maps at ε =3.23% in samples with different aging times. They showed that the strain map of the sample with T6 condition had fewer localized strains. In order to quantitatively investigate their observation, Ganesan et al. [8] studied the strain probability distribution of the tested samples (Figure 23). They showed that it is $\sim$ 3.84 more probable to find a localized strain of 4 $\times$ (Average applied strain) in underaged samples compared to the T6 condition.

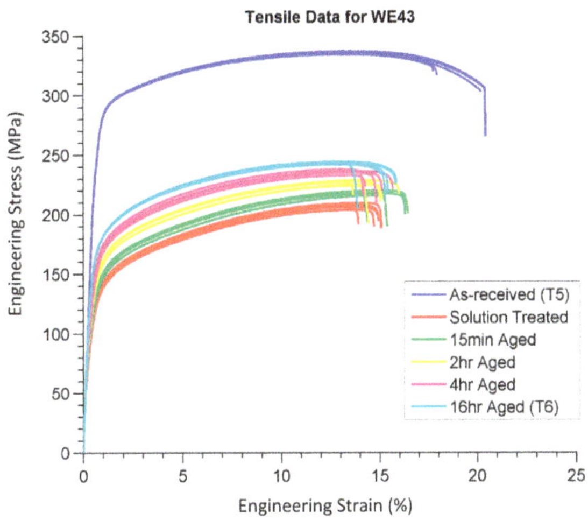

Figure 21. The stress-strain response of Mg alloy WE43 with different heat treatment conditions during uniaxial tension along the rolling direction (After Ganesan et al. [8]).

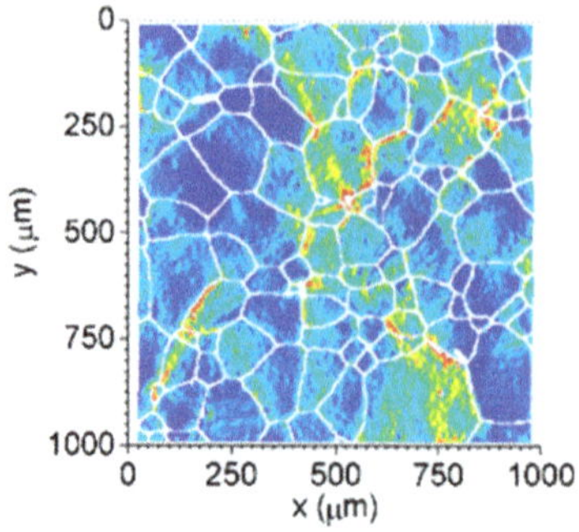

(**a**) Solution treated

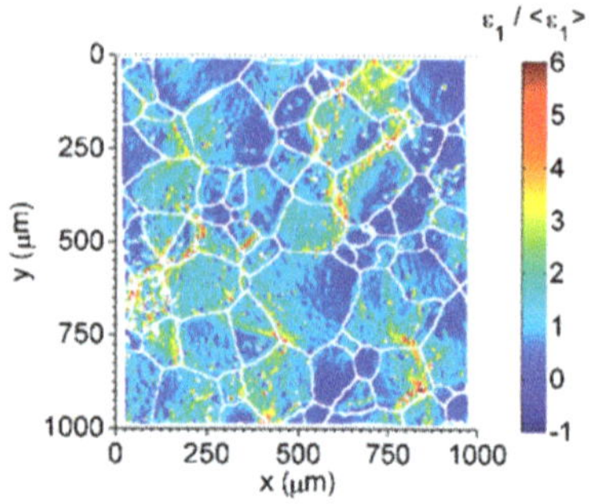

(**b**) 15 min aged

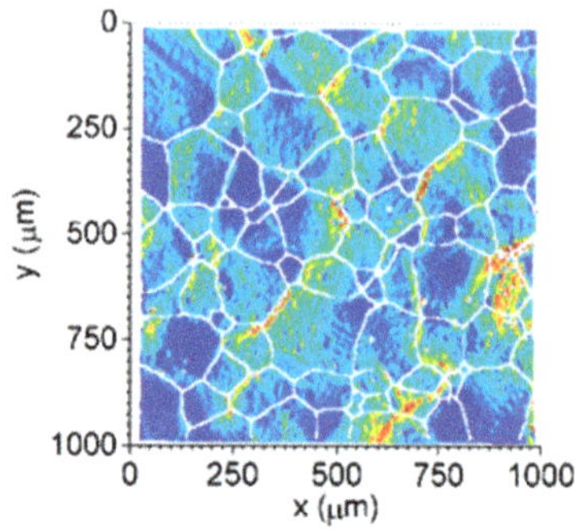

(**c**) 2 h aged

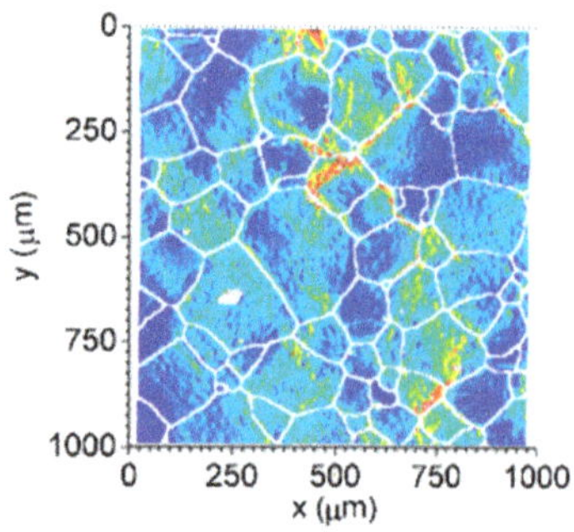

(**d**) 4 h aged

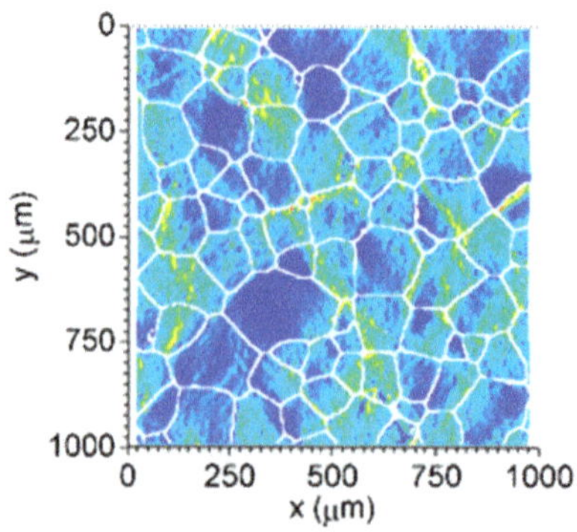

(**e**) 16 h aged (T6 condition)

Figure 22. The effect of heat treatment condition on normalized maximum principal strain map subjected to a uniaxial tension along the rolling direction at a strain of $\varepsilon = -3.23\%$ for different heat treatment conditions of: (**a**) Solution treated, (**b**) 15 min aged, (**c**) 2 h aged, (**d**) 4 h aged, (**e**) 16 h aged (T6 condition) (After Ganesan et al. [8]).

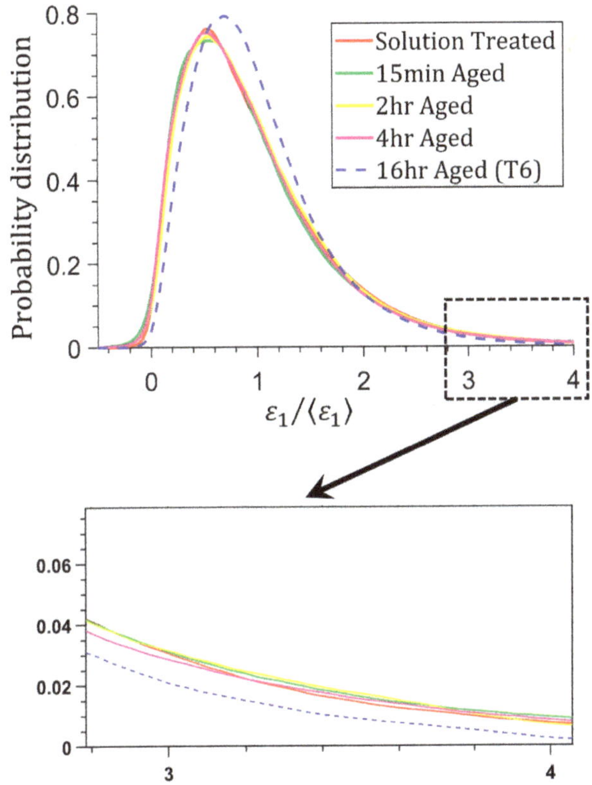

Figure 23. The probability distribution of normalized maximum principal strain $\varepsilon_1/\langle\varepsilon_1\rangle$ for WE43 Mg alloy subjected to different heat treatment conditions after solution treatment at the uniaxial tensile strain of $\varepsilon = -3.23\%$ along the rolling direction (After Ganesan et al. [8]).

They further investigated the effect of heat treatment using the CPFE simulation. They validated the CPFE framework and showed that the simulation can capture the stress-strain response of samples with different heat treatment conditions (Figure 24). They further validated the simulation by predicting the strain map of WE43-T6 at different tensile strains (Figure 25). The results showed that not only can the CPFE capture the macro responses, but it can also accurately model the local field maps. After validation of the CPFE framework, Ganesan et al. [8] used the CPFE simulation to investigate the deformation mechanisms involved in the response of T5 and T6 conditions. Accordingly, they compared the contribution of each deformation mode for each of these heat treatment conditions of T5 and T6 during uniaxial tension and compression (Figure 26). They showed that the heat treatment condition alters the contribution of different deformation modes. A sample with a T5 condition has finer grains, which alter the CRSS of deformation modes. For instance, they obtained the ratio of $CRSS_{\text{prismatic}}/CRSS_{\text{pyramidal}\langle a\rangle} = 1.02$ in the T5 condition, while this ratio was 0.88 for the T6 condition. They stated that this was the underlying mechanism for more active pyramidal $\langle a\rangle$ in the T5 condition compared to the activity of prismatic, while pyramidal $\langle a\rangle$ slip mode was nearly inactive in the T6 condition. Finally, Ganesan et al. [8] used the CPFE simulation results to capture the Hall–Petch constant for different deformation modes in WE43 Mg alloy, which were 4.53, 11.33, 8.9, 11.18, 5.1 MPa $\times$ mm$^{1/2}$ for different deformation modes of basal, prismatic, pyramidal $\langle a\rangle$, pyramidal $\langle c + a\rangle$, and extension twinning, respectively.

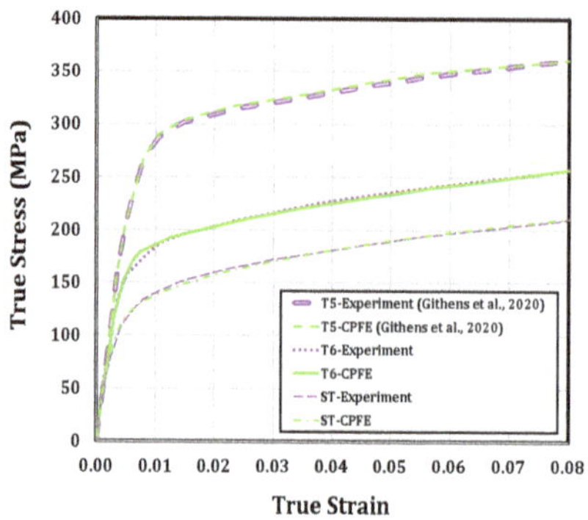

Figure 24. Comparison of the stress-strain curves of WE43 with different heat treatment conditions of ST, T5, and T6 subjected to uniaxial tension along the rolling direction: CPFE simulation versus the experimental results (After Ganesan et al. [8]).

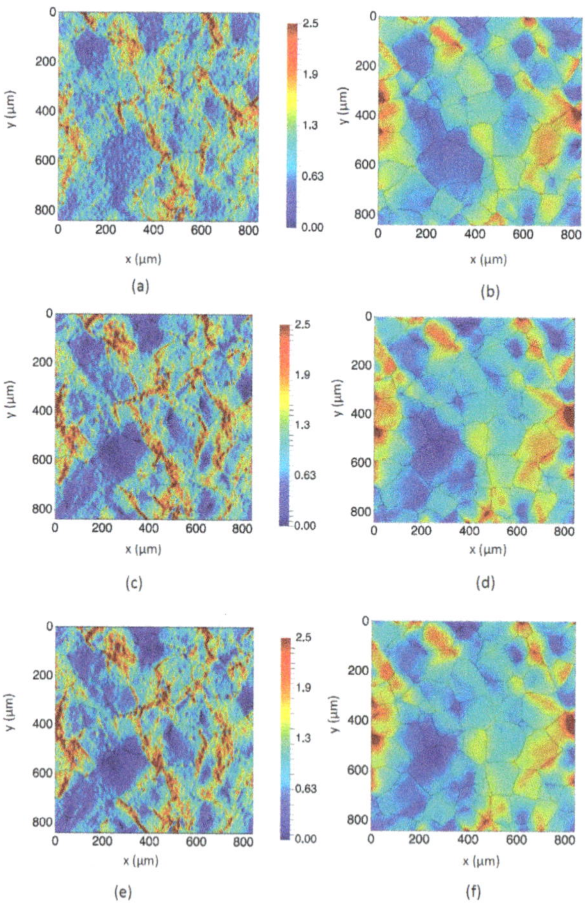

Figure 25. The strain maps obtained by the SEM-DIC experiment compared to the CPFE prediction in WE-43 T6 sample during uniaxial tension at different strain values: (**a**) $\varepsilon = 0.76\%$ (SEM-DIC), (**b**) $\varepsilon = 0.76\%$ (Simulation), (**c**) $\varepsilon = 4.83\%$ (SEM-DIC), (**d**) $\varepsilon = 4.83\%$ (Simulation), (**e**) $\varepsilon = 8.15\%$ (SEM-DIC), (**f**) $\varepsilon = 8.15\%$ (Simulation) (After Ganesan et al. [8]).

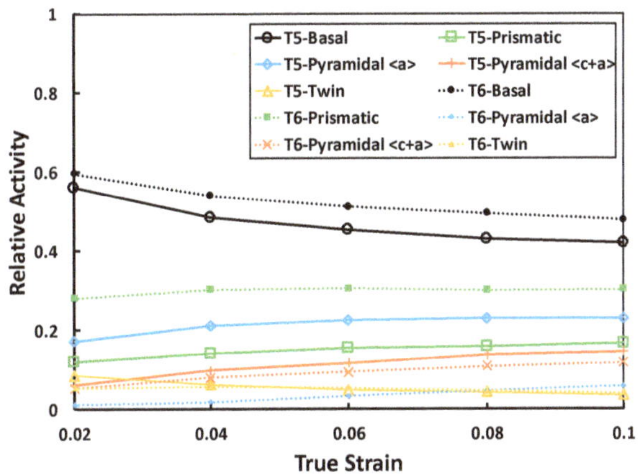

(a) Uniaxial tension

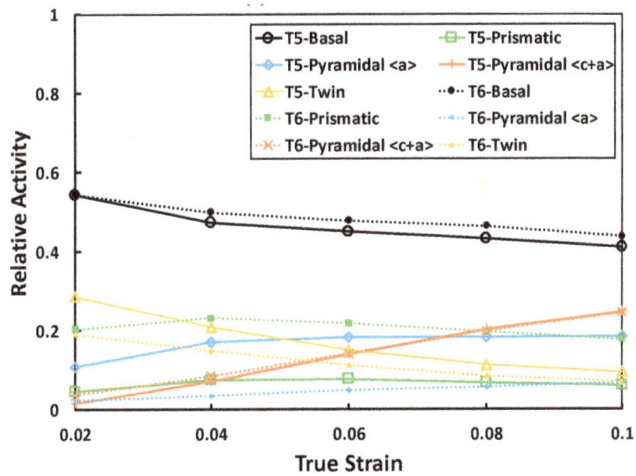

(b) Uniaxial compression

Figure 26. The comparison of deformation modes relative activity in T5 and T6 heat treatment conditions of WE43 Mg alloy during uniaxial loadings along the rolling direction obtained by the CPFE simulation: (**a**) Uniaxial tension, (**b**) uniaxial compression (After Ganesan et al. [8]).

3.2. *In-Situ Synchrotron X-ray Techniques*

3.2.1. Micromechanics of Twinning

Abdolvand et al. [27] used a combination of CPFE and in-situ 3DXRD experiments to investigate deformation twinning in the AZ31B Mg alloy. They applied the uniaxial tension on the sample with the texture favored for twinning during the loading and used the in-situ 3DXRD to study the microstructural evolution. They calibrated the CPFE model to capture the stress-strain response, as shown in Figure 27. They also studied the effect of twinning on texture. Accordingly, they measured the $\{0002\}$ pole figures at a tensile strain of $\varepsilon = 0\%$ and $\varepsilon = 0.4\%$. In the case of $\varepsilon = 0\%$, as shown in Figure 28a, basal poles were observed to be aligned almost parallel to the normal direction. However, in the case of $\varepsilon = 0.4\%$, as shown in Figure 28b, the deformation twinning reorients the $\{0002\}$ poles, which were observed towards the rolling direction and transverse direction on the outer edges of the pole figure.

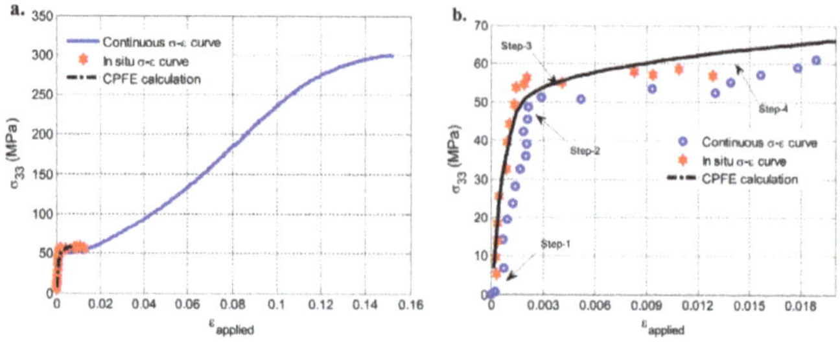

Figure 27. Stress-strain curves for AZ31B Mg alloy during uniaxial tension along normal direction obtained by continuous deformation, in-situ 3DXRD experiment, and CPFE simulation. The steps of 1 to 4 correspond to the strain values of $\varepsilon = 0\%, 0.2\%,\ 0.4\%$, and 1.4% at which the measurements of diffraction were performed. (**b**) is a magnified version of (**a**) in the strain range of $\varepsilon = 0\% - 2\%$ (After Abdolvand et al. [27]).

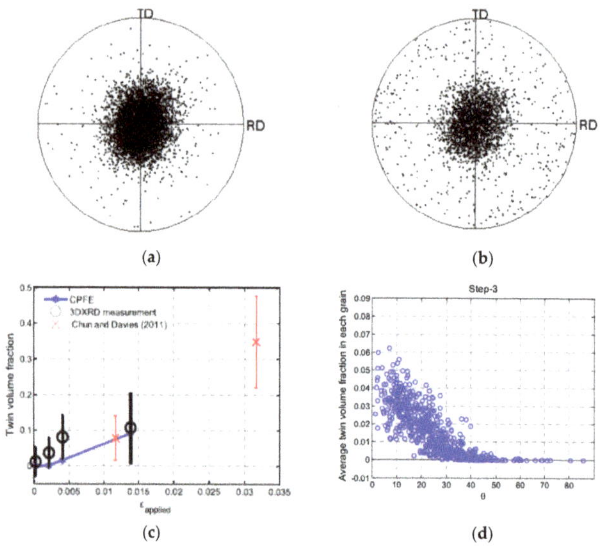

Figure 28. Effect of twinning on texture: (**a**) The $\{0002\}$ pole figures at a tensile strain of $\varepsilon = 0\%$ (Step-1), (**b**) the $\{0002\}$ pole figures at a tensile strain of $\varepsilon = 0.4\%$ (Step-3), (**c**) twin volume computed by CPFE versus the measured twin fraction, (**d**) variation of predicted twin volume versus misorientation between normal to the basal plane of the grain and loading direction predicted by CPFE at a tensile strain of $\varepsilon = 0.4\%$ (After Abdolvand et al. [27]).

Abdolvand et al. [27] also predicted the twin volume fraction using the CPFE simulation, which agrees well with the measured twin volume fraction (Figure 28c). They further investigated the CPFE simulation results by studying the variation of predicted twin volume versus misorientation angle (θ) between normal to the basal plane of the grain and the loading direction predicted by CPFE at a tensile strain of $\varepsilon = 0.4\%$ (Figure 28d). In the case of grains with $\theta > 40^{\circ}$, no twinning was predicted by CPFE. Furthermore, the scatter of twin volume for grains with similar θ showed that global θ was not the only governing factor of twinning volume as the effect of the local neighborhood and load sharing considerably varied the predicted twin volume fraction.

Another important subject Abdolvand et al. [27] addressed was the stress inside the twinned children and parent grains using the in-situ 3DXRD results (Figure 29). They selected 15 parent grains along 20 twinned children nucleated within those parent grains.

Figure 29i shows different components of the stress presented in the results summary. Different stress components were the normal stress σ_{33}, stress along the c-axis of the untwinned parent σ_{cp}, and normal and shear stress according to the twinned child plane, i.e., σ_n and τ_{rs}, respectively. The mean value of each stress was represented by a line. Abdolvand et al. [27] observed that the mean normal stress σ_{33} and σ_{cp} almost followed the applied stress trend. Because of the texture (c-axis is almost aligned with the loading direction), the values of σ_{33} and σ_{cp} were close. The values of σ_n in untwinned parents and twinned children grains were close, which were in the region of 23−30 MPa, which did not vary considerably during the loading. Interestingly, Abdolvand et al. [27] observed that the shear stress, which was resolved on the twinning plane, i.e., τ_{rs}, was considerably different in untwinned parent and twinned children grains. They reported that the average value of τ_{rs} in untwinned parent grains was 22 MPa higher than that of the twinned children grains. To further analyze the experimental results, Abdolvand et al. [27] focused on an untwinned parent grain with its two twinned children grains, as shown in Figure 29e–h, which were nucleated due to the activation of the twin variants 1 and 6. The twin variant 1 was nucleated at the normal stress of σ_n ~20 MPa, which was close to that of the parent, as shown in Figure 29a–d. The twin variant 6 appeared at the strain value of 0.4%, in which the shear stress τ_{rs} is ~19 MPa in the untwinned parent grain and ~ −10 MPa for the twinned grain. The results showed that the shear stress, which was resolved on the twinning plane, i.e., τ_{rs}, was smaller for both twin variants compared to the parent grains, which was in line with the results shown in Figure 29a–d.

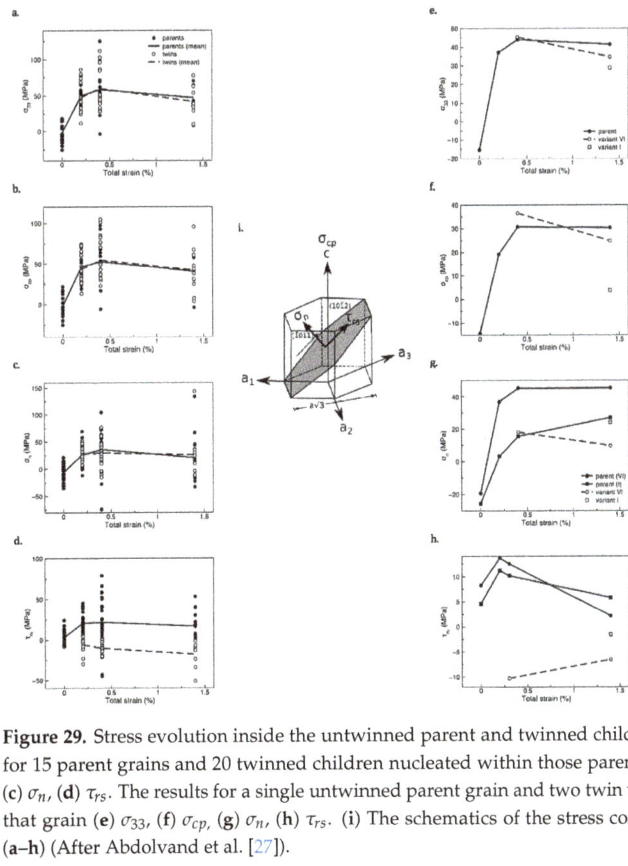

Figure 29. Stress evolution inside the untwinned parent and twinned children grains: The results for 15 parent grains and 20 twinned children nucleated within those parent grains (**a**) σ_{33}, (**b**) σ_{cp}, (**c**) σ_n, (**d**) τ_{rs}. The results for a single untwinned parent grain and two twin variants nucleated from that grain (**e**) σ_{33}, (**f**) σ_{cp}, (**g**) σ_n, (**h**) τ_{rs}. (**i**) The schematics of the stress components presented in (**a–h**) (After Abdolvand et al. [27]).

3.2.2. Detwinning

The deformation slip and twinning mechanisms were enough to define the response of the Mg alloys during simple monotonic loadings. However, in the case of complex loading paths, such as cyclic loadings, this was not the case. One should also define the detwinning mechanisms in these cases, as described in Section 2.3. In-situ Synchrotron X-ray techniques have contributed to investigate the detwinning mechanisms. In the case of Mg and its alloys, Murphy-Leonard et al. [7] incorporated the HEDM technique to investigate the twinning-detwinning in pure Mg. They applied cyclic loading with three different applied strains of 0.4%, 0.52%, and 0.75%. The sample was extruded, and loading was applied along the extrusion direction. The c-axis was mostly perpendicular to the extrusion direction. Accordingly, extension twinning was a favored mechanism during compression loading. To monitor the twin volume, Murphy-Leonard et al. [7] measured the X-Ray diffraction peaks of the basal $\{0\,0\,0\,2\}$ planes. In other words, initially, there was no basal $\{0\,0\,0\,2\}$ peak intensity along the loading direction. As a twinned child is nucleated and grows, a 86.3° reorientation of the basal pole occurs within a parent grain, and the peak intensity starts increasing. Accordingly, Murphy-Leonard et al. [7] introduced the formulation for the twin volume fraction ϕ based on the HEDM data as follows:

$$\phi = \frac{I_{LD}}{I^0_{ND}} \tag{16}$$

where I_{LD} is the basal $\{0\,0\,0\,2\}$ peak intensity along the loading direction, and I^0_{ND} is the initial basal $\{0\,0\,0\,2\}$ peak intensity along the normal direction.

Figure 30 shows the cyclic response of the sample at different loading cycles, including the stress-strain loops, basal $\{0\,0\,0\,2\}$ peak intensity along the loading direction, and basal $\{0\,0\,0\,2\}$ peak intensity along the normal direction. In cycle 1, Figure 30b shows that the basal $\{0\,0\,0\,2\}$ peak intensity along the loading direction remained zero during the tensile part of the loading, which means there no twin nucleation during the initial tensile loading. The basal $\{0\,0\,0\,2\}$ peak intensity along the loading direction started increasing at the C-1 point in compression and kept increasing until point B, which was the maximum compression loading. At the same time, at point C-1, Figure 30c shows that the basal $\{0\,0\,0\,2\}$ peak intensity along the normal direction started dropping. This was in fact, the reorientation of the c-axis along the normal direction towards the loading direction due to twinning.

As soon as unloading starts, the basal $\{0\,0\,0\,2\}$ peak intensity along the loading direction starts decreasing, while the basal $\{0\,0\,0\,2\}$ peak intensity along the normal direction starts increasing. In other words, the previously twinned children during compression loading were reoriented back to the parent grains. The twin exhaustion occurred at the point T-2 during the second loading cycle, as shown in Figure 30d–f. The variation of basal $\{0\,0\,0\,2\}$ peak intensity in the second cycle followed the same trend similar to the first cycle. However, the twinning initiates at smaller compressive strains C-2 compared to C-1 in the first cycle. In both cycles 1 and 2, the twin was fully exhausted in the tensile loading. However, as shown in Figure 30g–i, all the twinned children were not fully detwinned as the minimum twin volume fraction can be observed at the maximum tension strain A. This twin volume content was named residual twins. Murphy-Leonard et al. [7] showed that the residual twins increased by the number of cycles. One can compare the residual twins in loading cycle 500 at maximum tensile strain (point A) to that of loading cycle 200 to observe the increase in the residual twin content as the number of loading cycle increases.

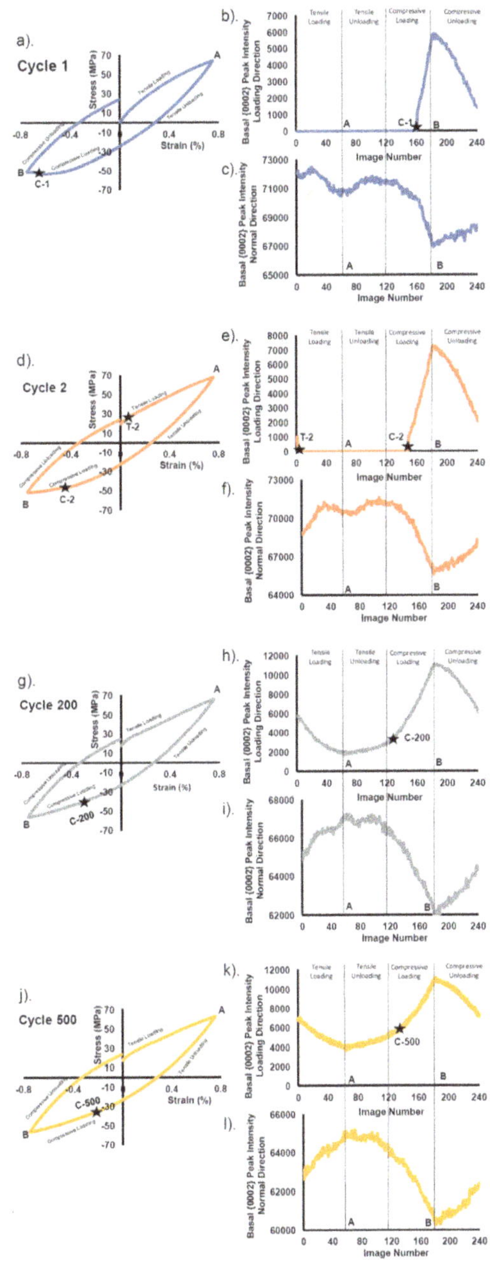

Figure 30. The cyclic response of pure Mg samples during the cyclic loading with a strain amplitude of 0.75% along the extrusion direction, including the stress-strain loops and variations of basal {0 0 0 2} peak intensity along with loading and normal directions during the cyclic loading (**a–c**) Cycle 1 (**d–f**) Cycle 2, (**g–i**) Cycle 200 (**j–l**) Cycle 500 (After Murphy-Leonard et al. [7]).

Murphy-Leonard et al. [7] also investigated the variation of twin initiation stress and detwinning exhaustion stress versus the number of cycles, as shown in Figure 31, in the case of cyclic loadings with the strain amplitude of 0.75% and 0.52%. The results showed that as the number of cycles increased, the magnitude of twinning initiation stress decreased while the magnitude of the twin exhaustion stress increased. They further investigated the results and analyzed the variation of twin intensity at maximum tensile strain, which can be inferred as the twin residuals, versus the number of cycles in the case of cyclic loadings with the strain amplitude of 0.75% and 0.52%, as shown in Figure 32. The results showed that before the loading cycle 100, there was no twinning intensity at maximum tensile strain independent of the strain amplitude, i.e., there were no residual twins. In the cases of loading cycles higher than 100, the residual twin content increased as the number of cycles increased. Furthermore, the residual twin content of samples subjected to the strain amplitude of 0.75% was larger than those subjected to the strain amplitude of 0.52%. In other words, the residual twin content increased by the strain amplitude. An open-source CPFE software PRISMS-Plasticity [103] was later used to capture the cyclic response of the sample, as shown in Figure 33. The results showed that the CPFE simulation can successfully capture the experimental cyclic stress-strain response.

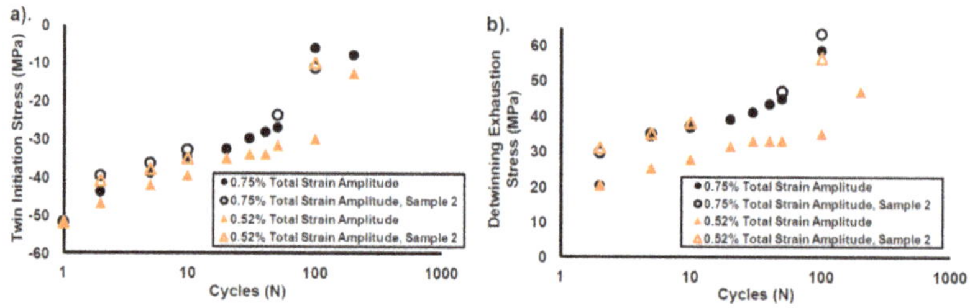

Figure 31. The variation of (a) twin initiation stress and (b) detwinning exhaustion stress versus the number of cycles in the case of cyclic loadings with the strain amplitude of 0.75% and 0.52% (After Murphy-Leonard et al. [7]).

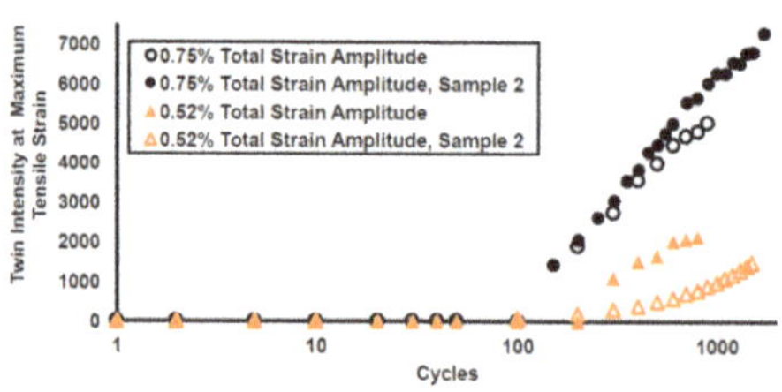

Figure 32. The variation of twin intensity at maximum tensile strain versus the number of cycles in the case of cyclic loadings with the strain amplitude of 0.75% and 0.52% (After Murphy-Leonard et al. [7]).

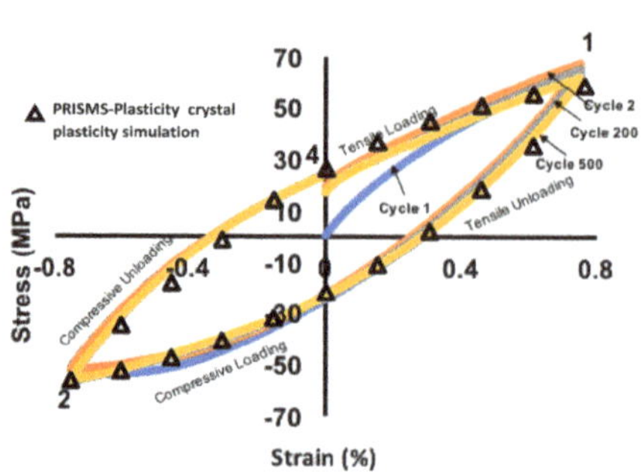

Figure 33. The stress-strain response of pure Mg sample during cyclic loading along the extrusion direction: CPFE simulation results for cycle 2 (Δ symbol) are compared versus the experimental results (After Aagesen et al. [108]).

3.3. In-Situ Neutron Diffraction

Micromechanics of Twinning

Brown et al. [34] incorporated the in-situ neutron diffraction to investigate the evolution of internal strain and texture during extension twinning in the AZ31B Mg alloy. They applied different loading paths that include in-plane compression (IPC), in-plane tension (IPT), and through-thickness compression (TTC), where the c-axis of most of the grains were perpendicular to the loading direction. Among these three loading paths, the extension twinning was the governing deformation mechanism in the IPC, which was the main focus of their study [34]. They used the VPSC framework to simulate the sample response during in-plane compression. Figure 34 presents the stress-strain responses of AZ31B Mg alloy during TTC and IPC. The circles on the IPC curve represent the strains at which the in-situ neutron diffraction measurements were conducted. The VPSC results were also presented as the dashed line for the IPC loading. The stress-strain response of IPC loading yields at the stress value of ∼60 MPa, which was followed by a plateau with a small hardening rate. This plateau is commonly attributed to the role of extension twinning [34]. The hardening rate starts increasing at the inflection point that occurs at the strain of ∼5%. For strain values larger than ∼8%, the stress during IPC becomes equal or larger than that of the TTC loading, and the rate of work hardening decreases again.

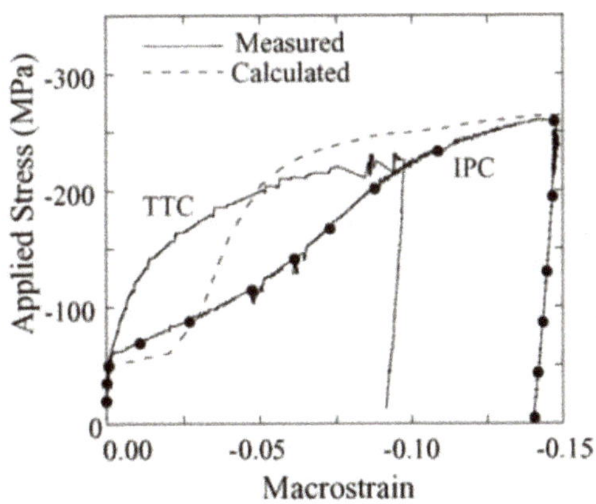

Figure 34. The stress-strain responses of AZ31B Mg alloy during thorough-thickness compression (TTC) and in-plane compression (IPC). The circles on IPC curve represent the strains at which the in-situ neutron diffraction measurements were conducted. The VPSC results are also presented as the dashed line for the IPC loading (After Brown et al. [34]).

Brown et al. [34] investigated the diffraction pattern of the AZ31B Mg alloy subjected to the IPC loading path in parallel and perpendicular detector banks at different stress values, as shown in Figure 35. Initially, the parallel detector banker does not detect any diffraction peaks from {0 0 0 2}, while the transverse detector bank shows very strong diffraction peaks from {0 0 0 2}. As the applied compression increases, the diffraction peaks intensity detected by the parallel detector banker increases, while the one detected by the transverse detector bank decreases. Brown et al. [34] stated that the increase in the diffraction peaks intensity of {0 0 0 2} detected by the parallel detector banker manifests the development of twinned grains. Accordingly, they measured the twinning variation and showed that the twinning initiates at the strain ∼1%. It was observed that the twinning volume linearly increases with the applied strain up to a strain ∼6%. The twinning is then saturated at the strain ∼8%, with the maximum twinning volume fraction of 80%. Brown et al. [34] mentioned that the variation of twin volume fraction is almost proportional to the increase in the diffraction peaks intensity of {0 0 0 2} detected by the parallel detector banker, with the slope of 0.145 per unit plastic strain. The VPSC predicted faster twin formation, which saturates at the strain ~4%. The increased hardening rate at larger strain observed in Figure 34 can be attributed to the twin saturation and initiation of pyramidal slip in the twinned regions [34].

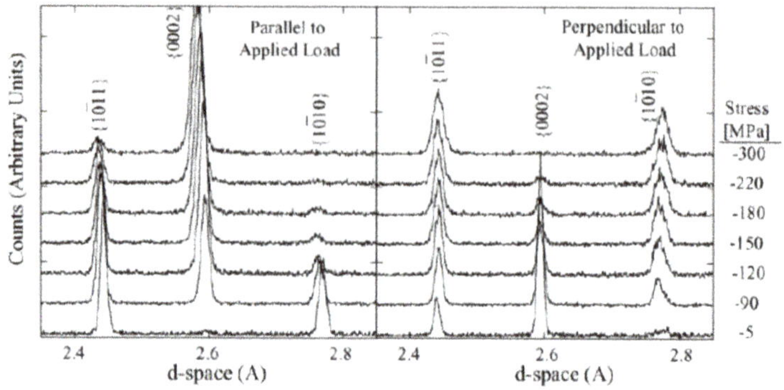

Figure 35. The diffraction pattern of the AZ31B Mg alloy subjected to in-plane compression loading in parallel and perpendicular detector banks at different stress values (After Brown et al. [34]).

Figure 36 shows the variation of lattice strain parallel and perpendicular to the loading axis in untwinned parent grains and daughter twinned grains versus the applied load [34]. Brown et al. [34] obtained the lattice strain for a grain orientation contributing to a given $(h\,k\,l)$ diffraction peak as follows:

$$\varepsilon = \frac{d^{hkl} - d_0^{hkl}}{d_0^{hkl}} \tag{17}$$

where d_0 denotes the unstrained interatomic spacing, which can be approximated by the initial interplanar spacing. Figure 36a shows the internal strain in the untwinned parent grain parallel and normal to the loading axis versus the applied load. The theoretical linear elastic relation is shown as dashed lines. Prior to the applied compressive stress of 70 MPa, the untwinned parent grains behave according to the linear elasticity. Afterward, the lattice strain becomes less than the elastic prediction as other grains with different orientations, including the daughter grains, which sustain stresses larger than average applied stress.

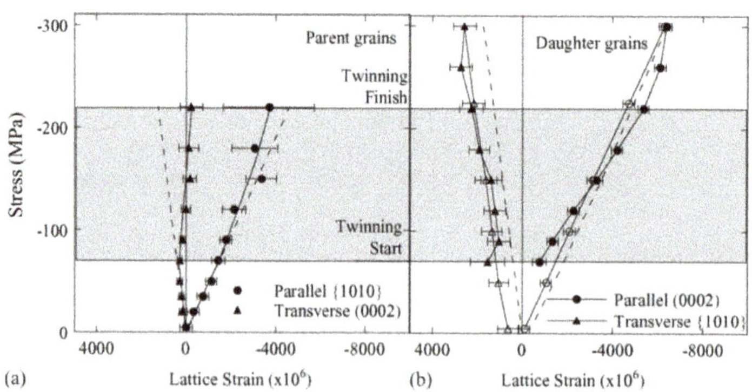

Figure 36. The variation of lattice strain parallel and perpendicular to the loading axis versus the applied load: (**a**) Untwinned Parent grains, (**b**) daughter twinned grains (After Brown et al. [34]). The dashed line represents the theoretical linear elastic relation. Closed markers denote the loading path, while the open markers represent the unloading path.

Figure 36b shows the internal strain in the daughter twinned grains parallel and normal to the loading axis versus the applied load. The lattice strain in daughter grains can be first identified at the compressive applied stress of 70 MPa, where twin reorientation occurs. Brown et al. [34] reported an interesting observation in which the lattice strain of newly twinned daughters was $-700\ \mu\varepsilon$, which was considerably less than the untwinned parent and the surrounding grains. In other words, the daughter twinned grains were in a relaxed state compared to the surrounding grains. In the region of applied compressive loading of 70 MPa − 220 MPa, the increase in lattice strain was much larger than the elastic prediction. The stress/lattice strain slope was reported as 32 GPa, which was smaller than the elastic slope of 48 GPa calculated for grains with basal pole parallel to the loading axis. This can be attributed to the fact that the daughter twinned grains were hard orientations and sustain more elastic strain compared to the untwinned parent and surrounding grains, which have soft orientations and deform inelastically [34]. For applied stress greater than 220 MPa, one can see an inflection point in the variation of lattice strain versus the applied loading as the variation of strain sustained by the daughter grain is less than the elastic prediction.

Brown et al. [34] calculated the CRSS required for twin formation using the in-situ neutron diffraction experimental data. They observed that the twin formation occurs at the stress in the region of 50 MPa − 70 MPa. In the case of the tested sample, the maximum Schmid factor is 0.499, which leads to the twinning CRSS of 25 to 35 MPa for AZ31B Mg alloy. In the last step, Brown et al. [34] tried to derive how much strain is sustained by twinning. They derived the contribution of twin to the total plastic strain as follows:

$$\frac{\Delta\varepsilon_{\text{twin}}}{\Delta\varepsilon_{\text{plastic}}} = 0.32S\frac{\Delta v}{\Delta\varepsilon_{\text{plastic}}} \tag{18}$$

where the coefficient 0.32 is obtained based on the texture of tested AZ31B Mg alloy, S is the characteristic twin shear strain defined in Equation (10), and v is the twin volume fraction. They estimated that 61% of the plastic deformation is sustained by twinning mechanisms at the strain of 5%. Afterward, the twinning rate versus the strain decreases in which it saturates at the strain of 8%. At this strain, 42% of the plastic deformation is sustained by the twinning mechanism.

4. Conclusions and Future Works

In the present study, the crystal plasticity models, which have been incorporated for Mg and its alloys, are reviewed. These models include different deformation mechanisms such as plastic slip, twinning, and detwinning. The recent experimental frameworks, such as in-situ neutron diffraction, 3D high energy synchrotron X-ray techniques, and digital image correlation under scanning electron microscopy, which have been incorporated along crystal plasticity models to investigate the properties of Mg and its alloys, are also reviewed. Although many studies have tried to model the behavior of Mg and its alloys using crystal plasticity, which was often coupled by the advanced in-situ techniques, there are still some critical challenges that need to be addressed. A summary is presented for the future challenges as below:

- Real-time crystal plasticity simulation coupled to in-situ experiments to guide identification of outliers that can in-turn improve crystal plasticity theories.
- A general map to include the effect of alloying for a variety of Mg alloys using crystal plasticity models along with synchrotron X-ray techniques in a consistent framework.
- Using machine learning techniques to learn the crystal plasticity models and generate surrogate models which can be used to design specific Mg alloy loading paths to achieve target properties.
- Developing a crystal plasticity model with a physically based twinning and detwinning model, which include the correct isotropic and kinematic hardenings to capture the appropriate cyclic response of Mg alloy. This is extremely important in the prediction of fatigue simulation using crystal plasticity simulation.

- Developing an integrated framework of crystal plasticity models and phase field simulation to better capture the twin morphology in Mg alloys.
- The interaction of slip modes and twinning and detwinning mechanisms, which is typically reflected as latent hardening in crystal plasticity models.
- Improved modeling of slip/twin and grain boundary interactions: Effects of grain size in different Mg alloys using crystal plasticity models, specifically via micro-Hall Petch models whose parameters can be inferred through experiments [98,102] and including the effect of grain boundary on twin nucleation and growth in crystal plasticity models.
- Integrating the crystal plasticity models of Mg and its alloys with the PRISMS-Fatigue framework [109] to investigate the effects of texture, grain morphology, sample size, multiaxial strain, and strain amplitude on their fatigue response.
- Coupling the crystal plasticity models with phase-field simulations to address the effect of deformation mechanisms such as plastic slip and twinning on the dynamic recrystallization of Mg alloys.

Author Contributions: Conceptualization, M.Y., G.Z.V., and V.S.; methodology, M.Y., G.Z.V., and V.S.; formal analysis, M.Y.; investigation, M.Y., G.Z.V., and V.S.; resources, G.Z.V.; data curation, M.Y.; writing—original draft preparation, M.Y.; writing—review and editing, M.Y., G.Z.V., and V.S.; visualization, M.Y.; supervision, G.Z.V. and V.S.; project administration, M.Y., G.Z.V., and V.S.; funding acquisition, G.Z.V., M.Y., and V.S. All authors have read and agreed to the published version of the manuscript.

Funding: The current work is partially funded by the NSF EPSCoR CIMM project under award #OIA-1541079. In addition, this work is partially supported by the U.S. Department of Energy, Office of Basic Energy Sciences, Division of Materials Sciences and Engineering under Award#DE-SC0008637 as part of the Center for Predictive Integrated Structural Materials Science (PRISMS Center) at the University of Michigan. We also acknowledge the financial cost-share support of the University of Michigan College of Engineering and Office of the Vice President for Research.

Acknowledgments: George Z. Voyiadjis acknowledges the funding from NSF EPSCoR CIMM project under award #OIA-1541079. Mohammadreza Yaghoobi and Veera Sundararaghavan acknowledge the funding from U.S. Department of Energy, Office of Basic Energy Sciences, Division of Materials Sciences and Engineering under Award#DE-SC0008637 as part of the Center for Predictive Integrated Structural Materials Science (PRISMS Center) at the University of Michigan. Mohammadreza Yaghoobi and Veera Sundararaghavan also acknowledge the financial cost-share support of the University of Michigan College of Engineering and Office of the Vice President for Research.

Conflicts of Interest: The authors declare no conflict of interest.

References

1. Cole, G.S. Summary of "Magnesium vision 2020: A North American automotive strategic vision for magnesium". In *Essential Readings in Magnesium Technology*; Springer: Cham, Switzerland, 2016; pp. 35–40.
2. Begum, S.; Chen, D.; Xu, S.; Luo, A.A. Low cycle fatigue properties of an extruded AZ31 magnesium alloy. *Int. J. Fatigue* **2009**, *31*, 726–735. [CrossRef]
3. Beyerlein, I.J.; Capolungo, L.; Marshall, P.E.; McCabe, R.J.; Tome, C.N. Statistical analyses of deformation twinning in magnesium. *Philos. Mag.* **2010**, *90*, 2161–2190. [CrossRef]
4. Mirza, F.A.; Chen, D.; Li, D.; Zeng, X. Low cycle fatigue of a rare-earth containing extruded magnesium alloy. *Mater. Sci. Eng. A* **2013**, *575*, 65–73. [CrossRef]
5. Mokdad, F.; Chen, D. Strain-controlled low cycle fatigue properties of a rare-earth containing ZEK100 magnesium alloy. *Mater. Des.* **2015**, *67*, 436–447. [CrossRef]
6. Yu, Q.; Jiang, Y.; Wang, J. Cyclic deformation and fatigue damage in single-crystal magnesium under fully reversed strain-controlled tension–compression in the [1 0$\bar{1}$0] direction. *Scr. Mater.* **2015**, *96*, 41–44. [CrossRef]
7. Murphy-Leonard, A.D.; Pagan, D.C.; Beaudoin, A.; Miller, M.P.; Allison, J.E. Quantification of cyclic twinning-detwinning behavior during low-cycle fatigue of pure magnesium using high energy X-ray diffraction. *Int. J. Fatigue* **2019**, *125*, 314–323. [CrossRef]
8. Ganesan, S.; Yaghoobi, M.; Githens, A.; Chen, Z.; Daly, S.; Allison, J.E.; Sundararaghavan, V. The effects of heat treatment on the response of WE43 Mg alloy: Crystal plasticity finite element simulation and SEM-DIC experiment. *Int. J. Plast.* **2021**, *137*, 102917. [CrossRef]

9. Yaghoobi, M.; Allison, J.E.; Sundararaghavan, V. Multiscale modeling of twinning and detwinning behavior of HCP polycrystals. *Int. J. Plast.* **2020**, *127*, 102653. [CrossRef]
10. Capolungo, L.; Beyerlein, I.J. Nucleation and stability of twins in hcp metals. *Phys. Rev. B* **2008**, *78*, 024117. [CrossRef]
11. Wang, J.; Hirth, J.; Tomé, C. (1¯012) Twinning nucleation mechanisms in hexagonal-close-packed crystals. *Acta Mater.* **2009**, *57*, 5521–5530. [CrossRef]
12. Beyerlein, I.J.; Tomé, C.N. A probabilistic twin nucleation model for HCP polycrystalline metals. *Proc. R. Soc. A Math. Phys. Eng. Sci.* **2010**, *466*, 2517–2544. [CrossRef]
13. Beyerlein, I.J.; McCabe, R.J.; Tomé, C.N. Effect of microstructure on the nucleation of deformation twins in polycrys-talline high-purity magnesium: A multi-scale modeling study. *J. Mech. Phys. Solids* **2011**, *59*, 988–1003. [CrossRef]
14. Zheng, S.; Beyerlein, I.; Wang, J.; Carpenter, J.; Han, W.; Mara, N. Deformation twinning mechanisms from bimetal interfaces as revealed by in situ straining in the TEM. *Acta Mater.* **2012**, *60*, 5858–5866. [CrossRef]
15. Barrett, C.D.; el Kadiri, H. The roles of grain boundary dislocations and disclinations in the nucleation of {1 0 1 2} twinning. *Acta Mater.* **2014**, *63*, 1–15. [CrossRef]
16. Beyerlein, I.J.; Kumar, M.A. The Stochastic Nature of Deformation Twinning: Application to HCP Materials. In *Handbook of Materials Modeling: Methods: Theory and Modeling*; Andreoni, W., Yip, S., Eds.; Springer International Publishing: Cham, Switzerland, 2018; pp. 1–39.
17. Wu, L.; Jain, A.; Brown, D.; Stoica, G.; Agnew, S.; Clausen, B.; Fielden, D.; Liaw, P. Twinning–detwinning behavior during the strain-controlled low-cycle fatigue testing of a wrought magnesium alloy, ZK60A. *Acta Mater.* **2008**, *56*, 688–695. [CrossRef]
18. Koike, J.; Fujiyama, N.; Ando, D.; Sutou, Y. Roles of deformation twinning and dislocation slip in the fatigue failure mechanism of AZ31 Mg alloys. *Scr. Mater.* **2010**, *63*, 747–750. [CrossRef]
19. Yu, Q.; Zhang, J.; Jiang, Y. Fatigue damage development in pure polycrystalline magnesium under cyclic tension-compression loading. *Mater. Sci. Eng. A* **2011**, *528*, 7816–7826. [CrossRef]
20. Zhang, J.; Yu, Q.; Jiang, Y.; Li, Q. An experimental study of cyclic deformation of extruded AZ61A magnesium alloy. *Int. J. Plast.* **2011**, *27*, 768–787. [CrossRef]
21. Wu, W.; Lee, S.Y.; Paradowska, A.M.; Gao, Y.; Liaw, P.K. Twinning–detwinning behavior during fatigue-crack propagation in a wrought magnesium alloy AZ31B. *Mater. Sci. Eng. A* **2012**, *556*, 278–286. [CrossRef]
22. Zhang, H.; Jérusalem, A.; Salvati, E.; Papadaki, C.; Fong, K.S.; Song, X.; Korsunsky, A.M. Multi-scale mechanisms of twinning-detwinning in magnesium alloy AZ31B simulated by crystal plasticity modeling and validated via in situ synchrotron XRD and in situ SEM-EBSD. *Int. J. Plast.* **2019**, *119*, 43–56. [CrossRef]
23. Aydıner, C.C.; Telemez, M.A. Multiscale deformation heterogeneity in twinning magnesium investigated with in situ image correlation. *Int. J. Plast.* **2014**, *56*, 203–218. [CrossRef]
24. Paudel, Y.; Indeck, J.; Hazeli, K.; Priddy, M.W.; Inal, K.; Rhee, H.; Barrett, C.D.; Whittington, W.R.; Limmer, K.R.; el Kadiri, H.; et al. Characterization and modeling of {101¯ 2} twin banding in magnesium. *Acta Mater.* **2020**, *183*, 438–451. [CrossRef]
25. Githens, A.; Ganesan, S.; Chen, Z.; Allison, J.; Sundararaghavan, V.; Daly, S. Characterizing microscale deformation mechanisms and macroscopic tensile properties of a high strength magnesium rare-earth alloy: A combined experimental and crystal plasticity approach. *Acta Mater.* **2020**, *186*, 77–94. [CrossRef]
26. Aydıner, C.C.; Bernier, J.V.; Clausen, B.; Lienert, U.; Tomé, C.N.; Brown, D.W. Evolution of stress in individual grains and twins in a magnesium alloy aggregate. *Phys. Rev. B* **2009**, *80*, 024113. [CrossRef]
27. Abdolvand, H.; Majkut, M.; Oddershede, J.; Schmidt, S.; Lienert, U.; Diak, B.J.; Withers, P.J.; Daymondd, M.R. On the deformation twinning of Mg AZ31B: A three-dimensional synchrotron X-ray diffraction experiment and crystal plasticity finite element model. *Int. J. Plast.* **2015**, *70*, 77–97. [CrossRef]
28. Lentz, M.; Klaus, M.; Wagner, M.; Fahrenson, C.; Beyerlein, I.J.; Zecevic, M.; Reimers, W.; Knezevic, M. Effect of age hardening on the deformation behavior of an Mg–Y–Nd alloy: In-situ X-ray diffraction and crystal plasticity modeling. *Mater. Sci. Eng. A* **2015**, *628*, 396–409. [CrossRef]
29. Kada, S.R.; Lynch, P.A.; Kimpton, J.A.; Barnett, M.R. In-situ X-ray diffraction studies of slip and twinning in the presence of precipitates in AZ91 alloy. *Acta Mater.* **2016**, *119*, 145–156. [CrossRef]
30. Greeley, D.; Yaghoobi, M.; Pagan, D.; Sundararaghavan, V.; Allison, J. Using synchrotron radiation to improve under-standing of deformation of polycrystalline metals by measuring, modelling and publishing 4D information. *Iop Conf. Ser. Mater. Sci. Eng.* **2019**, *580*, 012017. [CrossRef]
31. Li, L.; Yuanzhi, W.; Jie, W.; Xiao, L.C. In situ analysis of deformation twins within a magnesium polycrystal: (II) twin growth. *Micron* **2019**, *119*, 8–16. [CrossRef] [PubMed]
32. Zhang, N.; Zhang, Y.; Chen, S.; Li, Z.; Xie, H.; Lu, L.; Yao, X.; Luo, S. Onset of detwinning in Mg-3Al-1Zn alloy: A synchrotron-based X-ray diffraction study. *Scr. Mater.* **2021**, *190*, 113–117. [CrossRef]
33. Gharghouri, G.C.W.M.A.; Gharghouri, M.A.; Weatherly, G.C.; Embury, J.D.; Root, J. Study of the mechanical properties of Mg-7.7at.% Al by in-situ neutron diffraction. *Philos. Mag. A* **1999**, *79*, 1671–1695. [CrossRef]
34. Brown, D.; Agnew, S.; Bourke, M.; Holden, T.; Vogel, S.; Tomé, C. Internal strain and texture evolution during deformation twinning in magnesium. *Mater. Sci. Eng. A* **2005**, *399*, 1–12. [CrossRef]

35. Wu, L.; Agnew, S.; Brown, D.; Stoica, G.; Clausen, B.; Jain, A.; Fielden, D.; Liaw, P. Internal stress relaxation and load redistribution during the twinning–detwinning-dominated cyclic deformation of a wrought magnesium alloy, ZK60A. *Acta Mater.* **2008**, *56*, 3699–3707. [CrossRef]
36. Muránsky, O.; Carr, D.; Šittner, P.; Oliver, E. In situ neutron diffraction investigation of deformation twinning and pseudoelastic-like behaviour of extruded AZ31 magnesium alloy. *Int. J. Plast.* **2009**, *25*, 1107–1127. [CrossRef]
37. Wu, L.; Agnew, S.; Ren, Y.; Brown, D.; Clausen, B.; Stoica, G.; Wenk, H.; Liaw, P. The effects of texture and extension twinning on the low-cycle fatigue behavior of a rolled magnesium alloy, AZ31B. *Mater. Sci. Eng. A* **2010**, *527*, 7057–7067. [CrossRef]
38. Wu, W.; Qiao, H.; An, K.; Guo, X.; Wu, P.; Liaw, P.K. Investigation of deformation dynamics in a wrought magnesium alloy. *Int. J. Plast.* **2014**, *62*, 105–120. [CrossRef]
39. Wu, W.; Liaw, P.K.; An, K. Unraveling cyclic deformation mechanisms of a rolled magnesium alloy using in situ neutron diffraction. *Acta Mater.* **2015**, *85*, 343–353. [CrossRef]
40. Namakian, R.; Voyiadjis, G.Z. An atomic displacive model for 101¯21¯011 twinning in hexagonal close packed metals with the emphasis on the role of partial stacking faults in formation of {101¯2} twins. *Acta Mater.* **2018**, *150*, 381–393. [CrossRef]
41. Namakian, R.; Voyiadjis, G.Z.; Kwaśniak, P. On the slip and twinning mechanisms on first order pyramidal plane of magnesium: Molecular dynamics simulations and first principal studies. *Mater. Des.* **2020**, *191*, 108648. [CrossRef]
42. Liu, X.-Y.; Ohotnicky, P.P.; Adams, J.B.; Rohrer, C.L.; Hyland, R.W., Jr. Anisotropic surface segregation in Al − Mg alloys. *Surf. Sci.* **1997**, *373*, 357–370. [CrossRef]
43. Wu, Z.; Francis, M.F.; Curtin, W.A. Magnesium interatomic potential for simulating plasticity and fracture phenomena. *Model. Simul. Mater. Sci. Eng.* **2015**, *23*, 015004. [CrossRef]
44. Yasi, J.A.; Hector, L.G.; Trinkle, D.R. First-principles data for solid-solution strengthening of magnesium: From geometry and chemistry to properties. *Acta Mater.* **2010**, *58*, 5704–5713. [CrossRef]
45. Ghazisaeidi, M.; Hector, L.; Curtin, W. Solute strengthening of twinning dislocations in Mg alloys. *Acta Mater.* **2014**, *80*, 278–287. [CrossRef]
46. Olmsted, D.L.; HectorJr, L.G.; A Curtin, W.; Clifton, R.J. Atomistic simulations of dislocation mobility in Al, Ni and Al/Mg alloys. *Model. Simul. Mater. Sci. Eng.* **2005**, *13*, 371–388. [CrossRef]
47. Wang, J.; Hoagland, R.G.; Hirth, J.P.; Misra, A. Atomistic modeling of the interaction of glide dislocations with "weak" interfaces. *Acta Mater.* **2008**, *56*, 5685–5693. [CrossRef]
48. Yaghoobi, M.; Voyiadjis, G.Z. Effect of boundary conditions on the MD simulation of nanoindentation. *Comput. Mater. Sci.* **2014**, *95*, 626–636. [CrossRef]
49. Voyiadjis, G.Z.; Yaghoobi, M. Large scale atomistic simulation of size effects during nanoindentation: Dislocation length and hardness. *Mater. Sci. Eng. A* **2015**, *634*, 20–31. [CrossRef]
50. Yaghoobi, M.; Voyiadjis, G.Z. Atomistic simulation of size effects in single-crystalline metals of confined volumes during nanoindentation. *Comput. Mater. Sci.* **2016**, *111*, 64–73. [CrossRef]
51. Yaghoobi, M.; Voyiadjis, G.Z. Size effects in fcc crystals during the high rate compression test. *Acta Mater.* **2016**, *121*, 190–201. [CrossRef]
52. Voyiadjis, G.Z.; Yaghoobi, M. Size and strain rate effects in metallic samples of confined volumes: Dislocation length distribution. *Scr. Mater.* **2017**, *130*, 182–186. [CrossRef]
53. Yaghoobi, M.; Voyiadjis, G.Z. Microstructural investigation of the hardening mechanism in fcc crystals during high rate deformations. *Comput. Mater. Sci.* **2017**, *138*, 10–15. [CrossRef]
54. Yaghoobi, M.; Voyiadjis, G.Z. The effects of temperature and strain rate in fcc and bcc metals during extreme defor-mation rates. *Acta Mater.* **2018**, *151*, 1–10. [CrossRef]
55. Shahbeyk, S.; Voyiadjis, G.Z.; Habibi, V.; Astaneh, S.H.; Yaghoobi, M. Review of Size Effects during Micropillar Com-pression Test: Experiments and Atomistic Simulations. *Crystals* **2019**, *9*, 591. [CrossRef]
56. Saffarini, M.H.; Voyiadjis, G.Z.; Ruestes, C.J.; Yaghoobi, M. Ligament size dependency of strain hardening and ductil-ity in nanoporous gold. *Comput. Mater. Sci.* **2021**, *186*, 109920. [CrossRef]
57. Fan, H.; Aubry, S.; Arsenlis, A.; El-Awady, J.A. The role of twinning deformation on the hardening response of poly-crystalline magnesium from discrete dislocation dynamics simulations. *Acta Mater.* **2015**, *92*, 126–139. [CrossRef]
58. Wei, D.; Zaiser, M.; Feng, Z.; Kang, G.; Fan, H.; Zhang, X. Effects of twin boundary orientation on plasticity of bicrystalline copper micropillars: A discrete dislocation dynamics simulation study. *Acta Mater.* **2019**, *176*, 289–296. [CrossRef]
59. Voyiadjis, G.Z.; Yaghoobi, M. *Size Effects in Plasticity: From Macro to Nano*; Academic Press: Cambridge, MA, USA, 2019.
60. Houtte, P. Simulation of the rolling and shear texture of brass by the Taylor theory adapted for mechanical twinning. *Acta Met.* **1978**, *26*, 591–604. [CrossRef]
61. Tomé, C.; Lebensohn, R.; Kocks, U. A model for texture development dominated by deformation twinning: Application to zirconium alloys. *Acta Met. Mater.* **1991**, *39*, 2667–2680. [CrossRef]
62. Kalidindi, S.R. Incorporation of deformation twinning in crystal plasticity models. *J. Mech. Phys. Solids* **1998**, *46*, 267–290. [CrossRef]
63. Staroselsky, A.; Anand, L. Inelastic deformation of polycrystalline face centered cubic materials by slip and twinning. *J. Mech. Phys. Solids* **1998**, *46*, 671–673. [CrossRef]

64. Staroselsky, A.; Anand, L. A constitutive model for hcp materials deforming by slip and twinning: Application to magnesium alloy AZ31B. *Int. J. Plast.* **2003**, *19*, 1843–1864. [CrossRef]
65. Abdolvand, H.; Daymond, M.R.; Mareau, C. Incorporation of twinning into a crystal plasticity finite element model: Evolution of lattice strains and texture in Zircaloy-2. *Int. J. Plast.* **2011**, *27*, 1721–1738. [CrossRef]
66. Fernández, A.; Prado, M.T.P.; Wei, Y.; Jérusalem, A. Continuum modeling of the response of a Mg alloy AZ31 rolled sheet during uniaxial deformation. *Int. J. Plast.* **2011**, *27*, 1739–1757. [CrossRef]
67. Zhang, J.; Joshi, S.P. Phenomenological crystal plasticity modeling and detailed micromechanical investigations of pure magnesium. *J. Mech. Phys. Solids* **2012**, *60*, 945–972. [CrossRef]
68. Fernández, A.; Jérusalem, A.; Gutiérrez-Urrutia, I.; Pérez-Prado, M.T. Three-dimensional investigation of grain boundary-twin interactions in a Mg AZ31 alloy by electron backscatter diffraction and continuum modeling. *Acta Mater.* **2013**, *61*, 7679–7692. [CrossRef]
69. Liu, Y.; Wei, Y. A polycrystal based numerical investigation on the temperature dependence of slip resistance and texture evolution in magnesium alloy AZ31B. *Int. J. Plast.* **2014**, *55*, 80–93. [CrossRef]
70. Abdolvand, H.; Daymond, M.R. Internal strain and texture development during twinning: Comparing neutron dif-fraction measurements with crystal plasticity finite-element approaches. *Acta Mater.* **2012**, *60*, 2240–2248. [CrossRef]
71. Abdolvand, H.; Daymond, M.R. Multi-scale modeling and experimental study of twin inception and propagation in hexagonal close-packed materials using a crystal plasticity finite element approach; part II: Local behavior. *J. Mech. Phys. Solids* **2013**, *61*, 803–818. [CrossRef]
72. Abdolvand, H.; Majkut, M.; Oddershede, J.; Wright, J.P.; Daymond, M.R. Study of 3-D stress development in parent and twin pairs of a hexagonal close-packed polycrystal: Part II—Crystal plasticity finite element modeling. *Acta Mater.* **2015**, *93*, 235–245. [CrossRef]
73. Qiao, H.; Barnett, M.; Wu, P. Modeling of twin formation, propagation and growth in a Mg single crystal based on crystal plasticity finite element method. *Int. J. Plast.* **2016**, *86*, 70–92. [CrossRef]
74. Hama, T.; Tanaka, Y.; Uratani, M.; Takuda, H. Deformation behavior upon two-step loading in a magnesium alloy sheet. *Int. J. Plast.* **2016**, *82*, 283–304. [CrossRef]
75. Prasad, N.S.; Narasimhan, R.; Suwas, S. Effects of lattice orientation and crack tip constraint on ductile fracture initiation in Mg single crystals. *Int. J. Plast.* **2017**, *97*, 222–245. [CrossRef]
76. Lebensohn, R.; Tomé, C. A self-consistent anisotropic approach for the simulation of plastic deformation and texture development of polycrystals: Application to zirconium alloys. *Acta Met. Mater.* **1993**, *41*, 2611–2624. [CrossRef]
77. Agnew, S.; Yoo, M.; Tomé, C. Application of texture simulation to understanding mechanical behavior of Mg and solid solution alloys containing Li or Y. *Acta Mater.* **2001**, *49*, 4277–4289. [CrossRef]
78. Kumar, M.A.; Kanjarla, A.K.; Niezgoda, S.R.; Lebensohn, R.A.; Tomé, C.N. Numerical study of the stress state of a deformation twin in magnesium. *Acta Mater.* **2015**, *84*, 349–358. [CrossRef]
79. Lévesque, J.; Mohammadi, M.; Mishra, R.K.; Inal, K. An extended Taylor model to simulate localized deformation phenomena in magnesium alloys. *Int. J. Plast.* **2016**, *78*, 203–222. [CrossRef]
80. Segurado, J.; Lebensohn, R.A.; Llorca, J.; Tomé, C.N. Multiscale modeling of plasticity based on embedding the viscoplastic selfconsistent formulation in implicit finite elements. *Int. J. Plast.* **2012**, *28*, 124–140. [CrossRef]
81. Knezevic, M.; McCabe, R.J.; Lebensohn, R.A.; Tomé, C.N.; Liu, C.; Lovato, M.L.; Mihaila, B. Integration of self-consistent polycrystal plasticity with dislocation density based hardening laws within an implicit finite element framework: Application to low-symmetry metals. *J. Mech. Phys. Solids* **2013**, *61*, 2034–2046. [CrossRef]
82. Prakash, A.; Nöhring, W.G.; Lebensohn, R.A.; Höppel, H.W.; Bitzek, E. A multiscale simulation framework of the accumulative roll bonding process accounting for texture evolution. *Mater. Sci. Eng. A* **2015**, *631*, 104–119. [CrossRef]
83. Ardeljan, M.; Beyerlein, I.J.; McWilliams, B.A.; Knezevic, M. Strain rate and temperature sensitive multi-level crystal plasticity model for large plastic deformation behavior: Application to AZ31 magnesium alloy. *Int. J. Plast.* **2016**, *83*, 90–109. [CrossRef]
84. Feather, W.G.; Ghorbanpour, S.; Savage, D.J.; Ardeljan, M.; Jahedi, M.; McWilliams, B.A.; Gupta, N.; Xiang, C.; Vogel, S.C.; Knezevic, M. Mechanical response, twinning, and texture evolution of WE43 magnesium-rare earth alloy as a function of strain rate: Experiments and multi-level crystal plasticity modeling. *Int. J. Plast.* **2019**, *120*, 180–204. [CrossRef]
85. Kondo, R.; Tadano, Y.; Shizawa, K. A phase-field model of twinning and detwinning coupled with dislocation-based crystal plasticity for HCP metals. *Comput. Mater. Sci.* **2014**, *95*, 672–683. [CrossRef]
86. Liu, C.; Shanthraj, P.; Diehl, M.; Roters, F.; Dong, S.; Dong, J.; Ding, W.; Raabe, D. An integrated crystal plasticity–phase field model for spatially resolved twin nucleation, propagation, and growth in hexagonal materials. *Int. J. Plast.* **2018**, *106*, 203–227. [CrossRef]
87. Proust, G.; Tomé, C.; Kaschner, G. Modeling texture, twinning and hardening evolution during deformation of hexagonal materials. *Acta Mater.* **2007**, *55*, 2137–2148. [CrossRef]
88. Proust, G.; Tomé, C.N.; Jain, A.; Agnew, S.R. Modeling the effect of twinning and detwinning during strain-path changes of magnesium alloy AZ31. *Int. J. Plast.* **2009**, *25*, 861–880. [CrossRef]
89. Guillemer, C.; Clavel, M.; Cailletaud, G. Cyclic behavior of extruded magnesium: Experimental, microstructural and numerical approach. *Int. J. Plast.* **2011**, *27*, 2068–2084. [CrossRef]

90. Wang, H.; Wu, P.; Tomé, C.; Wang, J. A constitutive model of twinning and detwinning for hexagonal close packed polycrystals. *Mater. Sci. Eng. A* **2012**, *555*, 93–98. [CrossRef]
91. Wang, H.; Wu, P.; Wang, J. Modeling inelastic behavior of magnesium alloys during cyclic loading–unloading. *Int. J. Plast.* **2013**, *47*, 49–64. [CrossRef]
92. Wang, H.; Wu, P.D.; Wang, J.; Tomé, C.N. A crystal plasticity model for hexagonal close packed (HCP) crystals including twinning and de-twinning mechanisms. *Int. J. Plast.* **2013**, *49*, 36–52. [CrossRef]
93. Qiao, H.; Agnew, S.; Wu, P. Modeling twinning and detwinning behavior of Mg alloy ZK60A during monotonic and cyclic loading. *Int. J. Plast.* **2015**, *65*, 61–84. [CrossRef]
94. Hama, T.; Takuda, H. Crystal plasticity finite-element simulation of deformation behavior in a magnesium alloy sheet considering detwinning. *Steel Res. Int.* **2012**, 1115–1118.
95. Hama, T.; Kitamura, N.; Takuda, H. Effect of twinning and detwinning on inelastic behavior during unloading in a magnesium alloy sheet. *Mater. Sci. Eng. A* **2013**, *583*, 232–241. [CrossRef]
96. Hama, T.; Kobuki, A.; Takuda, H. Crystal-plasticity finite-element analysis of anisotropic deformation behavior in a commercially pure titanium Grade 1 sheet. *Int. J. Plast.* **2017**, *91*, 77–108. [CrossRef]
97. Briffod, F.; Shiraiwa, T.; Enoki, M. Numerical investigation of the influence of twinning/detwinning on fatigue crack initiation in AZ31 magnesium alloy. *Mater. Sci. Eng. A* **2019**, *753*, 79–90. [CrossRef]
98. Andani, M.T.; Lakshmanan, A.; Karamooz-Ravari, M.; Sundararaghavan, V.; Allison, J.; Misra, A. A quantitative study of stress fields ahead of a slip band blocked by a grain boundary in unalloyed magnesium. *Sci. Rep.* **2020**, *10*, 1–8. [CrossRef] [PubMed]
99. Armstrong, R.; Codd, I.; Douthwaite, R.M.; Petch, N.J. The plastic deformation of polycrystalline aggregates. *Philos. Mag.* **1962**, *7*, 45–58. [CrossRef]
100. Weng, G. A micromechanical theory of grain-size dependence in metal plasticity. *J. Mech. Phys. Solids* **1983**, *31*, 193–203. [CrossRef]
101. Sun, S.; Sundararaghavan, V. A probabilistic crystal plasticity model for modeling grain shape effects based on slip geometry. *Acta Mater.* **2012**, *60*, 5233–5244. [CrossRef]
102. Andani, M.T.; Lakshmanan, A.; Sundararaghavan, V.; Allison, J.; Misra, A. Quantitative study of the effect of grain boundary parameters on the slip system level Hall-Petch slope for basal slip system in Mg-4Al. *Acta Mater.* **2020**, *200*, 148–161. [CrossRef]
103. Yaghoobi, M.; Ganesan, S.; Sundar, S.; Lakshmanan, A.; Rudraraju, S.; Allison, J.E.; Sundararaghavan, V. PRISMS-Plasticity: An open-source crystal plasticity finite element software. *Comput. Mater. Sci.* **2019**, *169*, 109078. [CrossRef]
104. Christian, J.; Mahajan, S. Deformation twinning. *Prog. Mater. Sci.* **1995**, *39*, 1–157. [CrossRef]
105. Wu, P.D.; Guo, X.Q.; Qiao, H.; Lloyd, D.J. A constitutive model of twin nucleation, propagation and growth in magnesium crystals. *Mater. Sci. Eng. A* **2015**, *625*, 140–145. [CrossRef]
106. Lou, X.; Li, M.; Boger, R.; Agnew, S.; Wagoner, R. Hardening evolution of AZ31B Mg sheet. *Int. J. Plast.* **2007**, *23*, 44–86. [CrossRef]
107. Wu, L. Mechanical Behavior and the Role of Deformation Twinning in Wrought Magnesium Alloys Investigated Using Neutron and Synchrotron X-ray Diffraction. Ph.D. Thesis, University of Tennessee, Knoxville, TN, USA, 2009.
108. Aagesen, L.K.; Adams, J.F.; Allison, J.E.; Andrews, W.B.; Araullo-Peters, V.; Berman, T.; Chen, Z.; Daly, S.; Das, S.; de Wittet, S.; et al. PRISMS: An Integrated, Open-Source Framework for Accelerating Predictive Structural Materials Science. *JOM* **2018**, *70*, 2298–2314. [CrossRef]
109. Yaghoobi, M.; Stopka, K.S.; Lakshmanan, A.; Sundararaghavan, V.; Allison, J.E.; McDowell, D.L. PRISMS-Fatigue computational framework for fatigue analysis in polycrystalline metals and alloys. *Npj Comput. Mater.* **2021**, *7*, 1–12. [CrossRef]

crystals

Article

Influence of Non-Metallic Inclusions on Local Deformation and Damage Behavior of Modified 16MnCrS5 Steel

Faisal Qayyum [1,*], Muhammad Umar [1,2], Vladislav Elagin [1], Markus Kirschner [1], Frank Hoffmann [1], Sergey Guk [1] and Ulrich Prahl [1]

1 Institut für Metallformung, Technische Universität Bergakademie Freiberg, 09599 Freiberg, Germany; umar.matrix@gmail.com (M.U.); vladislav.elagin@student.tu-freiberg.de (V.E.); markus.kirschner@imf.tu-freiberg.de (M.K.); frank.hoffmann@imf.tu-freiberg.de (F.H.); sergey.guk@imf.tu-freiberg.de (S.G.); ulrich.prahl@imf.tu-freiberg.de (U.P.)

2 Department of Mechanical Engineering, Khwaja Fareed University of Engineering and Information Technology, Rahim Yar Khan 64200, Pakistan

* Correspondence: faisal.qayyum@imf.tu-freiberg.de; Tel.: +49-373-139-4073

Abstract: This work investigates a ferrite matrix with multiple non-metallic inclusions to evaluate their influence on the global and local deformation and damage behavior of modified 16MnCrS5 steel. For this purpose, appropriate specimens are prepared and polished. The EBSD technique is used to record local phase and orientation data, then analyze and identify the size and type of inclusions present in the material. The EBSD data are then used to run full phase crystal plasticity simulations using DAMASK-calibrated material model parameters. The qualitative and quantitative analysis of these full phase simulations provides a detailed insight into how the distribution of non-metallic inclusions within the ferrite matrix affects the local stress, strain, and damage behavior. In situ tensile tests are carried out on specially prepared miniature dog-bone-shaped notched specimens in ZEISS Gemini 450 scanning electron microscope with a Kammrath and Weiss tensile test stage. By adopting an optimized scheme, tensile tests are carried out, and local images around one large and several small MnS inclusions are taken at incremental strain values. These images are then processed using VEDDAC, a digital image correlation-based microstrain measurement tool. The damage initiation around several inclusions is recorded during the in situ tensile tests, and damage initiation, propagation, and strain localization are analyzed. The experimental results validate the simulation outcomes, providing deeper insight into the experimentally observed trends.

Keywords: damage mechanics; crystal plasticity; numerical simulation; local deformation behavior; in situ tensile test; VEDDAC; DAMASK; digital image correlation; non-metallic inclusions

Citation: Qayyum, F.; Umar, M.; Elagin, V.; Kirschner, M.; Hoffmann, F.; Guk, S.; Prahl, U. Influence of Non-Metallic Inclusions on Local Deformation and Damage Behavior of Modified 16MnCrS5 Steel. *Crystals* **2022**, *12*, 281. https://doi.org/10.3390/cryst12020281

Academic Editor: Mingyi Zheng

Received: 31 January 2022
Accepted: 15 February 2022
Published: 18 February 2022

1. Introduction

The steel industry is playing an important role in providing about 45% of the raw material to the automotive sector [1]. The advancement in the steel parts used in an automobile is inevitable to competitively improve the strength to weight ratio, fuel economy, and crash safety of cars. Researchers have studied various multi-phase steels to increase the ultimate tensile strength (UTS) and percentage elongation of steels using different thermo-mechanical methods and phase combinations [2,3].

The ever-increasing demand for lightweight materials for the mobility sector has promoted the utilization of the multi-phase phenomenon in steel [4]. For instance, quenched and partitioned (Q&P) steels have shown a UTS up to 1.4 GPa with a uniform elongation of up to 20% [5]. This remarkable combination is desirable during steel formation into sheets and during individual component making with application-based mechanical strength capabilities. Small-sized hard phase inclusions added into the soft ferrite matrix during steel making act as reinforcement participants [6–8]. The common inclusions in the ferrite matrix—i.e., alumina (Al_2O_3), cementite (Fe_3C), manganese sulfide (MnS), and pyrite (Fe_2S)—are very small in size, in the micron range [9–11]. Alloy steel thus formed with this microstructure, called gear steel, has good machinability and improved hardenability [12,13]. Post-carburizing-quenching case-hardened parts are used in an automobile where a core tensile strength of 800–1100 MPa is needed. Their good wear resistance and low-temperature impact toughness make them usable for piston bolts, camshaft levers, gears, worms, seals, and other sleeve parts [14].

The particle size, distribution, and morphology have a great effect on the overall mechanical properties of these steels [15,16]. MnS with better plasticity and major inclusions in desulphurized alloy steels affects the mechanical anisotropy after being rolled into elongated strips [6]. The deoxidation of the molten steel produces Al_2O_3; being hard, it can act to initiate microcracks and as a propagation source during the application of a load on the component [17]. Moreover, Al_2O_3 can also cause excessive wear of the tool during the machining of the component. Therefore, understanding, controlling, and utilizing the limiting effects of these two inclusions can assist in the material forming processes and producing application-based materials [18–20].

A crystal plasticity-based microstructural approach for investigating material behavior is comparatively more accurate than empirical and phenomenological studies [21–23]. The large-scale crystal plasticity finite element method (CPFEM), with multiple calculation points in a single orientation, can evaluate the average mechanical response of a polycrystalline material. An added advantage of this type of modeling and simulation is the analysis of the local stress and strain behavior to mitigate the concentrations and prolong the global elongation process [24,25].

Although large-scale CPFEM methods are considered important for the true behavior recognition of multi-phase materials, the literature has limited results that can be applied satisfactorily to the mass production of parts [26–31]. Recently, DP steel (martensite in a ferrite matrix) has been studied using the relaxed grain cluster homogenization technique using the Düsseldorf Advanced Material Simulation Kit (DAMASK) [32–34]. The results gave an interesting insight into microstructural evolution during the deformation of heterogeneous multi-phase materials. However, there is still a need for extensive study based on individual phase constituents in the steel matrix to reach a point where industrially required process maps can be generated and utilized for microstructurally informed material production.

This study is a unique combination of various techniques to evaluate the micromechanical response of polycrystalline multi-phase material. Industrially produced modified 16MnCrS5 with alumina, pyrite, and cementite as precipitate phase particles spread all over the ferrite matrix has been used as a starting material. An EBSD map has been generated, and in situ tensile deformation has been performed while recording the local material state at various global stress values. A crystal plasticity-based numerical simulation model built on finite strain theory has been used to study slip-based plastic deformation in a ferrite matrix. The correlation of simulation results with the experimental obtained local material behavior is studied and presented.

The manuscript has been categorized so that Section 1 presents the introduction and background of the study. Section 2 has very brief information about the experimental procedures. The modeling technique with the current numerical simulation procedure with boundary conditions is mentioned in Section 3, while the results are detailed in Section 4. Finally, Section 5 gives a complete discussion of the results compared to the already published literature, while the study's conclusions are presented in Section 6.

2. Methodology

In this work, in situ tensile tests were carried out on specially prepared specimens inside the SEM chamber in a deformation stage. The local EBSD data from the same specimen were used to run a full phase crystal, plasticity-based numerical simulation model. The material's local deformation and damage behavior, especially around an MnS inclusion, was analyzed and compared with the numerical simulation results. This section contains a complete methodology of the in situ tensile tests, EBSD data collection, and data processing that was carried out.

2.1. In Situ Test Setup and Data Collection

For in situ investigations, a specialized method was adopted based on previously published work by other researchers [35–39]. For in situ testing, tensile specimens according to DIN EN ISO 6892-1 with a length of 50 mm and a thickness of 0.6 mm with notches for strain concentration were prepared as shown in Figure 1c. The samples were cut from the discs using a water jet cutting machine to avoid material wastage and residual stress accumulation on the surfaces during conventional machining. The prepared samples were finely milled to obtain a better edge finish and produce small u-shaped notches in the middle of the neck, as shown in Figure 1c. They were prepared by metallographic polishing. Since the specimens were very thin, a special embedding compound was used during the polishing process that could be dissolved in acetone when heated to 30 °C after the polishing process was complete. As the test specimens' deformed surface was later to be analyzed with digital image correlation software, the test specimens were etched for 5 s in a 3% Nital solution to achieve the required surface contrast. The visual inspection of specimens before testing revealed that only non-deformed inclusions were found on the surface of the specimens, and a few MnS inclusions with an elongated shape were present in the middle of the specimen. Further details about specimen preparation and the achievement of the correct contrast of micrographs using image processing techniques are provided in Appendix A for interested readers.

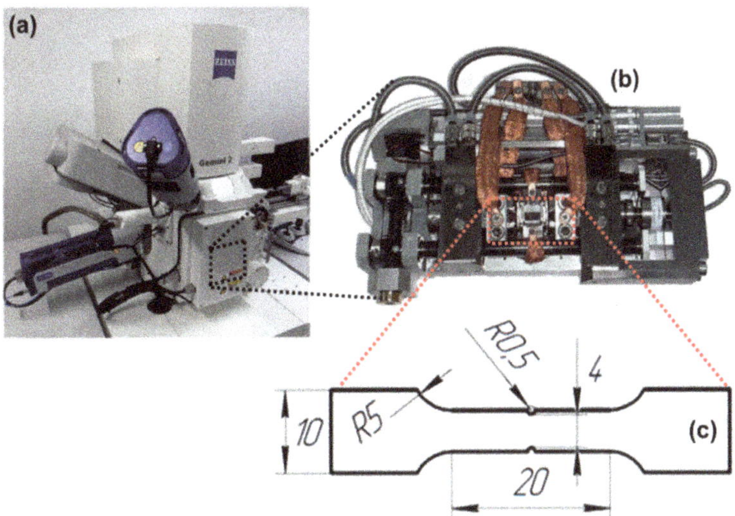

Figure 1. (**a**) ZEISS Gemini SEM 450 scanning electron microscope used (**b**) and the in situ tensile model. (**c**) Geometry of the in situ specimens with notches for the stress concentration, where all the dimensions are in mm.

The overall chemical composition of the current material is shown in Table 1 in comparison with the standard chemical composition of the alloy. It is modified 16MnCrS5 with 50% lower carbon content and 20% more sulfur. Therefore, the strength of this modified material is slightly lower; due to higher sulfur content and less carbon, it contains more MnS inclusions, with some being predominantly large (as discussed later in Sections 2.3 and 2.4).

Table 1. Chemical composition of the investigated modified 16MnCrS5 steel in weight percentage in comparison with standard non-modified 16MnCrS5 steel.

	Element	C	Si	Mn	P	S	Cr	Al	Cu
% Wt.	non-modified	0.16	0.25	1.15	<0.01	0.035	1.00	<0.01	0.03
	modified	0.081	0.038	1.07	<0.01	0.131	1.06	<0.01	0.03

The in situ tensile tests were carried out with a ZEISS Gemini SEM 450 scanning electron microscope using a Kammrath and Weiss tensile test stage, as shown in Figure 1a,b. This module allows in situ data collection during tensile or compressive loading using SE, EBSD, and EDS in the Gemini SEM450 of Carl Zeiss AG. For the experiments, different points on the tensile stress–strain curve were targeted at a forming speed of 8 μm/s, and the specimen was held at various values of stress and strain. At the same time, analyses of the microstructure and chemistry were performed. A series of images was taken during tensile testing. All images were taken with a resolution of 2048 × 1536 pixels. In addition, all photos in the area of interest were taken with a magnification of 11k and WD (working distance) of 16.6 mm with the "InLens" module.

2.2. Selection of Area for Full Phase Simulations

For the crystal structure analysis, the specimens were placed in the Gemini SEM450 from Carl Zeiss AG. The EBSD analysis was performed with an accelerating voltage of 20 kV, a working distance of 11 mm, and a specimen holder pre-tilted by 70°. The magnification was 10,000×. For the EBSD analysis, the Symmetry S2 detector from Oxford Instruments PLC was used. The analysis area was 95.4 μm × 71.8 μm, with a step size of 0.2 μm. The advanced Symmetry S2 EBSD detector utilizes a unique combination of speed (4500 pixels/second), sensitivity (CMOS technology sensor), and diffraction pattern details (1244 × 1024 pixel resolution) to efficiently produce high-quality EBSD data for the selected area, which can then be post-processed to obtain information in the desired format. Modified 16MnCrS5 steel contains inclusions that are quantitatively analyzed and recoded using EDS Ultim MaxX from Oxford Instruments PLC.

The multi-phase steel material Representative Volume Element (RVE) was obtained after EBSD analysis. The data were initially raw and required cleaning for unindexed points due to crystallographic noise at the grain boundaries. This noise was reduced using an intelligent algorithm implemented on the MTEX toolbox with MATLAB [40]. The procedural details of the EBSD data cleaning and the developed algorithm are published elsewhere [41]. Therefore, readers are encouraged to refer to that work for further details.

The ferrite matrix distribution is shown in Figure 2a, and the distributions of alumina (Al_2O_3), cementite (Fe_3C), and manganese sulfide (MnS) concerning the grain boundaries are displayed in Figure 2b. Most of the inclusions detected in this area of the specimens are present on the grain boundaries. This kind of microstructure with appropriately large second phase particles distributed homogeneously on the grain boundaries of adequately large ferrite grains appears to be due to the adopted manufacturing strategy.

2.3. Statistical Analysis of the EBSD Data

A quantitative comparison of the grain shape (aspect ratio) and particle size of the ferrite matrix and inclusions is presented in Figures 3 and 4, respectively. In Figure 3, the aspect ratio of 1 corresponds to almost perfectly round shapes, whereas the higher aspect ratio corresponds to the adequately elongated grain or inclusion. Considering this criterion, the aspect ratio of ferrite grains in Figure 3a predominantly lies between 1–2, except for some grains with a high degree of elongation of >4, which is the normal ferrite microstructure reported in previous literature [42,43]. On the other hand, the inclusions in Figure 3b–d are largely round (aspect ratio = 1), with some particles slightly elongated. There can be two reasons why the manufacturing process yields such inclusion shapes, as reported previously [44], or it can be due to the limitations of the measurement and magnification chosen in this work. During EBSD analysis, only these slightly large inclusions with round profiles were detected due to the chosen magnification and the set step size. In contrast, very thin and long inclusions were missed.

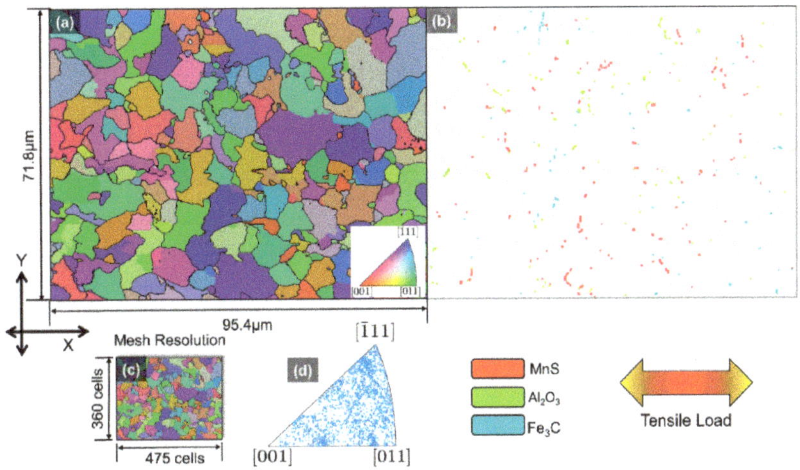

Figure 2. Multi-phase steel specimen. (**a**) IPF for ferrite matrix with BCC crystal structure, with different colors showing different crystallographic orientations. (**b**) Three types of inclusions spread all over the ferrite matrix and (**c**) the mesh resolution showing the grid size for calculation points. (**d**) IPF figure showing orientations of ferrite grain orientations with reference directions.

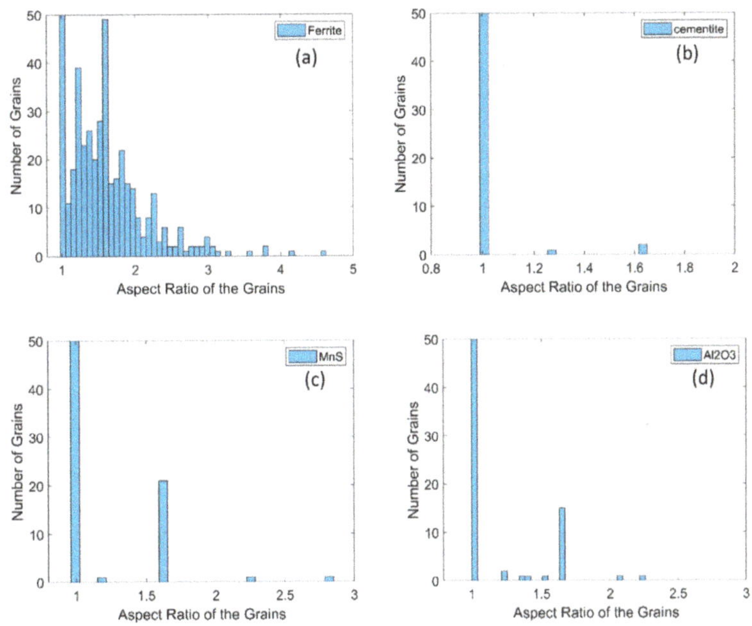

Figure 3. Distribution of aspect ratios of (**a**) ferrite grains, which are relatively larger and have a wider spread; (**b**) cementite inclusions, which are close to 1, with few inclusions at 1.3 and 1.7; (**c**) MnS inclusions are also usually round, with a few with a spread of 2.8; (**d**) Al_2O_3 inclusions are also usually round with an aspect ratio of 1, with some inclusions reaching up to 2.3.

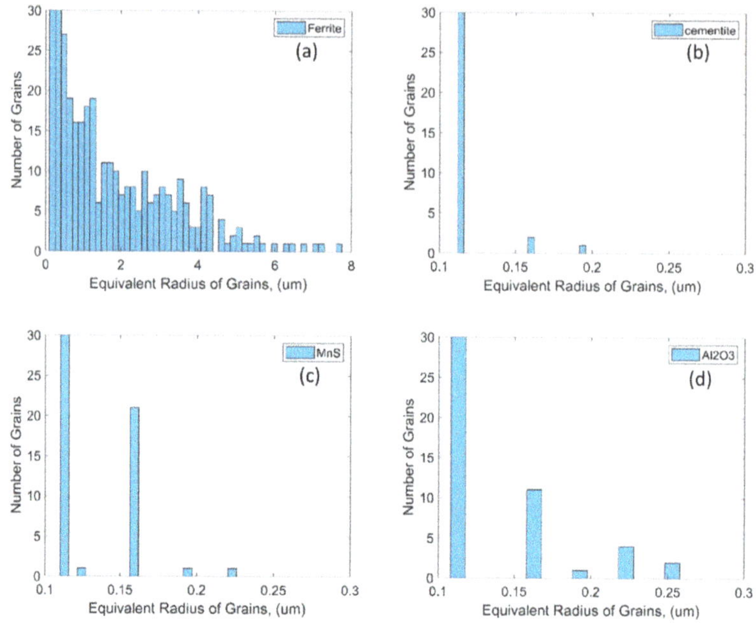

Figure 4. Distribution of grain size in terms of equivalent radius (ER) for (**a**) ferrite, which is biased towards small grains but being widespread reaches up to 8 µm; (**b**) cementite inclusions, which are generally small with sizes close to 0.1 µm, with a few inclusions up to 0.18 µm ER; (**c**) MnS inclusions, which although usually small are in a large number with 0.16 µm ER and a few as large as 0.23 µm. (**d**) The Al_2O_3 inclusions have a large spread with many having sizes close to 0.1 µm and then having a downslope spread up to 0.25 µm ER.

The grain size in Figure 4 is measured in terms of the mean sphere diameter. A significant number of ferrite grains in the selected area for EBSD analysis are below 4 µm, as observed in Figure 4a, and some grains are larger than 6 µm. As shown in Figure 4b–d, the inclusions are comparatively very small. Most inclusions are around 0.1 µm in size, whereas very few are larger than 0.2 µm. Even smaller inclusions may be present within the microstructure but were not detected due to the selected magnification and EBSD measurement step size. This lack of smaller inclusions negligibly alters the material chemistry or local phase distribution and hence is assumed to have almost no effect on the simulation results.

Second phase inclusions have been talked about a great deal in this section. The elemental analysis of an MnS inclusion using EDS analysis is presented in Figure 5 for reference. All the other inclusions—i.e., Al_2O_3 and Fe_3C—have also been identified using the same process.

2.4. Selection of Area, Tools Used, and Methodology Adopted for In Situ Strain Measurement

During the in situ tensile tests, a large, elongated sulfide inclusion (MnS) was analyzed. The inclusion, along with the chemical analysis confirming it to be MnS, is shown in Table 2. The length of the inclusion together with the cavity is about 42.5 µm. The elongated inclusion aligned with the tensile loading axis (horizontal to the shown micrograph) was expected to be prone to severe fracture and was therefore of great interest.

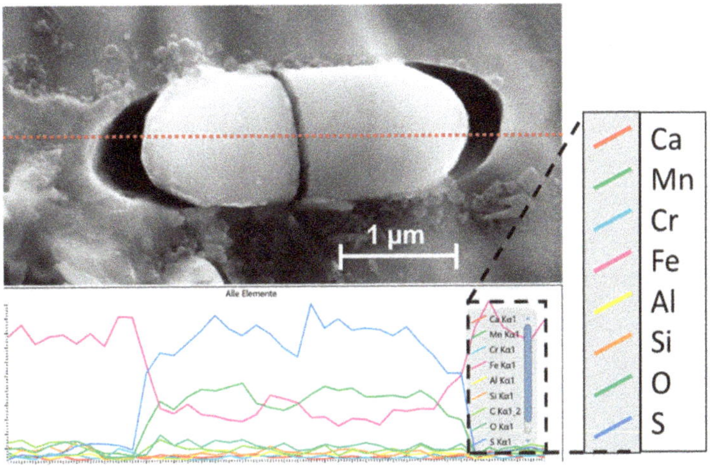

Figure 5. Elemental analysis of the modified 16MnCrS5 steel specimen using EDS to identify the inclusions (the red dotted line represents the line of measurement). The inclusion shows a major concentration of Mn and S; therefore, it is classified as an MnS inclusion.

Table 2. Chemical analysis of inclusion at different points, confirming MnS. The units of the measurement are in mass, %.

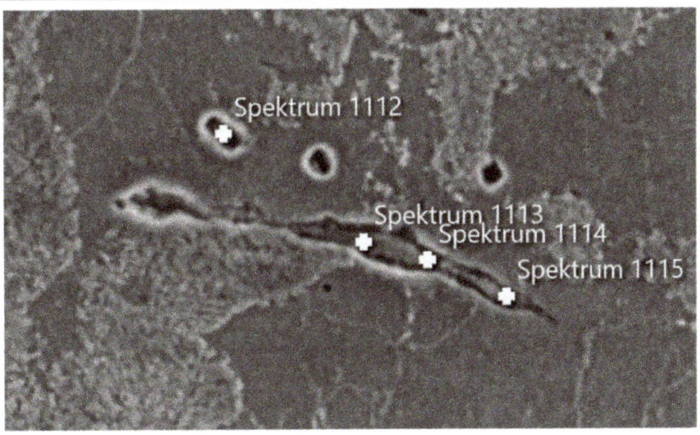

Element	Spectrum 1112	Spectrum 1113	Spectrum 1114	Spectrum 1115
S	10.50	6.46	10.34	14.65
Cr	1.19	1.70	0.95	0.95
Mn	17.06	11.61	17.24	23.48
Fe	69.56	75.98	70.66	60.16
Others	1.69	4.25	0.81	0.76
Total	100.00	100.00	100.00	100.00

The in situ tests come with their own challenge of image acquisition [35–37]. The images taken during the test should have a specific contrast and clear feature identification possibility [45]. To ensure that the images taken during the current tests are of acceptable quality and grayscale distribution, an analysis was performed that has not been included in the main part of the body but is presented in Appendix A. The effect of different settings and attributes is discussed and can be interesting for readers interested in performing such tests themselves. Appendix A also contains detailed information about the local and global strain measurement during experimentation.

The digital image correlation software VEDDAC from Chemnitzer Werkstoffmechanik GmbH was used to process the captured image series. When processing the images, images with strong contrast or sharpness fluctuations were removed from the calculations, and the final set contained images. Two calculation areas were selected during DIC. The first calculation area is shown in Figure 6a, which consists of the matrix around the inclusions and the cavities. The second calculation is the inclusion area and cavities around it, as shown in Figure 6b. The optimal grid sizes for a particular series of images were 9 × 9 pixels (coarse) for the matrix and 5 × 5 pixels for the inclusion (fine). More information about the local strain measurement methodology is provided in Appendix B.

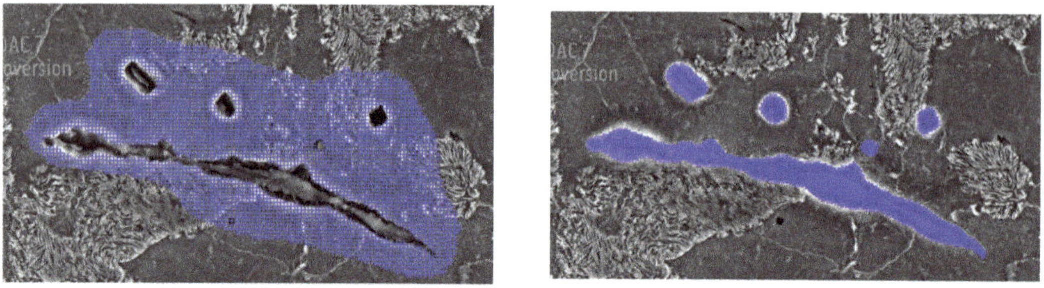

(**a**) (**b**)

Figure 6. Calculation network for digital image correlation (**a**) around non-metallic inclusions (**b**) in non-metallic inclusions and cavity area. In this work, a grid of 9 × 9 pixels (coarse) for the matrix and 5 × 5 pixels for the inclusion (fine) was chosen to get accurate and fully resolved results.

3. Numerical Simulation Model Setup

The EBSD data and the information of each phase's chemical and crystallographic structure were used to model and run a full phase crystal plasticity-based numerical model with calibrated material models. These simulations were used to provide detailed information about the local deformation and damage behavior of this material and how different inclusions, morphologies, and distributions affect the local deformation and damage behaviors. This insight is useful to understand the data from limited in situ tests.

First developed by Hutchinson [46] and later extended by Kalidindi [47], the phenomenological model implemented in the DAMASK framework [33] calculates the plastic deformation by slip planes based on the initial and saturated slip resistance—S_0, S_s, respectively (refer to Table 3). This resistance value for slip systems = 1,2,3 … N_{slip} increases from the initial to saturation value depending on the crystal structure specifications and the critical shear value on the slip plane during deformation.

This model uses a mathematical description to correlate the initial deformation gradient (F) with the resulting first Piola–Kirchoff stress tensor (P), as shown in Equations (1) and (2). The elastic part is simplified in the form of the generalized Hook's law, and for the plastic part, it is given with the help of the plastic velocity gradient given in Equation (3).

$$\dot{\overline{F}}_{ij} = \begin{bmatrix} 1 & 0 & 0 \\ 0 & * & 0 \\ 0 & 0 & * \end{bmatrix} \times 10^{-3}\text{s}^{-1}$$
$$\overline{P}_{ij} = \begin{bmatrix} * & * & * \\ * & 0 & * \\ * & * & 0 \end{bmatrix} Pa \quad (1)$$

$$S = CE^{e} \quad (2)$$

where

S = second Piola–Kirchoff stress tensor;

C = fourth-order elastic stiffness Tensor;
E^e = second-order Langrangian strain tensor.

$$L_p = \sum_{\alpha=1}^{\mathrm{Nslip}} \dot{\gamma}^{\alpha} \left(\mathrm{m}^{\alpha} \otimes \mathrm{n}^{\alpha}\right) \tag{3}$$

The value of the velocity plasticity gradient L_p is dependent on the shear strain γ (Equation (4)), which is a function of the resolved shear stress τ^{α} and slip-resistance on the α slip plane S^{α}.

$$\dot{\gamma}^{\alpha} = \dot{\gamma}_0 \left|\frac{\tau^{\alpha}}{S^{\alpha}}\right|^{n} \mathrm{sgn}(\tau^{\alpha}) \tag{4}$$

and

$\alpha = 1, 2, 3 \cdots, N_{slip}$ (For BCC Ferrite N_{slip} = 24)

A non-conserved damage field φ is governed by the continuous release of the stored mechanical energy density from undamaged to fully damaged conditions in a region; i.e., $\varphi = 1$ to $\varphi = 0$. In the ductile damage criterion implemented in DAMASK, plastic energy dissipation at a material point is the driving force here for damage flux f_{φ}. Therefore, plastic energy dissipation in the form of the following equation is used:

$$w_{plastic} = \frac{1}{2}\varphi^2 \int M_p.L_p dt. \tag{5}$$

Consequently, the minimization of the total free energy density is responsible for force to drive damage, and it is given as follows:

$$f_{\varphi} = -\left[\frac{\partial w_{plastic}}{\partial \varphi} + \frac{\partial w_{surface}}{\partial \varphi}\right] = \frac{G}{l_c} - \varphi \int M_p.L_p dt. \tag{6}$$

where

G = surface tension of the newly generated damage surface;
l_c = length scale of the diffused surface;
L_p = plastic velocity gradient.

The plastic governing law changes to the following when damage evolution is coupled with dissipated plastic energy:

$$L_p = f\left(\frac{M_p}{\varphi^2}\right) \tag{7}$$

Readers are encouraged to read the pioneering work by the developers of DAMASK for further details and understanding of the model [33,34,48,49]. For instance, a strategy was published recently by Qayyum et al. [50] by choosing an optimal RVE that behaves isotopically yet is small enough to produce fast results for the calibration of the material model parameters of single-phase materials. The strategy comprises a 10 × 10 × 10 size RVE containing 1000 grains, where each point represents a different grain. The RVE adopted from that published methodology is shown in Figure 7a. This RVE was used to calibrate the fitting parameters in the material model by comparing simulation outcomes with the experimental observations, and the results are shown in Figure 7b.

It is observed that the simulation results match well with the experimental observations globally. The dips in the experimental data are positions where the test was stopped and pictorial data were recorded. The adopted and calibrated material model parameters used in this study are shown in Table 3. This set of parameters was used to run full phase simulations and then analyze the local material deformation and damage behavior.

The DAMASK framework is designed to get all the inputs in the form of specifically structured text files with a complete microstructural description of RVE with geometrical details, material behavior attributes, and boundary values. Elastic and plastic phase parameters for ferrite and only elastic parameters of all the inclusions are defined to observe the micromechanical deformation of ferrite caused by stiffness degradation. Ductile damage criteria have been incorporated in the ferrite matrix as well. The RVE generated by using the EBSD data shown in Figure 2 was subjected to quasi-static tensile load with a strain rate of 10^{-3} s^{-1} in the x-direction.

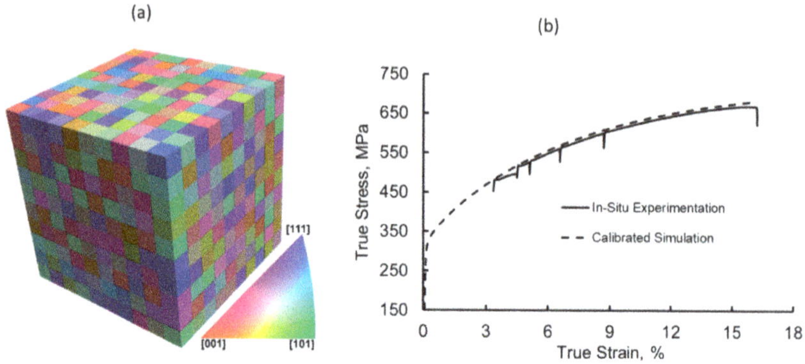

Figure 7. (**a**) 10 × 10 × 10 size RVE containing 1000 individual grain orientations, (**b**) comparison of experimental and simulation flow curves, showing a good match of results.

Table 3. Elastic parameters of second phase inclusions obtained from the literature [10,20,23,26,28,41,51,52] and calibrated parameters of ferrite adopted from the literature [23,26,28,41,52] and adjusted by comparing with the flow curve from in situ tests. The ductile damage parameters were calibrated by comparing them with the in situ test results.

	Elastic Parameters of All the Phases	
Parameter	**Value**	**Unit**
Ferrite-C_{11},C_{12},C_{44}	233.3, 235.5, 128.0	GPa
Fe_3C-C_{11},C_{12},C_{44} [17]	375.0, 161.0, 130.0	GPa
MnS-C_{11},C_{12},C_{44} [6]	177.3, 117.0, 25.2	GPa
Al_2O_3-C_{11},C_{12},C_{44} [10]	496, 109, 206 [18]	GPa
	Plastic Parameters of Ferrite Phase	
$\dot{\gamma}_0$	5.6×10^{-4}	ms^{-1}
S_0, [111]	95	MPa
S_s, [111]	222	MPa
S_0, [112]	96	MPa
S_s, [112]	412	MPa
h_o	1	GPa
$h_{\alpha\beta}$	1.0	GPa
n, w	3, 2.0	-
N_{slip}	12, 12	-
N_{twin}	0	-
	Ductile Damage Parameters	
Parameter Description	**Value for Ferrite**	**Unit**
Interface energy (g_0)	1.0	Jm^{-2}
Damage mobility coefficient (M)	0.001	s^{-1}
Critical plastic strain (ε_{crit})	0.5	-
Damage rate sensitivity coefficient (P)	10	-
Damage diffusion (D)	1.0	-

4. Results

4.1. Global Results of Numerical Simulation

The global results of the simulation in comparison with experimental observations are shown in Figure 8. The global results are calculated by taking an average of the results at each solution point for each increment. The data are combined in a meaningful way to produce observable trends. The stress and damage evolution with respect to the true global strain in the selected RVE is shown here in Figure 8.

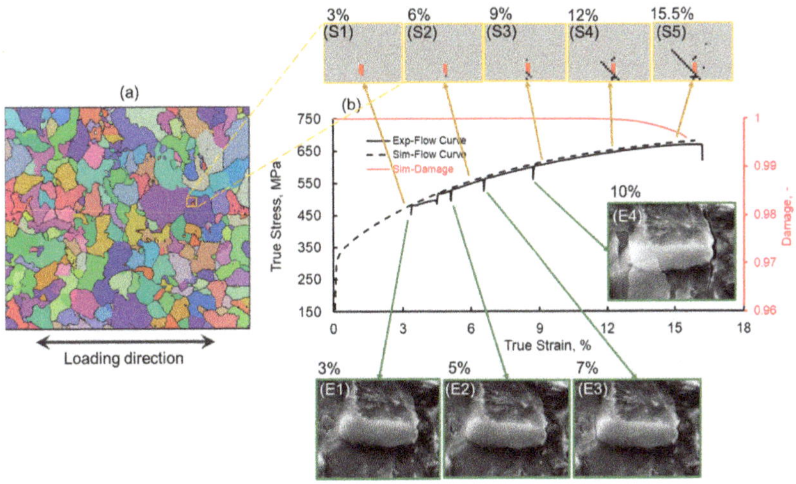

Figure 8. (**a**) RVE chosen to run full phase simulations in IPF colors; yellow box represents one of the crack initiation zones. (**b**) Outcome of the simulation flow curve in comparison with the in situ experimental test results, showing a good match; the dips in the experimental results are the points where the test was stopped to take local pictures of the inclusions for later DIC. E1–E4 show the damage evolution around an MnS inclusion, S1–S5 inset figures show the damage evolution around a comparable MnS inclusion during numerical simulation.

A good correlation between global experimental and numerical simulation results is observed with less than 3% difference of predictions. The inset figures in Figure 8 from S1–S5 show damage evolution around an MnS inclusion. Inset figures E1–E4 show damage evolution around a similar-sized MnS inclusion captured during in situ testing. Interestingly, not only do the global results match well, but the local damage evolution is also comparable. These local deformation and damage trends are discussed later in this article.

4.2. Local Results of Numerical Simulation

The local distributions of strains, stress, triaxiality, and damage at 3%, 9%, and 15.5% of global strain are shown in Figure 9. Due to the heterogeneous distribution of non-metallic inclusions and different Schmidt factors of each ferrite grain, the local distributions of each attribute are largely heterogeneous. Therefore, the grain boundaries have been overlaid in contrasting colors to clarify the distributions of attributes within individual grains.

It is observed that at 3% global strain, the local strain distribution is relatively homogeneous within the selected RVE. As the global strain increases and reaches a maximum of 15.5%, an extreme heterogeneous distribution of strain along the grain boundaries in thick channels aligned at 45 degrees to the loading axis is observed with intermediate areas of very low local strain. On the other hand, the stress distribution is high from the beginning in the ferrite grains with a high Schmidt factor and increases significantly within those grains. The value remains similar as the global strain reaches the maximum value.

Of course, the local stress and strain distribution is largely affected by the presence and distribution of non-metallic inclusions. The behavior is discussed in more detail in the next section.

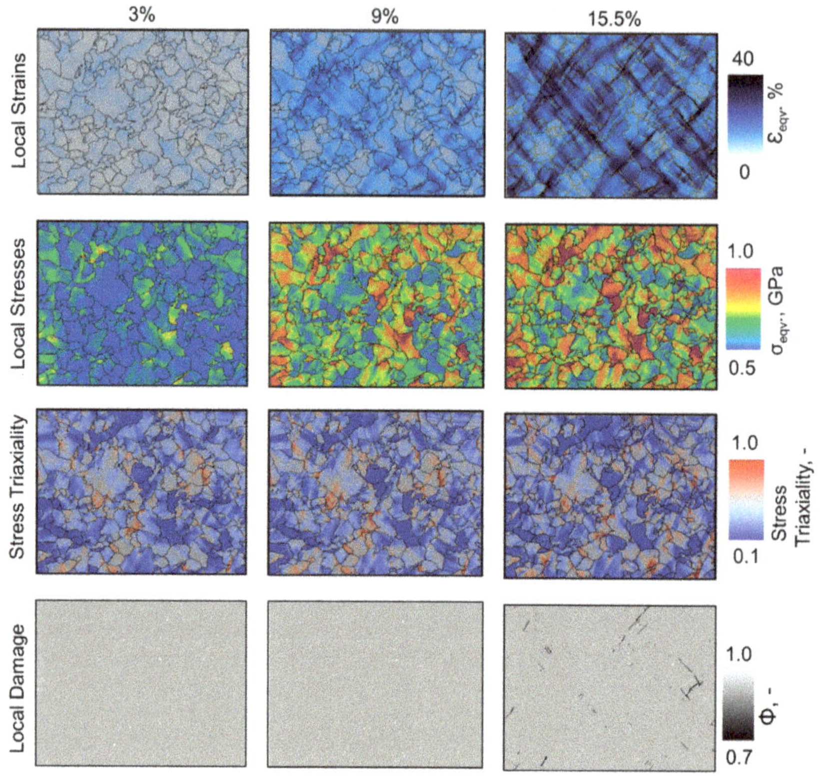

Figure 9. Evolution of local strain, local stress, stress triaxiality, and local damage in the selected RVE at 3%, 9%, and 15.5% global true strains, where the loading axis is horizontal to the micrographs.

Stress triaxiality is the relative degree of hydrostatic stress in each stress state, which is usually used to estimate the type of fracture that might take place in the material. In the current RVE, it is observed that the triaxiality measure in most of the grains is close to 0.1, which means that the stress (no matter how high) is predominantly hydrostatic. Therefore, the deviatoric component of stress in these grains is low and will not contribute significantly towards material flow and eventual damage. In addition, it is interesting to note that the triaxial stress remains consistent with the increase in strain and only slightly shifts to neighboring grains after the damage initiation and propagation.

Generally, it is observed that the damage initiates at the interface of the non-metallic inclusions, especially in the narrow zones locked between the cluster of inclusions. After initiations, the damage propagates at an oblique angle to the applied external load. As a result, the local voids coalesce and form larger cracks that grow faster, and material damage accelerates. It is important to mention here that the continuum mechanics simulations are intrinsically unstable due to the loss of stiffness at several solution points. Furthermore, it takes immense computing power to calculate the corresponding equilibrium for each increment after damage initiates in the RVE. Therefore, the simulations were only computed up to 15.5% of the global strain, and the results are shown in Figure 9.

Although the local results generally for the ferrite matrix are shown in Figure 9, one of the main objectives of this article is to investigate the effect of non-metallic inclusions on the material's local deformation and damage behavior. In Figure 10, the focus has specifically been shifted towards three different zones in the RVE of comparable size with different inclusion compositions and distributions. Zone-I comprises a single grain with a large MnS inclusion in the middle. Zone-II comprises several small ferrite grains with a clustered distribution of non-metallic inclusions primarily on the grain boundaries. Zone-III comprises relatively larger ferrite grains with a diffused distribution of non-metallic inclusions within the grains and on the grain boundaries. The distribution of local attributes in the three zones is selectively identified in the corresponding subplots Figure 10b–e. There is a relatively low stress concentration in zone-I and less strain due to almost nonexistent stress triaxiality, and therefore, no damage takes place in this zone. This can be due to the high Schmidt factor of this specific grain and the central position of a large hard MnS inclusion which restricts the slip planes from moving in this grain freely.

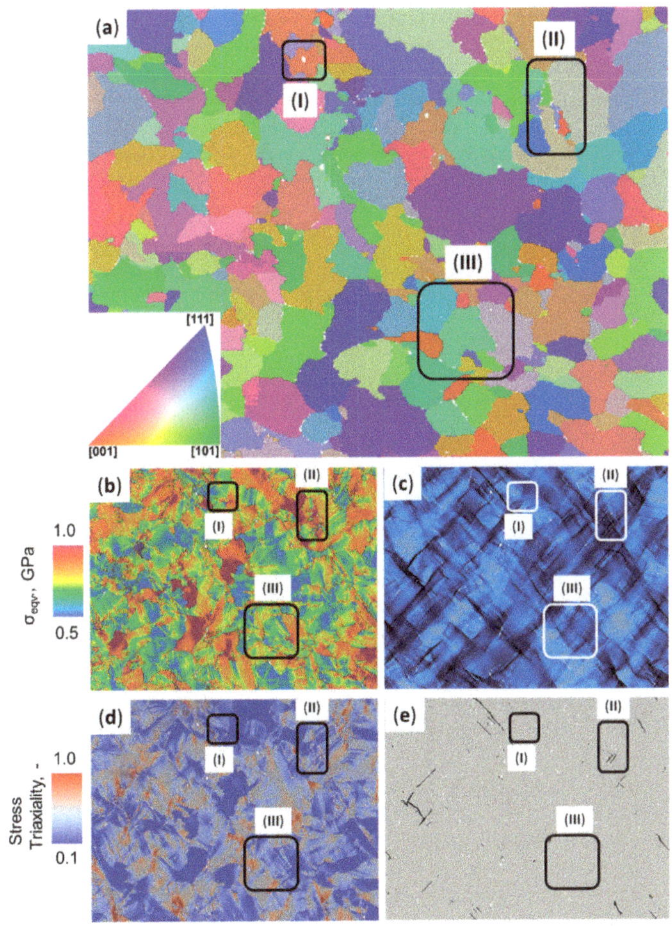

Figure 10. A construction figure with (**a**) showing the selected RVE with the identification of three zones. All non-metallic inclusions are displayed by white points. In other subplots, (**b**) true Mises stress, (**c**) true Mises strain, (**d**) stress triaxiality, and (**e**) damage propagation withing the ferrite matrix are shown. Zone-I has a ferrite grain with a large MnS inclusion in the middle, zone-II is a region of small ferrite grains with clustered non-metallic inclusions present on the grain boundaries, and zone-III has relatively large ferrite grains with a dispersed distribution of non-metallic inclusions within grains and on the grain boundaries. The loading axis is horizontal to the micrographs.

Small non-metallic inclusions distributed on the grain boundaries of small ferrite grains act as high-stress concentration sites with the large plastic flow in zone-II. Consequently, the stress triaxiality in this zone is relatively high, indicating a large plastic flow and evolution of ductile damage. This damage evolution due to the coalescence of small voids that form on the inclusion/matrix interface is visible in Figure 10e. It is observed that local damage incidents join together to form microcracks that propagate at maximum shear plane oriented 45 degrees to the loading axis. This can be due to two reasons: first, due to the 2D nature of the RVE, as has been pointed out in the previous work [52]; and second, due to the high amount of slip system activation in grains and areas with maximum critical resolved shear.

Zone-III comprises relatively larger ferrite grains with a relatively broader distribution of non-metallic inclusions within the grain and grain boundaries. In zone-III, although the triaxiality is comparable with zone-II, due to the lack of clustering on non-metallic inclusions, the stress distribution and strain distribution are lower. Hence, no damage occurs in the zone.

In Figure 10, a qualitative comparison of local attributes in three different zones shows how the distribution of non-metallic inclusions plays a role in affecting the local stress, strain, and damage. To quantitively compare three zones, a statistical normalized probability distribution was adopted to systematically compare the distribution of stress and strain within zone-I, zone-II, and zone-III, as shown in Figure 11.

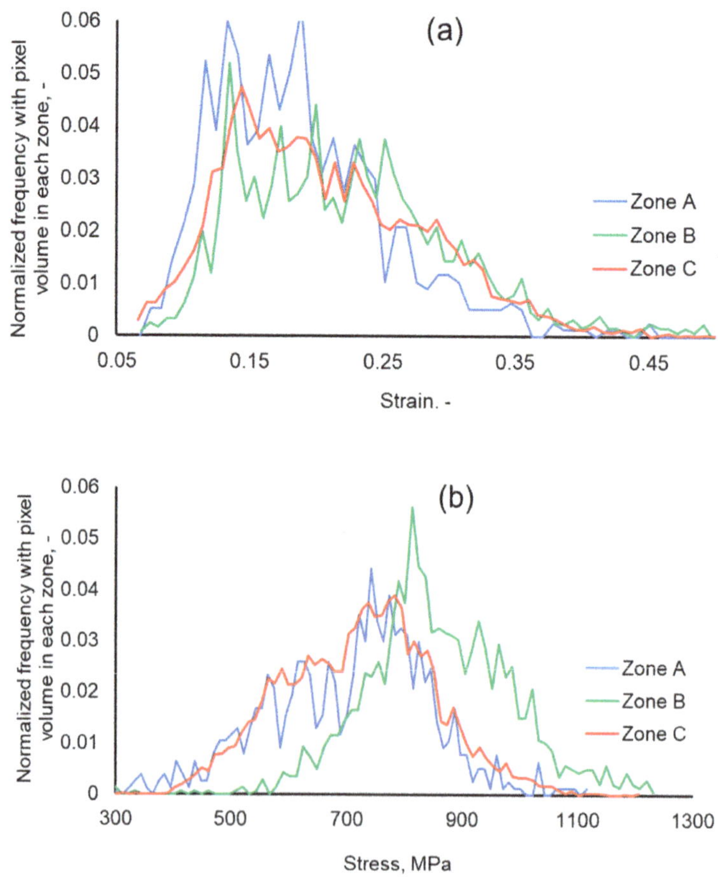

Figure 11. The normalized frequency of (**a**) strain and (**b**) stress for each zone identified in Figure 10.

To quantitatively compare the stress and strain distribution in zones I to III, a statistical distribution tool is used, and the results are presented in Figure 11. It is observed that the strain distribution in all three zones, as shown in Figure 11a, is very similar. It peaks around 0.15–0.18 and then linearly drops to a strain of 0.45. The strain in zone-I is slightly higher and that in zone-II is slightly lower due to the local orientation distribution. The stress distributions in these zones reach up to 1100 MPa.

The statistical stress distribution in the three zones is presented in Figure 11b and proves the already stated hypothesis. It is observed that the stress in zone-I and zone-III peaks around 750 MPa, with a slight bias towards the lower stress side. On the other hand, the stress distribution in zone-II, due to the close clustering of several non-metallic inclusions on the grain boundaries of small ferrite grains, peaks at 850 MPa (11% higher than the other zones), and its distribution is biased towards the higher-stress side. The distribution reaches 1250 MPa (12% higher than the other zones). This high local attribute distribution in zone-II makes it more prone to local damage initiation and propagation in high deformation regimes.

4.3. Local Results of the In Situ Tensile Test

The local results of the in situ tensile tests at 3%, 5%, and 8% of global strain are shown in Figure 12 The methodology and procedure of local strain measurement is provided in Appendix B. The complete set of results, due to its extensiveness, is provided in Appendix C of this article. Readers can refer to Appendix B and C for further details about collection and frame by frame evolution of local strain measurements.

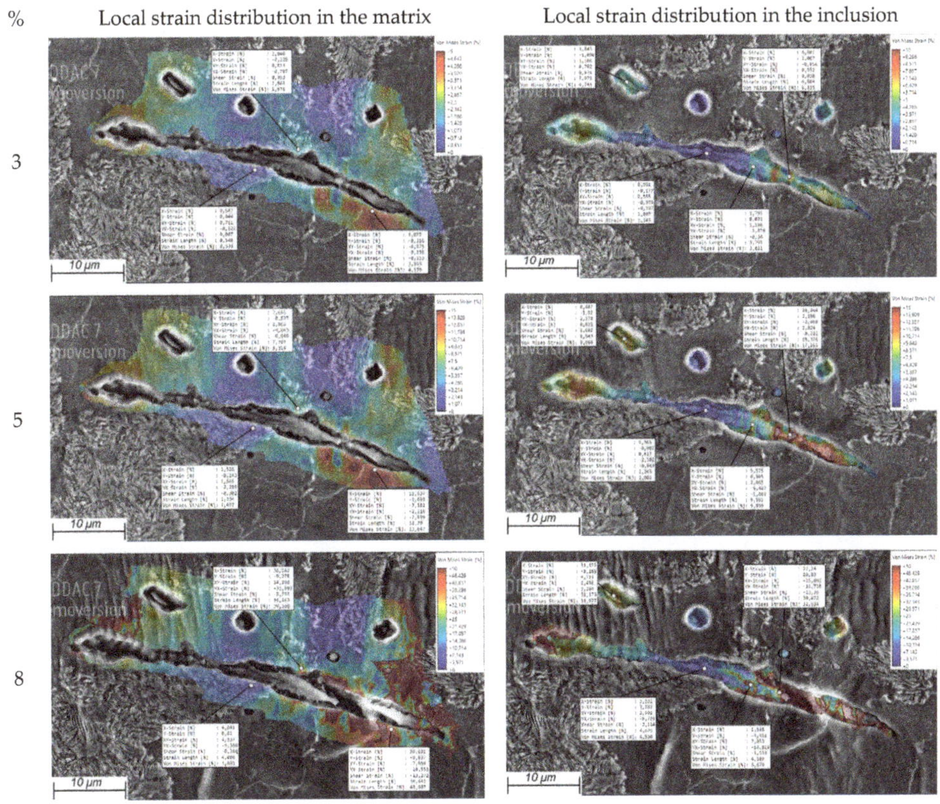

Figure 12. Local strain distribution in ferrite matrix and MnS inclusion at 3%, 5.5%, and 8% global strains.

In Figure 12, at a subsequent global strain of 3% and 400 MPa global stress, it can be observed that in the zones with increased elongation, the strain rate is much higher than in the zones where the strain rate has not increased much from the beginning of the test. The effect of the increase in elongation at the ferrite grain boundaries near the non-metallic inclusion is preserved. At this stress, two zones with a significant increase in strain can be seen in the middle of the inclusion. When comparing the entire image series, at a load of 450 MPa and global strain of 5%, the sulfide inclusion is cracked in the area of thinning. The inclusion in the left part of the crack is brighter due to its change in position, which significantly affects the strain values in this inclusion area, as the digital image correction software requires areas with contrast for accurate calculation. The breaking point itself is hidden at this point due to the topography.

The maximum values in the range of increased load are above 8.9%. It is observed that the two sulfide inclusions move further away from each other. The strain in the region increases, reaching a value of over 13–64%. Around the formed cavity of a small inclusion located on the upper right side of the elongated inclusion, the elongation near the pearlitic zone is greater than on the other side of the same grain. On the side of the ferrite grain around this cavity is a zone of increased deformation with a value of about 7–11%. There is a significant difference in elongation within and around the small cavities. In the middle cavity, the deformation is minimal and is between 2% and 3.5%, while in the left cavity, the deformation around and within the cavity is between 6.4% and 9.8%.

Due to further load increase and subsequent high strain values, the DIC program loses previously captured pixels for the calculation. By enlarging the gaps created, some pixels can significantly change their gray tone, causing the program to find a similar range of pixels elsewhere and greatly distorting the calculation values. At a maximum global strain of 8% (450 MPa), a large loss of pixels is observed (in areas beyond the specified deformation limit of dark red) because the topography changes too much, and the color range continues to change in many areas. In addition, the distance between the halves of the sulfide inclusion become significantly large. In the lower deformation zone in the matrix around the sulfide grain inclusion, the values are 43.6%. In contrast, the zone-I around the inclusion next to the perlite grain remains below the total deformation.

4.4. *Damage Evolution around Non-Metallic Inclusions*

The MnS inclusion tracked to analyze local deformation and damage behavior in Figure 12 is an exception. Usually, the inclusion size in the material user consideration is between 1–4 µm; please refer to Figure 4 for reference. Unfortunately, such small inclusions are hard to track and record local strain around due to high magnification, resulting in fast drifting during testing in the SEM chamber [53]. Despite this paradox, in the current work, during in situ testing, a few MnS inclusions of small size were tracked and recorded at different strains. The results are shown in Figures 13 and 14.

In Figures 13 and 14, several inclusions classified as MnS inclusions based on the EDS analysis were tracked at different external strains that were applied. The results shown here were at the beginning of the test when the inclusions were identified, at external load when the damage was initiated, and an external load when the damage started to evolve.

In Figure 13, it is observed that the damage initiates perpendicular to the line of action of the load at about 525 MPa global stress and propagates in that same direction. It is important to keep in mind that the inclusion is embedded in the matrix, and here only the surface is being observed. The surface matrix seems to be extremely deformed due to the internal propagation of the damage around the inclusion. It is also interesting to note that the damage initiates at the inclusion/matrix interface. In contrast, the inclusion remains free of any cracking, unlike the case of large inclusion, which was shown and discussed earlier in Figures 5 and 12. This is because the damage of the inclusion is a function of the size and distribution in the matrix, where small inclusions with an aspect ratio of less than 2.5 are generally not highly prone to damage.

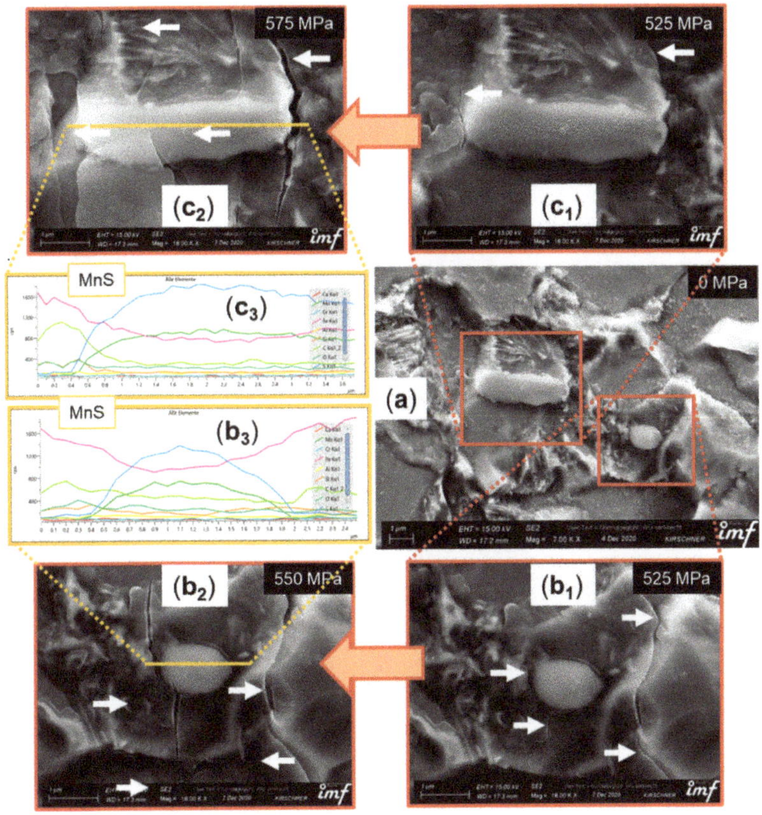

Figure 13. (**a**) Localized single inclusion identification, (**b**) represents a round inclusion at 525 MPa and 550 MPa, respectively. (**c**) represents an elliptical inclusion at 525 MPa and 550 MPa, respectively. EDS show the maximum concentration of manganese and sulfur along the measured length verifying both to be an MnS inclusions. The white arrows point to the local damage initiation and growth.

In Figure 14, two other inclusions are tracked, again identified as MnS based on the EDS analysis shown in the same figure. Both inclusions are completely embedded in small ferrite grains and resemble the case of zone-I in Figure 10.

In Figure 14, it is observed that the damage starts to initiate at 550 MPa of external load, which is slightly higher than the damage initiation load for other inclusions but is in a similar range. At 575 MPa of external load, damage on the inclusion matrix interface grows so extensive that it is easily identifiable and is marked with the help of white arrows. In this case, it is observed that voids initiate at an oblique angle to the applied external load and not perpendicular to the load as observed earlier. This position of the damage initiation depends on several factors: i.e., orientation, size, shape, and sharpness of the inclusion edges and the place of the inclusion (inside a large grain or on the grain boundary). It would also largely depend on the orientation of the surrounding matrix grains. As mentioned earlier, all these inclusions have 3D geometry, and the third dimension would also affect the results observed on the surface of the specimen.

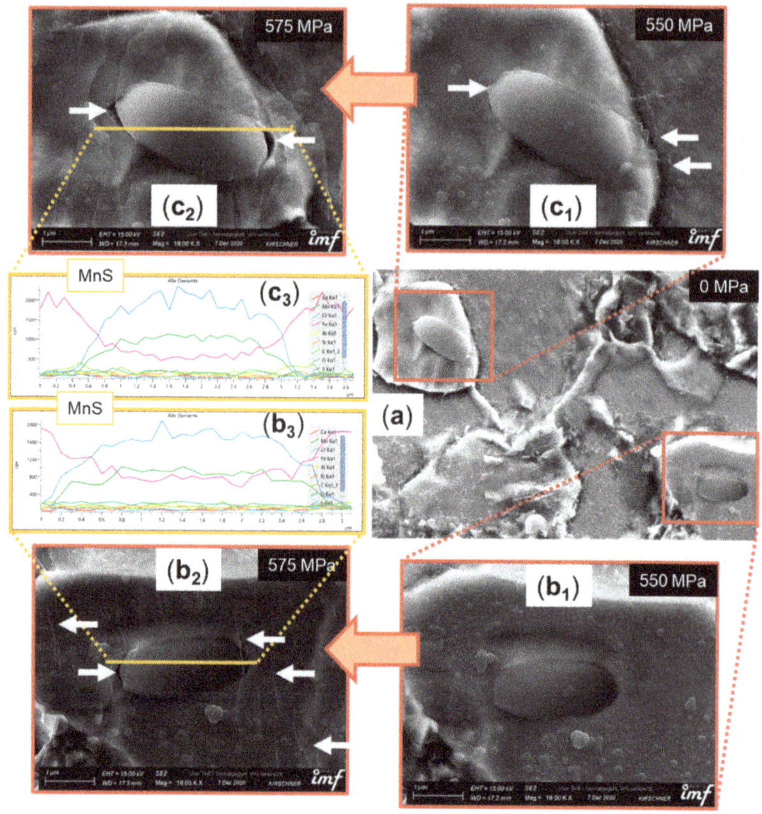

Figure 14. (**a**) Localized single inclusion identification, (**b**) represents an elliptical inclusion parallel to the applied load at 550 MPa and 575 MPa, respectively. (**c**) represents an elliptical inclusion placed 30 degree to the applied load at 550 MPa and 575 MPa, respectively. EDS show the maximum concentration of manganese and sulfur along the measured length verifying both to be an MnS inclusions. The white arrows point to the local damage initiation and growth.

5. Discussion

The global and local deformation and damage behavior of modified 16MnCrS5 steel (composition in Table 1) is analyzed. Before the in situ testing, the specimens were prepared using a special grinding and polishing technique (presented in detail in Appendix A), and detailed EBSD data were collected to understand the morphology of the matrix and the second phase inclusions. The outcome of the statistical analysis of the collected microstructural data is presented in Figures 3 and 4. It is evident that, generally, the non-metallic inclusions are quite small, and the MnS inclusions are slightly larger and more visible at the set magnification. Although the MnS inclusions were the focus of this work, other inclusions are also expected to behave similarly.

The parameters of the crystal plasticity model were calibrated and then incorporated in the full phase model employing the recorded and cleaned EBSD data. It is observed that the global results of the simulation match well with the experimental observations. Therefore, researchers can successfully use the identified parameters to model this material using DAMASK in the future. However, one can argue that this match of global results is because a similar data set was used to calibrate these parameters. To answer that critique, the damage pattern around a comparable inclusion in experiments and simulations was made at different strains, and a remarkable similarity in the trends was observed.

Apart from the global results, the local simulation results provide great insight into how the inclusion size, position within the matrix, and their distribution affect the local stress, strain, damage, and triaxiality. The trends of local strain distribution match well with the previously published data [28,54]. A qualitative and quantitative comparison of the three different zones was presented in Section 4.2. All three zones are intrinsically different based on the composition, inclusion size, and distribution. In addition, there is a clear difference in the damage behavior in these zones, and local stress distribution is observed. These local attributes account for the different local material behavior observed in the experimental part of this work.

Since the previous works have shown that the second phase inclusions are major players in defining the formability of the material [15,41,55], a special focus was given to them in this work. Previous work [51–56] shows that the morphology and distribution of the second phase inclusions dictate the formability limit and damage degradation in a material. In this work, in situ tests were carried out focusing on the strain evolution and damage evolution around and within these second phase inclusions. The second phase inclusion size distribution is observed to be close to 1 μm with a few larger inclusions in the specimen. The local strain around and within one of these exceptionally large MnS inclusions is studied in detail. MnS inclusions, due to their larger size, are easily identifiable at the magnification range selected in this study, and therefore they appear predominantly in this work. However, it is assumed that other inclusions behave similarly. This is also verified by the simulation results presented in Section 4.2, where all the inclusions behave almost indifferently under applied external load.

The results of the local strain measurement from the in situ tests are presented in Section 4.3. It is observed that the strain in the matrix is higher at some points and lower at some points, depending on the orientation of the corresponding ferrite grain. The inclusion is observed to have a low strain distribution until a brittle fracture appears, which damages the whole microstructure, and it becomes hard to track the local strain in the material further. Small MnS inclusions were also tracked in this work at increasing external loads. For these smaller inclusions, it is observed that matrix/inclusion interface decohesion starts to take place and grow at high strain regimes of >550 MPa. These findings match well with the previously published work of other researchers [4,6,19,44,56].

Although similar research with in situ tests and crystal plasticity simulations has been previously carried out by some researchers before [21,27,28,57,58], it was either pure simulation or pure experimental work. No correlation using both methods has been presented before concerning how the inclusion distribution, size, and morphology affect the material deformation and degradation. However, in this work, the findings of the developed crystal plasticity-based numerical simulation model have been validated. Researchers and industry can now use this tool to analyze and optimize the non-metallic inclusion size, distribution, and morphology to attain the desired material formability.

6. Conclusions

The global and local deformation and damage behavior of modified 16MnCrS5 steel using in situ experimental and crystal plasticity-based numerical simulation model is analyzed. Recorded EBSD data were used to evaluate the local microstructure of the material. The data were then adopted to run full phase crystal plasticity simulations in DAMASK with calibrated material model parameters. In situ tensile tests were carried out on specially prepared specimens with incremental data collection for different inclusions distributed within the ferrite matrix. A images were processed using digital image correlation-based tool to obtain microstrain measurements. The damage initiation, propagation, and strain localization around a large MnS inclusion were analyzed in detail. Other smaller inclusions tracked in this study helped in understanding the local material behavior. The conclusions of this extensive study are as follows:

1. The non-metallic inclusions are heterogeneously sized and heterogeneously distributed within the ferrite matrix. The inclusions are usually small (size ~2 μm)

and elliptical (aspect ratio < ~2), with the exceptions of some extremely large (>10 μm up to 50 μm) and elongated (aspect ratio > 5) inclusions that are also present in the matrix. These large and elongated inclusions play a critical role in defining the limiting formability of the steel under consideration.

2. The 2D full phase simulation model developed in the current work provides accurate information about the material's local damage initiation and propagation under consideration. Although the same areas were not compared quantitatively, the simulation results match the experimental observations of global stress–strain response and the local damage initiation around the inclusions. Hence, the model can be used by engineers and researchers for further material engineering and optimization with confidence.
3. The local stress and strain largely depend on the local composition and distribution of non-metallic inclusions and the size and orientation of the neighboring ferrite grains. The local stress in highly clustered zones is ~12% higher than the other material zones where the inclusions are more dispersed within the matrix.
4. The damage initiation and propagation also depend on the inclusion size and position. If the inclusion is very large, brittle fracture occurs at relatively lower applied external stress (~450 MPa), which results in the fast damage initiation and propagation in the matrix. For most small and relatively elliptical inclusions, the damage initiates on the matrix/inclusion interface at relatively high-stress regimes (~550 MPa). It propagates at an oblique angle to the applied load. These smaller and relatively elliptical inclusions are also less prone to brittle cracking and strengthen the matrix during deformation.
5. An adequate material manufacturing methodology should be employed for the material class under consideration, resulting in small, elliptical, homogeneously distributed inclusions within the ferrite matrix. This would result in a material with better formability and higher damage resistance.

Author Contributions: Conceptualization, F.Q. and M.U.; methodology, F.Q. and M.U.; software, F.Q. and M.U.; validation, F.Q., V.E., M.K. and S.G.; formal analysis, F.Q., M.U., V.E. and F.H.; investigation, V.E., M.K., F.H. and S.G.; resources, F.Q., V.E., M.K. and S.G.; data curation, F.Q. and M.U.; writing—original draft preparation, F.Q., M.U. and S.G.; writing—review and editing, F.Q., M.U., S.G. and U.P.; visualization, F.Q., M.U., S.G. and U.P.; supervision, U.P. and S.G.; project administration, S.G. and U.P.; funding acquisition, F.Q., S.G. and U.P. All authors have read and agreed to the published version of the manuscript.

Funding: This research was funded by Industrielle Gemeinschafts-Forschung (IGF) under project number 20429 BG, titled "Nichtmetallische Einschlüsse".

Institutional Review Board Statement: Not applicable.

Informed Consent Statement: Not applicable.

Data Availability Statement: The data are not publicly available but can be shared with interested researchers privately upon request.

Acknowledgments: The authors acknowledge the DAAD Faculty Development for Candidates (Balochistan), 2016 (57245990)-HRDI-UESTP's/UET's funding scheme in cooperation with the Higher Education Commission of Pakistan (HEC) for sponsoring the stay of Faisal Qayyum at IMF TU Freiberg. This work is conducted within the DFG-funded collaborative research group TRIP Matrix Composites (SFB 799). The authors gratefully acknowledge the German Research Foundation (DFG) for the financial support of the SFB 799. The authors also acknowledge the support of Martin Diehl and Franz Roters (MPIE, Düsseldorf) for their help regarding the functionality of DAMASK.

Conflicts of Interest: The authors declare no conflict of interest.

Nomenclature

Acronyms

Symbol	*Description*
SEM	Scanning electron microscope
SE	Secondary electron (detector)
BSE	Back scatter electron (detector)
EDS	Eenergy dispersive spectorscope
EBSD	Electron back scatter diffraction
DAMASK	Düsseldorf advanced material simulation kit
RVE	Representative volume element
NME	Non-metallic inclusions

Appendix A. Limitations and Challenges Associated with the In Situ Tensile Test Methodology

Before the in situ tensile tests, the specimens were metallographically polished (OPS). Since the deformed surface of the test specimens was later to be analyzed with digital image correlation software, the test specimens were etched for 5 s in a 3% Nital solution to achieve the required surface contrast. The experiment revealed that only non-deformed inclusions were found on the surface of the specimens, while non-metallic sulfide inclusions with an elongated shape were in the middle of the specimens examined. For further tests, the specimens were ground to a thickness of 0.6 mm and then polished and etched. Since the specimens were very thin, manual grinding was not possible because it was not possible to avoid deviations in the thickness of the specimens. A conventional embedding mass was not suitable for specimen preparation because there was a high risk of damaging the specimens when detached from the embedding mass therefore It was decided to use the embedding mass Technovit-5071 as shown in Figure A1. A special feature of this embedding mass was that the specimens could be dissolved after processing by dissolving the embedding mass in acetone when heated to 30 °C.

Figure A1. Specimens after polishing in Technovit-5071.

Within the scope of this work, an attempt was made to generate a special pattern [7] on the specimen surface for the digital image correlation program with the help of polystyrene latex beads with a diameter of 0.095 µm. However, due to the previous etching, it was difficult to achieve an even distribution on the surface, as all the beads accumulated in the resulting depressions (Figure A2). The available DIC systems, such as VEDDAC, GOM, and VIC-2D, have therefore not made a flat closed evaluation possible. Therefore, the procedure with balls was not pursued in further investigations. The focus was placed on a special etching of the specimen surface with the application of the "InLens" module.

The method for the deposition of SiO_2 nanoparticles on the specimen surface was as follows:

1. The surface of the specimen was metallographically prepared for colloidal silicon dioxide (OPS);
2. On a clean polished dip plate, 10–20 drops of OPS were placed along the diameters;
3. The specimen was pressed onto the cloth with 1–2 MPa pressure and turned for 1 s at 2 rpm;
4. With water flushing and 100 RPS plate speed, the specimen was rotated on the cloth at 1–2 MPa pressure at 2 RPS for 3 s;
5. The specimen surface was rinsed with ethanol and dried with a blower.

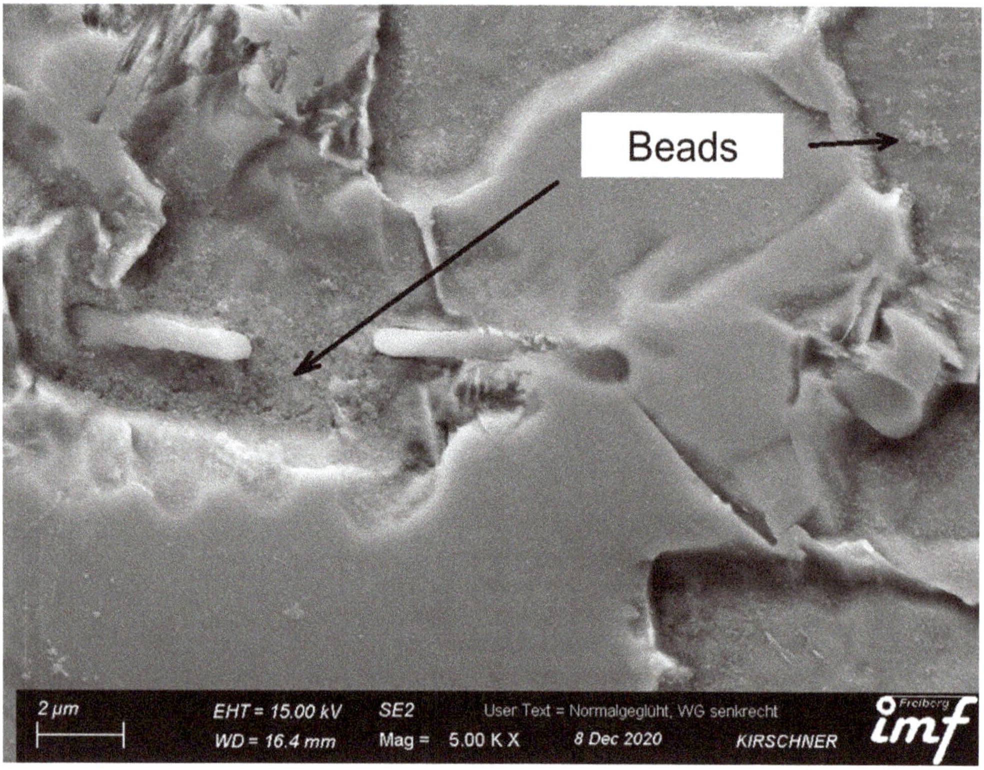

Figure A2. Example of the accumulation of polystyrene latex beads in the deep spots created by etching.

As part of this work, various tests were also carried out to create contrasts on the surface of the specimens. To achieve the desired contrast, different etching times of 2 to 15 s were used (Figure A3).

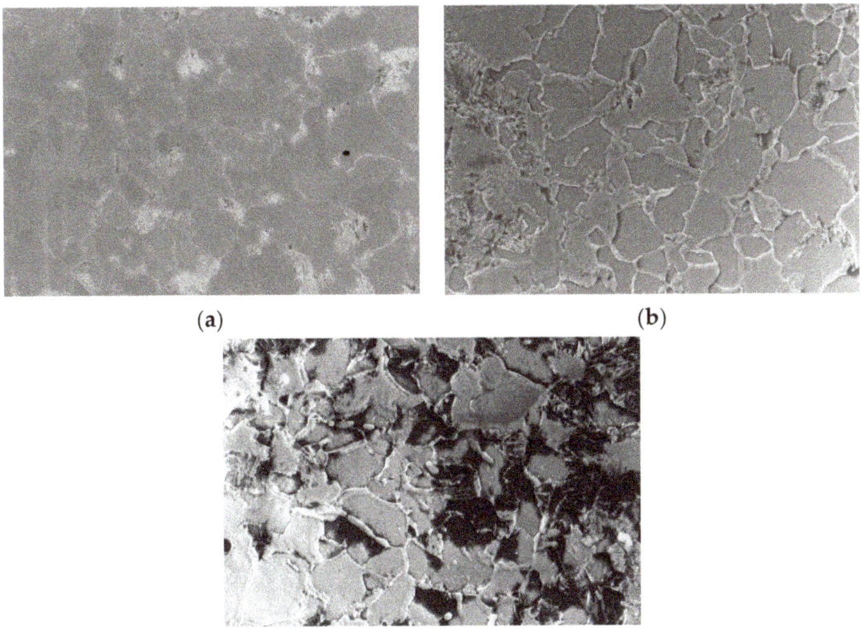

(a) (b) (c)

Figure A3. Example of a microstructure surface with different etching times: (**a**) 2 s, (**b**) 5 s, (**c**) 15 s.

Various scanning electron microscope modules such as SE (Figure A4a) and "InLens" (Figure A4b) were used.

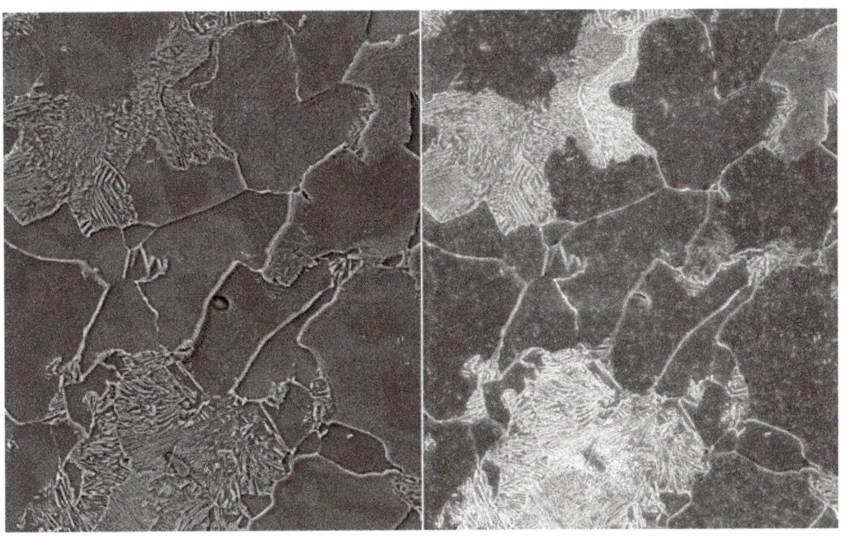

(a) (b)

Figure A4. Example of a microstructure surface using different modules: (**a**) SE, (**b**) "InLens".

Different contrast values were used in the settings of the electron beam microscope (Figure A5a contrast = 49%, Figure A5b contrast = 52%).

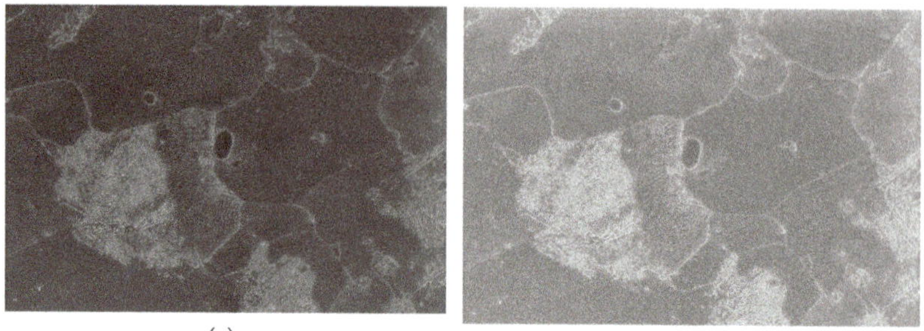

(**a**) (**b**)

Figure A5. Example of a microstructure surface at different contrasts of SEM: (**a**) contrast = 49%, (**b**) contrast = 52%.

Appendix B. Local Mises Strain Measurement

The von Mises strain at the meso plane was calculated using a scale in the areas studied (Figure A6). On each surface of the undeformed specimen, three points were selected so that one point was common; they formed two identical straights, and there was an angle of 90° between these straights.

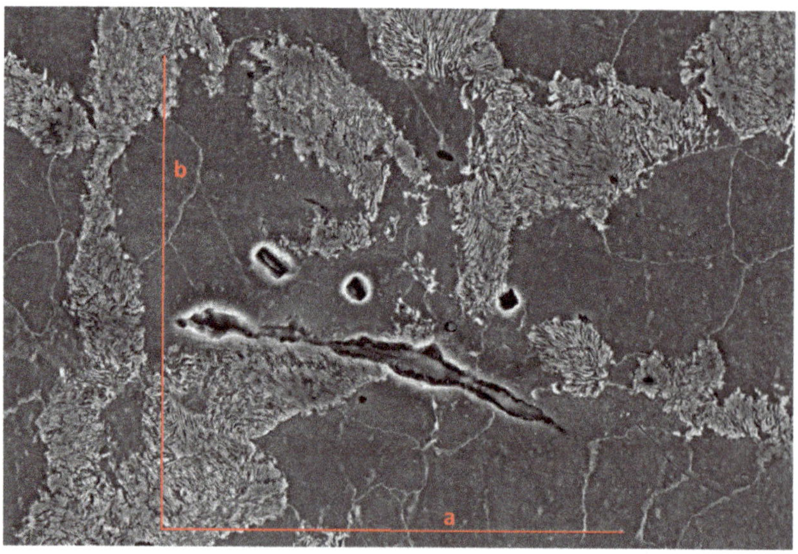

Figure A6. Account length of the selected area used to calculate mesoscale plane-strain for the tensile specimen (11K).

The strain was calculated using the following formula:

$$\varnothing_1 = ln\frac{a}{a_0} = \frac{1}{2}ln\frac{a_1^2 + b_1^2 + \sqrt{\left(a_1^2 + b_1^2\right)^2 - 4a_1^2 b_1^2 sin^2\delta}}{2a_0^2} \tag{A1}$$

$$\varnothing_2 = ln\frac{a}{a_0} = \frac{1}{2}ln\frac{a_1^2 + b_1^2 + \sqrt{\left(a_1^2 + b_1^2\right)^2 - 4a_1^2 b_1^2 sin^2\delta}}{2a_0^2} \tag{A2}$$

$$\varnothing_v = \frac{2}{\sqrt{3}}\sqrt{\varnothing_1^2 + \varnothing_2^2 + \varnothing_1\varnothing_2} \tag{A3}$$

As a result of the tests, the meso deformation in the long inclusion in the tensile specimen was calculated. The results of meso-scale deformation are presented in Table A1.

Table A1. Change in the length of the account of the selected area and angle at the meso plane for the tensile specimen.

MPa	A [μm]	B [μm]	Angle of Axis A [Degrees]	Angle of Axis B [Degrees]	Angle [Degree]
0	50	50	0	90	90
300	50.63	49.87	0.5	90	89.5
375	50.8	49.96	0.1	90.1	90
400	51.1	49.83	−0.19	90.4	90.59
425	51.52	49.66	−0.28	90.24	90.52
437	51.94	49.66	−0.37	90.19	90.56
450	52.41	49.62	−0.42	90.44	90.86
462	53.46	49.54	−0.54	90.73	91.27
450_2	57.01	49.89	−0.72	91.5	92.22

The tests yielded the stress–strain diagram shown in Figure A7. The diagram shows that the range of plastic deformation started at 300 MPa. The maximum tensile strength achieved was 462 Mpa, and the breakage of the specimen began at a stretch of 7.9%

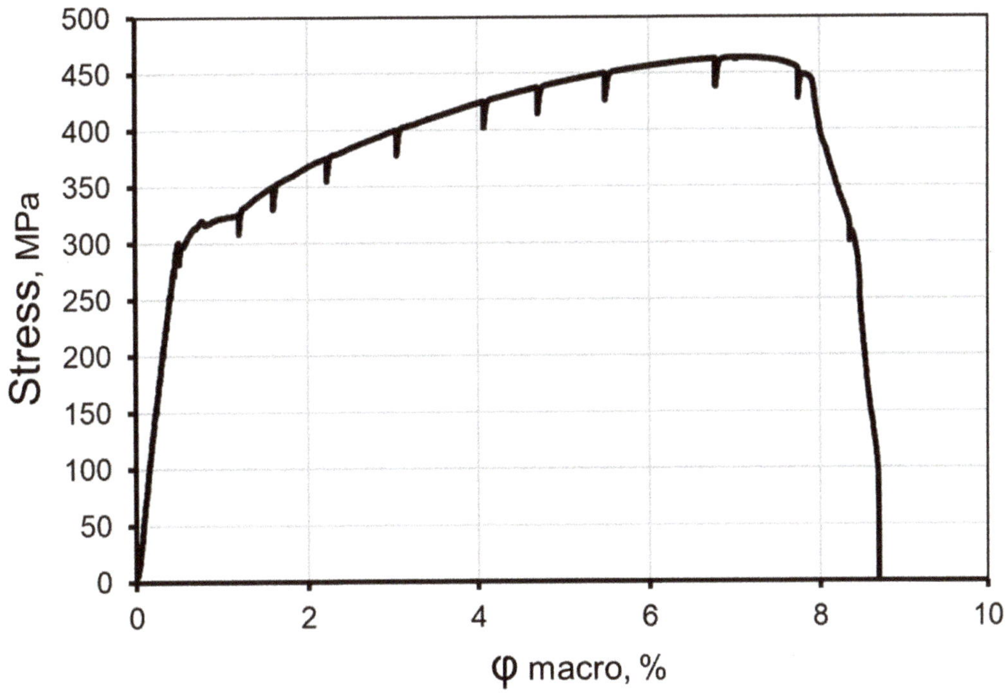

Figure A7. Stress–strain diagram of the tensile specimen.

Based on the data obtained, diagrams were created to compare the meso deformation and micro deformation in different areas of the study with the names of the selected areas (Figure A8). A diagram has also been created to compare macro and micro deformation (Figure A9).

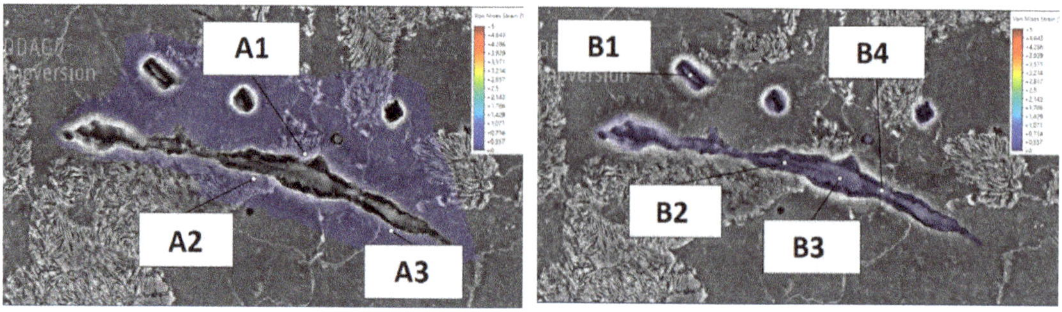

(a) (b)

Figure A8. Marking and numbering of measuring points (**a**) around non-metallic inclusions; (**b**) within non-metallic inclusions.

The diagram in Figure A9 shows that the intensity of micro deformation is significantly higher compared to macro deformation at the points far from the pearlite grains.

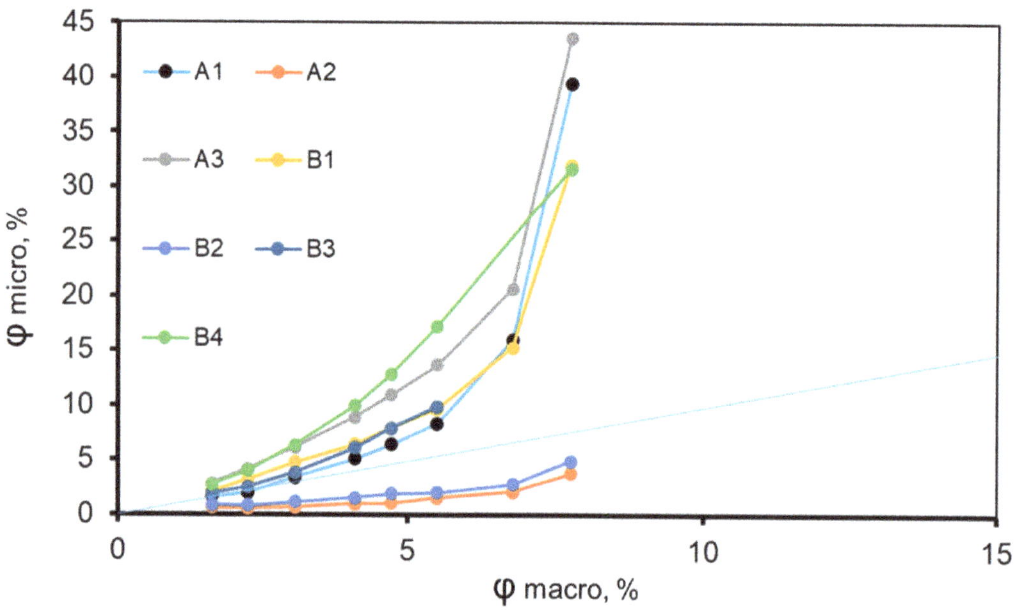

Figure A9. Comparison of meso deformation and micro deformation.

Appendix C. Local Strain Measurement Overlaid on SEM Micrographs

The frame by frame evolution of local strain distribution around (left) and within (right) the MnS inclusion is presented in Figure A10 from 1.6% to 7.75% global true strain.

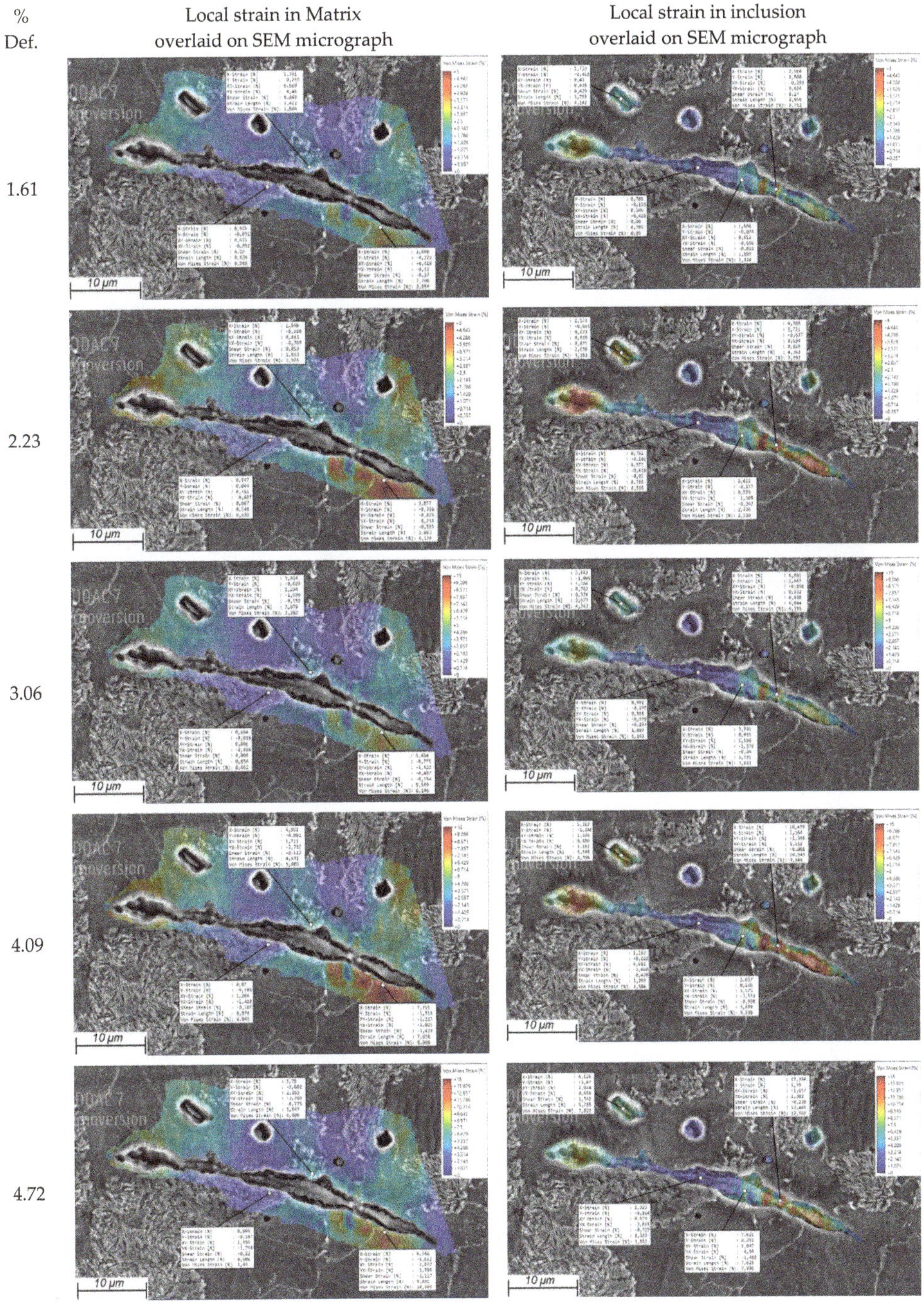

Figure A10. *Cont.*

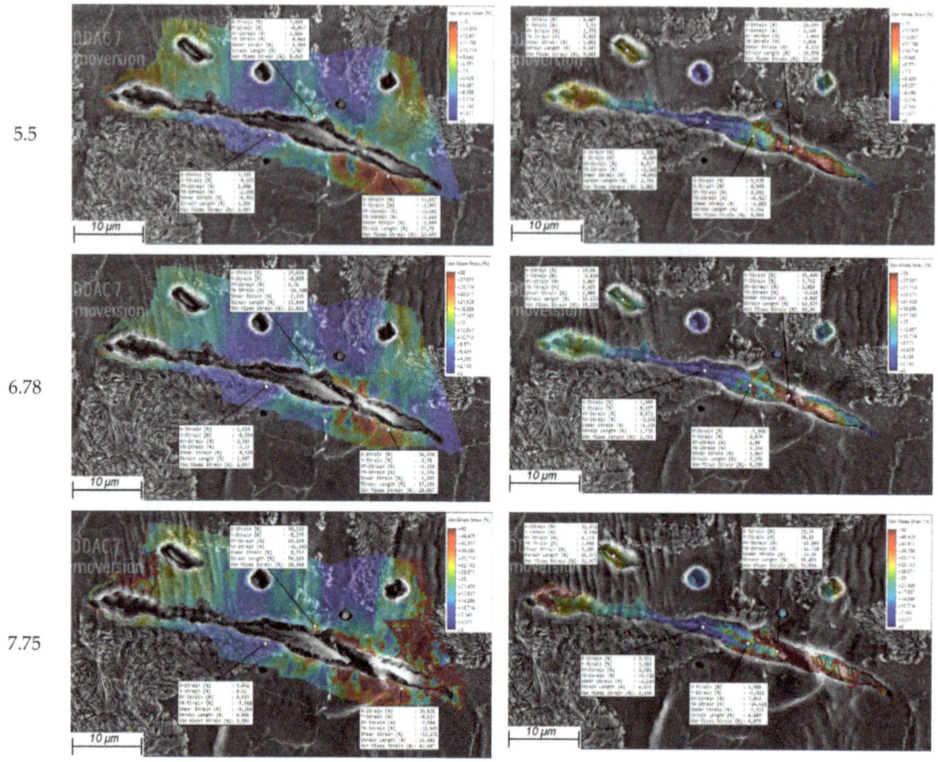

Figure A10. Local strain measurement overlaid on SEM micrographs.

References

1. Krauss, G. *Steels: Processing, Structure, and Performance*; ASM International: Materials Park, OH, USA, 2015.
2. Wang, J.; Shen, Y.; Liu, Y.; Wang, F.; Jia, N. Tailoring Strength and Ductility of a Cr-Containing High Carbon Steel by Cold-Working and Annealing. *Materials* **2019**, *12*, 4136. [CrossRef] [PubMed]
3. Das, T.; Li, J.; Painter, M.; Summerville, E. Evaluation of two AISI 4037 cold heading quality steel wires for improved tool life and product quality. *J. Mater. Eng. Perform.* **2002**, *11*, 86–91. [CrossRef]
4. Shen, P.; Fu, J. Morphology study on inclusion modifications using Mg–Ca treatment in resulfurized special steel. *Materials* **2019**, *12*, 197. [CrossRef] [PubMed]
5. Mohammed, B.; Park, T.; Pourboghrat, F.; Hu, J.; Esmaeilpour, R.; Abu-Farha, F. Multiscale crystal plasticity modeling of multi-phase advanced high strength steel. *Int. J. Solids Struct.* **2018**, *151*, 57–75. [CrossRef]
6. Hosseini, S.; Temmel, C.; Karlsson, B.; Ingesten, N.-G. An in-situ scanning electron microscopy study of the bonding between MnS inclusions and the matrix during tensile deformation of hot-rolled steels. *Metall. Mater. Trans. A* **2007**, *38*, 982–989. [CrossRef]
7. Mukhtar, F.; Qayyum, F.; Shah, M. Studying the effect of thermal fatigue on multiple cracks propagating in an SS316L thin flange on a shaft specimen using a multi-physics numerical simulation model. *Strojniški Vestnik-J. Mech. Eng.* **2019**, *65*, 565–573. [CrossRef]
8. Qayyum, F.; Shah, M.; Muqeet, A.; Afzal, J. The effect of anisotropy on the intermediate and final form in deep drawing of SS304L, with high draw ratios: Experimentation and numerical simulation. *IOP Conf. Ser. Mater. Sci. Eng.* **2016**, *146*, 012031. [CrossRef]
9. Miyamoto, G.; Usuki, H.; Li, Z.-D.; Furuhara, T. Effects of Mn, Si and Cr addition on reverse transformation at 1073 K from spheroidized cementite structure in Fe–0.6 mass% C alloy. *Acta Mater.* **2010**, *58*, 4492–4502. [CrossRef]
10. Khan, U.; Hussain, A.; Shah, M.; Shuaib, M.; Qayyum, F. Investigation of mechanical properties based on grain growth and microstructure evolution of alumina ceramics during two step sintering process. *IOP Conf. Ser. Mater. Sci. Eng.* **2016**, *146*, 012046. [CrossRef]
11. Qayyum, F.; Guk, S.; Kawalla, R.; Prahl, U. Experimental Investigations and Multiscale Modeling to Study the Effect of Sulfur Content on Formability of 16MnCr5 Alloy Steel. *Steel Res. Int.* **2018**, *90*, 1800369. [CrossRef]
12. Chen, H.; Zhou, X. Research progress of gear steel for automobiles. *J. Mater. Sci. Eng.* **2011**, *29*, 478.

13. Sheng, J.; La, P.; Su, J.; Ren, J.; Ma, J.; Shi, Y.; Li, Z.; Wang, J. In situ SEM analysis for deformation mechanism of micro/nanostructured 304 stainless steel with high strength and good plasticity. *Mod. Phys. Lett. B* **2018**, *32*, 1850182. [CrossRef]
14. Salawu, E.Y.; Ajayi, O.O.; Inegbenebor, A.; Popoola, A.; Uyor, U. Effects of Heat Treatment Techniques on the Fatigue Behaviour of Steel Gears: A Review. *J. Phys. Conf. Ser.* **2019**, *1378*, 042001. [CrossRef]
15. Umar, M.; Qayyum, F.; Farooq, M.U.; Khan, L.A.; Guk, S.; Prahl, U. Investigating the Effect of Cementite Particle Size and Distribution on Local Stress and Strain Evolution in Spheroidized Medium Carbon Steels using Crystal Plasticity-Based Numerical Simulations. *Steel Res. Int.* **2020**, *92*, 2000407. [CrossRef]
16. Guk, S.; Hofmann, K.; Beyer, S.; Kawalla, R. Anordnung und Verfahren zur Prüfung des Umformvermögens von Proben aus plastisch umformbaren Werkstoffen und/oder zur Bestimmung der tribologischen Eigenschaften einer vorbehandelten Oberfläche einer Probe. *Patent* **2016**, *10*, 225–530.
17. Kang, Y.-C.; Chan, S.L.-I. Tensile properties of nanometric Al_2O_3 particulate-reinforced aluminum matrix composites. *Mater. Chem. Phys.* **2004**, *85*, 438–443. [CrossRef]
18. Moehring, K.; Walther, F. Load Direction-Dependent Influence of Forming-Induced Initial Damage on the Fatigue Performance of 16MnCrS5 Steel. *Materials* **2020**, *13*, 2680. [CrossRef]
19. Kusche, C.F.; Gibson, J.S.-L.; Wollenweber, M.A.; Korte-Kerzel, S. On the mechanical properties and deformation mechanisms of manganese sulphide inclusions. *Mater. Des.* **2020**, *193*, 108801. [CrossRef]
20. Schowtjak, A.; Wang, S.; Hering, O.; Clausmeyer, T.; Lohmar, J.; Schulte, R.; Ostwald, R.; Hirt, G.; Tekkaya, A.E. Prediction and analysis of damage evolution during caliber rolling and subsequent cold forward extrusion. *Prod. Eng.* **2020**, *14*, 33–41. [CrossRef]
21. Diehl, M.; Groeber, M.; Haase, C.; Molodov, D.A.; Roters, F.; Raabe, D. Identifying Structure–Property Relationships Through DREAM.3D Representative Volume Elements and DAMASK Crystal Plasticity Simulations: An Integrated Computational Materials Engineering Approach. *JOM* **2017**, *69*, 848–855. [CrossRef]
22. Sedighiani, K.; Traka, K.; Roters, F.; Raabe, D.; Sietsma, J.; Diehl, M. Determination and analysis of the constitutive parameters of temperature-dependent dislocation-density-based crystal plasticity models. *Mech. Mater.* **2022**, *164*, 104117. [CrossRef]
23. Hussein, T.; Umar, M.; Qayyum, F.; Guk, S.; Prahl, U. Micromechanical Effect of Martensite Attributes on Forming Limits of Dual-Phase Steels Investigated by Crystal Plasticity-Based Numerical Simulations. *Crystals* **2022**, *12*, 155. [CrossRef]
24. Qayyum, F.; Guk, S.; Kawalla, R.; Prahl, U. On Attempting to Create a Virtual Laboratory for Application-Oriented Microstructural Optimization of Multi-Phase Materials. *Appl. Sci.* **2021**, *11*, 1506. [CrossRef]
25. Zhang, H.; Diehl, M.; Roters, F.; Raabe, D. A virtual laboratory using high resolution crystal plasticity simulations to determine the initial yield surface for sheet metal forming operations. *Int. J. Plast.* **2016**, *80*, 111–138. [CrossRef]
26. Diehl, M.; An, D.; Shanthraj, P.; Zaefferer, S.; Roters, F.; Raabe, D. Crystal plasticity study on stress and strain partitioning in a measured 3D dual phase steel microstructure. *Phys. Mesomech.* **2017**, *20*, 311–323. [CrossRef]
27. Diehl, M.; Wicke, M.; Shanthraj, P.; Roters, F.; Brueckner-Foit, A.; Raabe, D. Coupled Crystal Plasticity–Phase Field Fracture Simulation Study on Damage Evolution Around a Void: Pore Shape Versus Crystallographic Orientation. *JOM* **2017**, *69*, 872–878. [CrossRef]
28. Tasan, C.C.; Hoefnagels, J.P.M.; Diehl, M.; Yan, D.; Roters, F.; Raabe, D. Strain localization and damage in dual phase steels investigated by coupled in-situ deformation experiments and crystal plasticity simulations. *Int. J. Plast.* **2014**, *63*, 198–210. [CrossRef]
29. Wang, D.; Diehl, M.; Roters, F.; Raabe, D. On the role of the collinear dislocation interaction in deformation patterning and laminate formation in single crystal plasticity. *Mech. Mater.* **2018**, *125*, 70–79. [CrossRef]
30. Liu, C.; Shanthraj, P.; Diehl, M.; Roters, F.; Dong, S.; Dong, J.; Ding, W.; Raabe, D. An integrated crystal plasticity–phase field model for spatially resolved twin nucleation, propagation, and growth in hexagonal materials. *Int. J. Plast.* **2018**, *106*, 203–227. [CrossRef]
31. Cereceda, D.; Diehl, M.; Roters, F.; Raabe, D.; Perlado, J.M.; Marian, J. Unraveling the temperature dependence of the yield strength in single-crystal tungsten using atomistically-informed crystal plasticity calculations. *Int. J. Plast.* **2016**, *78*, 242–265. [CrossRef]
32. Shanthraj, P.; Eisenlohr, P.; Diehl, M.; Roters, F. Numerically robust spectral methods for crystal plasticity simulations of heterogeneous materials. *Int. J. Plast.* **2015**, *66*, 31–45. [CrossRef]
33. Roters, F.; Diehl, M.; Shanthraj, P.; Eisenlohr, P.; Reuber, C.; Wong, S.L.; Maiti, T.; Ebrahimi, A.; Hochrainer, T.; Fabritius, H.O.; et al. DAMASK—The Düsseldorf Advanced Material Simulation Kit for modeling multi-physics crystal plasticity, thermal, and damage phenomena from the single crystal up to the component scale. *Comput. Mater. Sci.* **2019**, *158*, 420–478. [CrossRef]
34. Roters, F.; Eisenlohr, P.; Kords, C.; Tjahjanto, D.; Diehl, M.; Raabe, D. DAMASK: The Düsseldorf Advanced MAterial Simulation Kit for studying crystal plasticity using an FE based or a spectral numerical solver. *Procedia IUTAM* **2012**, *3*, 3–10. [CrossRef]
35. Weidner, A.; Yanina, A.; Guk, S.; Kawalla, R.; Biermann, H. Microstructure and Local Strain Fields in a High-Alloyed Austenitic Cast Steel and a Steel-Matrix Composite Material after in situ Tensile and Cyclic Deformation. *Steel Res. Int.* **2011**, *82*, 990–997. [CrossRef]
36. Weidner, A.; Biermann, H. Combination of different in situ characterization techniques and scanning electron microscopy investigations for a comprehensive description of the tensile deformation behavior of a CrMnNi TRIP/TWIP steel. *JOM* **2015**, *67*, 1729–1747. [CrossRef]

37. Weidner, A.; Berek, H.; Segel, C.; Aneziris, C.G.; Biermann, H. In situ tensile deformation of TRIP steel/Mg-PSZ composites. *Mater. Sci. Forum* **2013**, *738–739*, 77–81. [CrossRef]
38. Agnew, S.R.; Brown, D.W.; Tomé, C.N. Validating a polycrystal model for the elastoplastic response of magnesium alloy AZ31 using in situ neutron diffraction. *Acta Mater.* **2006**, *54*, 4841–4852. [CrossRef]
39. Lu, W.; Zhang, D.; Zhang, X.; Wu, R.; Sakata, T.; Mori, H. Microstructural characterization of TiB in in situ synthesized titanium matrix composites prepared by common casting technique. *J. Alloys Compd.* **2001**, *327*, 240–247. [CrossRef]
40. Bachmann, F.; Hielscher, R.; Schaeben, H. Texture analysis with MTEX–free and open source software toolbox. *Solid State Phenom.* **2010**, *160*, 63–68. [CrossRef]
41. Umar, M.; Qayyum, F.; Farooq, M.U.; Guk, S.; Kirschner, M.; Korpala, G.; Prahl, U. Exploring the Structure–Property Relationship in Spheroidized C45EC Steel Using Full Phase Crystal Plasticity Numerical Simulations. *Steel Res. Int.* **2022**, *93*, 2100452. [CrossRef]
42. Pütz, F.; Henrich, M.; Roth, A.; Könemann, M.; Münstermann, S. Reconstruction of Microstructural and Morphological Parameters for RVE Simulations with Machine Learning. *Procedia Manuf.* **2020**, *47*, 629–635. [CrossRef]
43. Huang, Y.; Froyen, L. Quantitative analysis of microstructure in metals with computer assistance. *NDT Net* **2001**, *6*, 5.
44. Lipiński, T.; Wach, A. Dimensional structure of non-metallic inclusions in high-grade medium carbon steel melted in an electric furnace and subjected to desulfurization. *Solid State Phenom.* **2014**, *223*, 46–53. [CrossRef]
45. Ghadbeigi, H.; Pinna, C.; Celotto, S. Quantitative strain analysis of the large deformation at the scale of microstructure: Comparison between digital image correlation and microgrid techniques. *Exp. Mech.* **2012**, *52*, 1483–1492. [CrossRef]
46. Hutchinson, J.W. Bounds and self-consistent estimates for creep of polycrystalline materials. *Proc. R. Soc. Lond. A Math. Phys. Sci.* **1976**, *348*, 101–127.
47. Kalidindi, S.R. Incorporation of deformation twinning in crystal plasticity models. *J. Mech. Phys. Solids* **1998**, *46*, 267–290. [CrossRef]
48. Roters, F.; Eisenlohr, P.; Hantcherli, L.; Tjahjanto, D.D.; Bieler, T.R.; Raabe, D. Overview of constitutive laws, kinematics, homogenization and multiscale methods in crystal plasticity finite-element modeling: Theory, experiments, applications. *Acta Mater.* **2010**, *58*, 1152–1211. [CrossRef]
49. Eisenlohr, P.; Diehl, M.; Lebensohn, R.A.; Roters, F. A spectral method solution to crystal elasto-viscoplasticity at finite strains. *Int. J. Plast.* **2013**, *46*, 37–53. [CrossRef]
50. Qayyum, F.; Guk, S.; Prüger, S.; Schmidtchen, M.; Saenko, I.; Kiefer, B.; Kawalla, R.; Prahl, U. Investigating the local deformation and transformation behavior of sintered X3CrMnNi16-7-6 TRIP steel using a calibrated crystal plasticity-based numerical simulation model. *Int. J. Mater. Res.* **2020**, *111*, 392–404. [CrossRef]
51. Matsuno, F.; Nishikida, S.-i.; Ikesaki, H. Mechanical Properties of Manganese Sulphides in the Temperature Range between Room Temperature and 1000 deg. C. *Trans. Iron Steel Inst. Jpn.* **1985**, *25*, 989–998. [CrossRef]
52. Qayyum, F.; Umar, M.; Guk, S.; Schmidtchen, M.; Kawalla, R.; Prahl, U. Effect of the 3rd Dimension within the Representative Volume Element (RVE) on Damage Initiation and Propagation during Full-Phase Numerical Simulations of Single and Multi-Phase Steels. *Materials* **2021**, *14*, 42. [CrossRef]
53. Gussev, M.; Edmondson, P.; Leonard, K. Beam current effect as a potential challenge in SEM-EBSD in situ tensile testing. *Mater. Charact.* **2018**, *146*, 25–34. [CrossRef]
54. Tasan, C.; Hoefnagels, J.; Geers, M. Microstructural banding effects clarified through micrographic digital image correlation. *Scr. Mater.* **2010**, *62*, 835–838. [CrossRef]
55. Umar, M.; Qayyum, F.; Farooq, M.U.; Khan, L.A.; Guk, S.; Prahl, U. Analyzing the cementite particle size and distribution in heterogeneous microstructure of C45EC steel using crystal plasticity based DAMASK code. In Proceedings of the 2021 International Bhurban Conference on Applied Sciences and Technologies (IBCAST), Islamabad, Pakistan, 12–16 January 2021; pp. 15–20.
56. Needleman, A. A continuum model for void nucleation by inclusion debonding. *J. Appl. Mech.* **1987**, *54*, 525–531. [CrossRef]
57. Diehl, M.; Naunheim, Y.; Yan, D.; Morsdorf, L.; An, D.; Tasan, C.; Zaefferer, S.; Roters, F.; Raabe, D. Coupled Experimental-Numerical Analysis of Strain Partitioning in Metallic Microstructures: The Importance of Considering the 3D Morphology. In Proceedings of the BSSM 12th International Conference on Advances in Experimental Mechanics, Sheffield, UK, 29–31 August 2017; pp. 29–31.
58. Qayyum, F.; Guk, S.; Prahl, U. Studying the Damage Evolution and the Micro-Mechanical Response of X8CrMnNi16-6-6 TRIP Steel Matrix and 10% Zirconia Particle Composite Using a Calibrated Physics and Crystal-Plasticity-Based Numerical Simulation Model. *Crystals* **2021**, *11*, 759. [CrossRef]

crystals

MDPI

Article

Transformation of 2D RVE Local Stress and Strain Distributions to 3D Observations in Full Phase Crystal Plasticity Simulations of Dual-Phase Steels

Shaochen Tseng [1,2], Faisal Qayyum [1,*], Sergey Guk [1], Chingkong Chao [2] and Ulrich Prahl [1]

1 Institut für Metallformung, Technische Universität Bergakademie, 09599 Freiberg, Germany; shao-chen.tseng@extern.tu-freiberg.de (S.T.); sergey.guk@imf.tu-freiberg.de (S.G.); ulrich.prahl@imf.tu-freiberg.de (U.P.)

2 Department of Mechanical Engineering, National Taiwan University of Science and Technology, Taipei 106335, Taiwan; ckchao@mail.ntust.edu.tw

* Correspondence: faisal.qayyum@imf.tu-freiberg.de

Abstract: Crystal plasticity-based numerical simulations help understand the local deformation behavior of multiphase materials. It is known that in full phase simulations, the local 2-dimensional (2D) representative volume elements (RVEs) results are distinctly different from 3-dimensional (3D) RVEs. In this work, the difference in the results of 2D and 3D RVEs is investigated systematically, and the effect of magnification, total strain and composition are analyzed. The 3D RVEs of dual-phase (DP)-steel are generated using DREAM-3D. The 2D RVEs are the sliced surfaces of corresponding 3D RVEs for a direct pixel-to-pixel comparison of results. It is shown that the corresponding 3D distribution can be rapidly derived from the 2D result based on the alternative error and least square method. The interactive parameters for these processes are identified and analyzed for the ferrite phase, which provides information about the convergence. Examined by qualitative and quantitative statistical analysis, it is shown that the corresponding 2D distribution by the fourth iteration has a prominent similarity with the exact 3D distribution. The work presented here contributes toward solving the paradox of comparing local strain from 2D crystal plasticity (CP) simulations with the effective 3D specimen used for tests.

Keywords: crystal plasticity; DAMASK; representative volume element; least square method; alternative error method; dual-phase steel

Citation: Tseng, S.; Qayyum, F.; Guk, S.; Chao, C.; Prahl, U. Transformation of 2D RVE Local Stress and Strain Distributions to 3D Observations in Full Phase Crystal Plasticity Simulations of Dual-Phase Steels. *Crystals* **2022**, *12*, 955. https://doi.org/10.3390/cryst12070955

Academic Editors: Pavel Lukáč and Wojciech Polkowski

Received: 28 April 2022
Accepted: 6 July 2022
Published: 8 July 2022

1. Introduction

Multiple-phase composites possess admirable mechanical properties and service life. Combining softer matrix with harder islands [1], multiple-phase materials demonstrate the strong structure and ductility, such as in dual-phase steel (DP steel) [2], metal matrix composites, and other advanced steel [3]. The material properties of the multiphase materials depend on the microstructural attributes, such as the size, shape, composition, and distribution of the second phase within the matrix [4]. The effect of these attributes is interdependent and plays a key role in defining the deformation and damage behavior under varying loading conditions [5,6], i.e., strain rate, temperature, and loading direction. Numerical simulation models provide an interesting outlook for targeted material development by avoiding the expensive and time-consuming experimentation of every iterative modification in the material microstructure [7–9].

There are different numerical simulation models, i.e., empirical, analytical, data-driven, and hybrid [10,11]. Although they are useful in the general modeling of material deformation and damage behavior, they lack the fundamental dependence on the local microstructural attributes [12,13]. Fast Fourier transformation-based crystal plasticity models provide a comprehensive and accurate solution for modeling the multiphase material's dependence on the microstructural attributes [14,15]. The RVEs in such models comprise all the necessary phase compositions, size, orientation, and distribution that represent the actual material [16]. Therefore, the model's accuracy and applicability largely depend on the constructed RVE for such simulations.

Different simple and sophisticated models for accurate RVE development have been proposed previously, i.e., single-step Voronoi tessellation, multi-step Voronoi tessellation, and artificial neural networks yielding accurate results that are usually 3D. According to the RVE [17] method, both 2D (2-dimension) [18] and 3D (3-dimension) [19] microstructures have been constructed by random grain size, orientation, phase, and texture or measured from electron backscatter diffraction (EBSD) data [20]. Based on EBSD patterns from the serial-sectioning experiments, 3D polycrystalline microstructures were constructed by Groeber et al. [21,22]. Recently, DREAM-3D [23] was used to construct realistic RVEs from virtual or real statistical grain size, orientation, and texture data. Applying DREAM-3D and DAMASK [24], the performance of multilayer composites can be evaluated under mechanical loading by the crystal plasticity material model.

Compared to the simulation and experimental data for DP steels, Ramazani et al. [25] discovered that the 2D model displayed underestimated behavior, while the 3D result demonstrated a quantitative description of the flow curve in comparison to the experimental data. Qayyum et al. [26,27] studied the local deformation and transformation behavior of transformation-induced plasticity (TRIP) steel. In these articles, the simulation results of global stress and strain behavior were compared to the experimental result. Nevertheless, the validation of local distribution in 3D has a challenge to discuss, owing to the inconsistency between experiment and simulation results. Diehl et al. [28] analyzed the effect of "columnarity" and studied the influence of the nearby environment on stress and strain in DP microstructures. It was discovered that the local stress and strain distributions are strongly influenced by both the nearby grain shape and grain orientation. Due to this effect, the 2D simulations of heterogeneous microstructures could be definitely misleading for the damage prediction [29] of crack initiation and propagation.

The CP models are also largely dependent on several physical and a few fitting parameters. These parameters are obtained by comparing averaged numerical simulation results with experimental stress and strain curves [26,30]. Although a certain set of fitted parameters seems to accurately represent the overall deformation behavior, a slightly different set of fitted parameters is also expected to yield similar global results with a significant change in local results. This reliability of such models on the globally calibrated parameters can lead to incorrect local results. In the recent past, researchers have developed algorithms for automated sensitivity analyses and parametric identifications from global stress and strain curves [31,32]. A more reliable way of calibrating and validating the CP model parameters can be through direct local comparison of experimental in situ and simulation results.

The local strain measurement is experimentally possible now due to the advancement of in situ measurement and data processing tools, which can accurately capture local strain distribution and local microstructural attributes [33–35]. However, constructing an accurate 3D RVE of the same specimen is a paradox as 3D EBSD requires polishing, slicing, and measurement throughout the specimen. Moreover, different phases frequently cause false detection at the measurement surfaces and the grain boundaries [36].

Therefore, to carry out such a comparison, the 2D RVE results from CP simulation should be reliably transformed to keep the global results the same while transforming the local stress and strain distributions for comparison with the results of the 3D material surface. It is a mathematical challenge that can yield a model for transforming 2D RVE simulation results of local strain distribution for comparison with strain measurements on the surface of a 3D specimen. Recently, Qayyum et al. [37,38] investigated the stress and strain distribution between 2D and 3D RVEs with different total grain numbers. It was shown that the global distribution of the flow curve is similar across different RVEs. However, the local distribution showed obvious variation. The numerical finding of

Qayyum et al. [37,38] underlines the different stress and strain distribution in 2D and 3D via random distribution of grain size and orientation. In the current article, the previous study is carried one step further, and a corresponding local 3D stress and strain distribution on the RVE surface is achieved by transforming 2D stress and strain. The work is based on the alternative error and least square method. The transformation of the ferrite phase stress and strain from the 2D result is iteratively achieved. The iterative constants for different cases of magnification, strain levels, and volume fractions are presented and compared. The proposed new numerical approach is helpful in iteratively transforming the 2D stress and strain if the 3D outcome is known. To analyze and compare the derived 3D distribution with actual results, a statistical analysis of the obtained data is carried out.

The following Section 2 presents the detail of the numerical simulation model, including grain size, boundary condition, and material parameter of crystal plasticity. Then, the proposed iterative method, which is adopted in the current work, is explained in Section 3. Next, the derived 3D local stress and strain distribution calculated using the proposed method are displayed. Meanwhile, similarities between derived and exact 3D distribution are provided by statistical analysis in Section 4. Finally, the discussion and conclusion are presented in Sections 5 and 6.

2. Numerical Simulation Model

The material adopted in the current investigation is DP steel, which comprises two distinct phases, i.e., soft ferrite matrix and hard martensite island. This large local microstructural heterogeneity yields significant local strain contrast during deformation. The microstructural distribution, grain size, material properties, and grain orientation are adopted from previously published work [39–41]. This article investigates different parameters for varying the overall number of grains, strain levels, and volume fractions. Table 1 demonstrates the grain size distribution of ferrite and martensite, number of total grains, volume fractions, and strain levels for each model. For cases 1 to 3, the varying parameter is the total number of grains; for cases 4 to 6, the varying parameter is strain level; For cases 7 to 9, the varying parameter is volume fraction. Note that the unit of grains size is µm and total grain is a dimensionless value. Based on experimental observation [42], the smaller average grain sizes of ferrite and martensite were assigned as 6.35 and 4.6 in RVE-A, whereas the larger average grain sizes were 14.0 and 12.8 in RVE-C during the crystal plasticity simulation. Hence, the total grains of RVE-A are more than RVE-C.

Table 1. Grain size distribution, number of total grain, volume fraction, and strain level for the microstructure [37].

Case	RVE	Ferrite Grains			Martensite Grains			Total Grain	Strain Level [%]	Volume Fraction
		Min.	Max.	Avg.	Min.	Max.	Avg.			
1	A	5.1	7.6	6.35	3.5	5.7	4.6	8400	25	0.1
2	B	8.5	12.7	10.6	5.4	8.8	7.1	1900	25	0.1
3	C	11.2	16.8	14.0	9.8	15.8	12.8	700	25	0.1
4	D	8.5	12.7	10.6	5.4	8.8	7.1	1900	5	0.1
5	D	8.5	12.7	10.6	5.4	8.8	7.1	1900	15	0.1
6	D	8.5	12.7	10.6	5.4	8.8	7.1	1900	25	0.1
7	D	8.5	12.7	10.6	5.4	8.8	7.1	1900	15	0.1
8	E	8.5	12.7	10.6	5.4	8.8	7.1	1900	15	0.15
9	F	8.5	12.7	10.6	5.4	8.8	7.1	1900	15	0.2

A cubic equiaxed crystal structure and ellipsoid grain shape are assigned to both ferrite and martensite phases. $100 \times 100 \times 100$ voxels RVEs are systematically constructed using DREAM-3D for the sizes as shown in Table 1. Figure 1 shows a schematic diagram and flow chart in this article. The blue arrows indicate the steps involved in carrying out this work. In the beginning, 2D and 3D simulations for different RVEs are carried out. Both ferrite (F) and martensite (M) phases are then individually analyzed. Combined with the least square and alternative error method, the iterative constants (a_1, b_1, a_2, and b_2, ...) of the linear equation are calculated while transforming the 2D results into 3D.

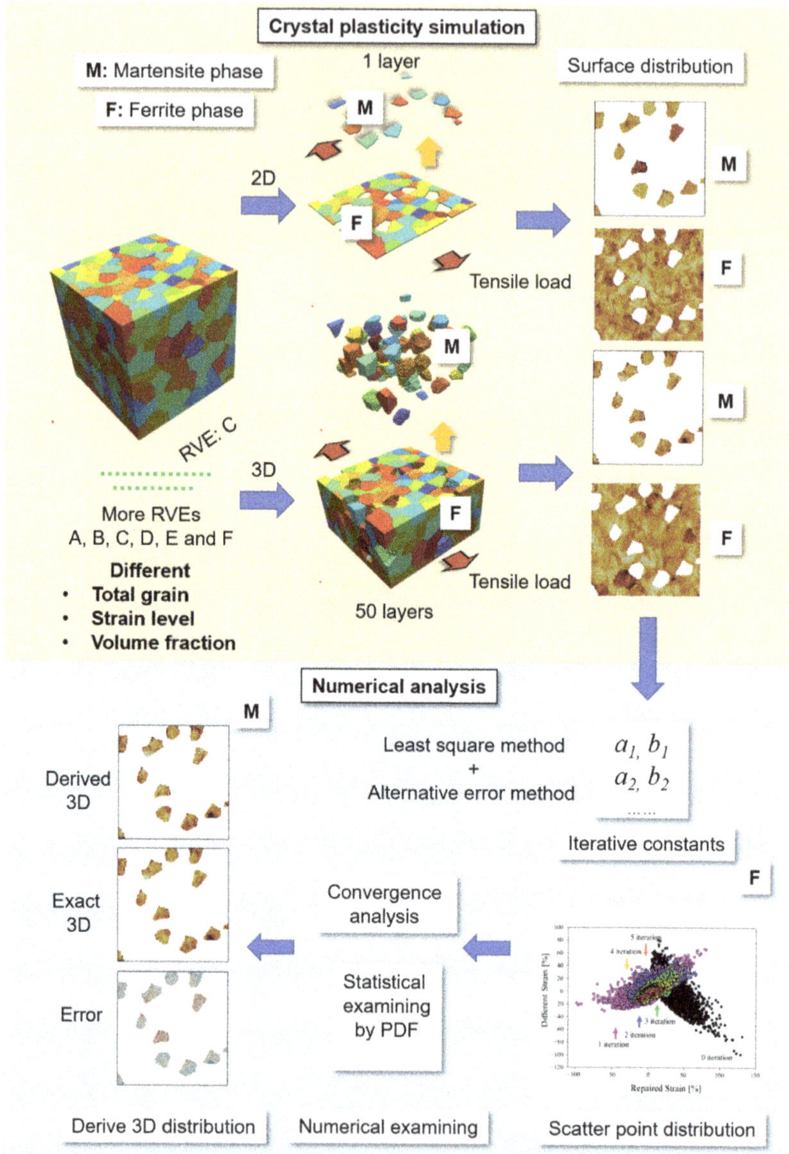

Figure 1. The schematic diagram represents the flow chart of the current work. The upper half (yellow background) indicates the flow chart of crystal simulation, and the bottom half (blue background) represents numerical analysis. The blue arrow indicates the flow of data. The red arrow represents tensile loading, and the yellow arrow separates the ferrite (F) and martensite (M) phases.

The principal aim of this ongoing study is to develop a robust methodology for deriving the 3D surface stress and strain distribution from the 2D RVE result for ferrite and martensite phases. This work brings us one step closer to that aim. As a representation,

RVE-C in inverse pole figure (IPF) colors is shown in Figure 2a (h = 1), where h indicates the length of z-direction normalized total length of z-direction. Following Qayyum et al. [37], 50- (h = 0.5) and 1-layer (h = 0.01) RVE were sliced from the initial microstructure and regarded as 3D and 2D RVE, respectively. The simulation results for 1- and 50-layers have been selected and defined as 2D and 3D results of the local stress and strain distribution. There are 10,000 Gaussian mesh elements in one layer, meaning 2D and 3D possess 10,000 and 500,000 elements, respectively. All the constructed RVEs were sliced into similar geometries with 1-layer (h = 0.01) RVE and shown in terms of texture style in Figure 2b.

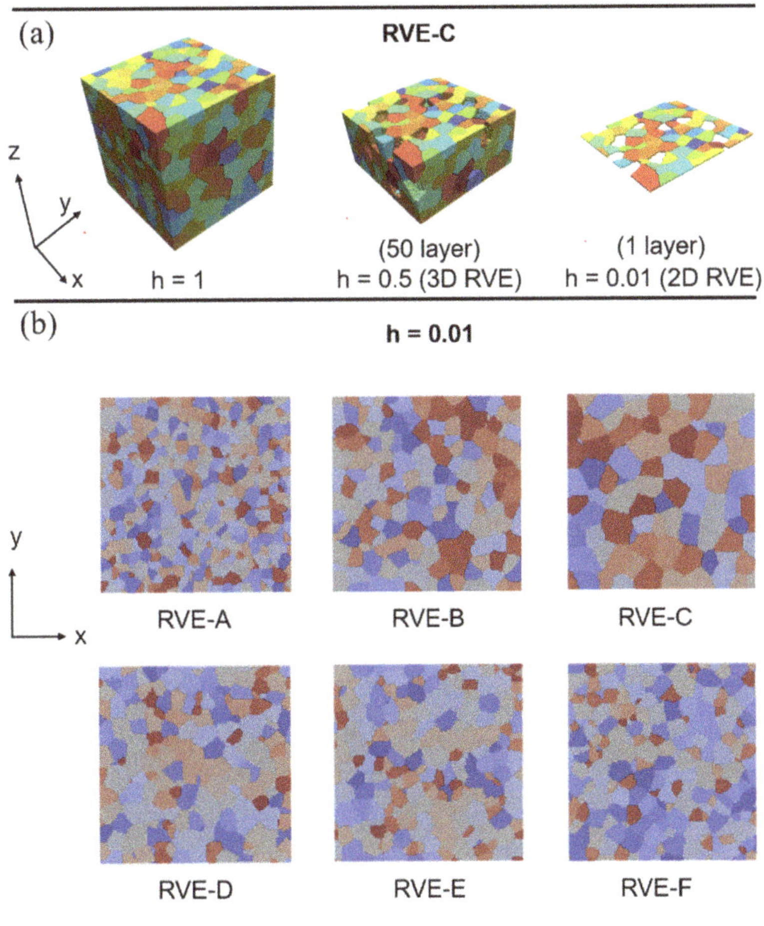

Figure 2. (**a**) The resulting RVE-C initial microstructure colored according to the inverse pole figure (IPF). (**b**) All RVE models (RVE-A to RVE-F) were sliced into geometries with 1-layer (h = 0.01) RVE in terms of texture style.

The ferrite and martensite are defined as elastic–viscoplastic deformable phases. The elastic stiffness, shear resistance, hardening behavior, and curve fitting parameter are adapted from previously published work [38] and presented in Table 2.

Table 2. Mechanical properties of multiphase (ferrite and martensite) were adopted from [37] for simulation modeling.

Parameter	Symbol	Ferrite	Martensite	Unit
First elastic stiffness constant with normal strain	C_{11}	233.3	417.4	GPa
Second elastic stiffness constant with normal strain	C_{12}	135.5	242.4	GPa
First elastic stiffness constant with shear strain	C_{44}	128.0	211.1	GPa
Initial shear resistance on [111]	S_O [111]	95	406	MPa
Saturation shear resistance on [111]	S_∞ [111]	222	873	MPa
Initial shear resistance on [112]	S_O [112]	96	457	MPa
Saturation shear resistance on [112]	S_∞ [112]	412	971	MPa
Slip hardening parameter	h_O	1.0	563	GPa
Interaction hardening parameter	$h_{\alpha,\beta}$	1.0	1.0	-
Stress exponent	n	20	20	-
Curve fitting parameter	w	2.0	2.0	-

The crystallographic orientation, mechanical properties, and phase of ferrite and martensite are included in the RVE geometry definition. A uniaxial load along the x-direction is defined using mixed boundary conditions as follows:

$$\dot{F}_{ij} = \begin{bmatrix} 1 & 0 & 0 \\ 0 & * & 0 \\ 0 & 0 & * \end{bmatrix} \times 10^{-3} {\cdot} s^{-1} \tag{1}$$

$$P_{ij} = \begin{bmatrix} * & * & * \\ * & 0 & * \\ * & * & 0 \end{bmatrix} Pa \tag{2}$$

where $\dot{F}_{ij}$ is the coefficients of the macroscopic rate of the deformation gradient, P_{ij} is the first Piola–Kirchhoff stress. $\dot{F}_{11} = 1/s$ indicates tensile condition, 0 is represented as restricted, and * is an arbitrary value during the simulation. It should be noted that the strain rate of all simulations is assumed to be $1 \times 10^{-3}/s$ in conjunction with periodic boundary conditions in all three directions. The simulations are performed in plain strain mode, where 2D RVE is interpreted as a columnar grain structure.

The spectral method via fast Fourier transform [43] is used to solve the continuum mechanics formulation mentioned above. After completing simulations, the 2D and 3D data are statistically analyzed. The 3D numerical results are used as a reference for iteratively modifying the 2D results by combining the least square and alternative error methods via Matlab programs (2019b, The MathWorks Inc., Natick, MA, USA).

3. Method

In the beginning, both the 50- and 1-layers were regarded as geometry models for crystal simulation and considered as 3D and 2D RVE models, as shown in Figure 2a. Note that stress/strain throughout this article means either stress or strain conditions. The results from the 2D RVE are compared with the surface of the corresponding 3D RVE to analyze the difference in stress/strain values. First, the stress/strain on the 2D layer result with respect to the difference in stress/strain between the 3D and 2D layer results is considered. Based on the least square method, the linear equation for this can be written as follows:

$$y_{1\mathrm{i}} = a_1 x_{1\mathrm{i}} + b_1, \quad \mathrm{i} = 1,2,3\ldots\ldots 10{,}000 \tag{3}$$

$$a_1 = \frac{\sum_{\mathrm{i}=1}^{10{,}000} (f_{1\mathrm{i}} - \overline{f_1})(f_{\mathrm{er0i}} - \overline{f_{\mathrm{er0}}})}{\sum_{\mathrm{i}=1}^{10{,}000} (f_{1\mathrm{i}} - \overline{f_1})^2}, \; b_1 = \overline{f_{\mathrm{er0}}} - a_1 \overline{f_1}, \; \mathrm{i} = 1,2,3\ldots\ldots 10{,}000 \tag{4}$$

where i is the number of elements on the top surface, and there are 100 × 100 voxels on both 2D and 3D layers. Where f_{1i} is stress/strain for 2D layer result; $\overline{f_1}$ is average stress/strain for 2D layer result; f_{50i} is stress/strain for 3D layer result; f_{er0i} is the difference in stress/strain between 3D and 2D layer results ($f_{er0i} = f_{50i} - f_{1i}$), which indicates the difference of non-revised (original 2D) stress/strain; and $\overline{f_{er0}}$ is the average difference in stress/strain between 3D and 2D layer results.

Through this method, the corresponding y_{1i}, a_1, and b_1 can be obtained. Hence, the first revised stress/strain (f_{r1i}) and the different first revised stress/strain (f_{er1i}) can be derived:

$$f_{r1i} = f_{1i} + y_{1i},\ i = 1,2,3\ldots\ldots 10,000 \tag{5}$$

$$f_{er1i} = f_{50i} - f_{r1i},\ i = 1,2,3\ldots\ldots 10,000 \tag{6}$$

where f_{er1i} can be calculated by the difference in stress/strain between the 3D layer and the first revised 2D layer result. However, if the error of the first revision is not convergent, the second iteration will be carried out.

Next, the difference in stress/strain between the 3D and 2D layer results is considered with respect to the different first revised stress/strain. Based on the least square method, the linear equation can be obtained as follows:

$$y_{2i} = a_2 x_{2i} + b_2,\ i = 1,2,3\ldots\ldots 10,000 \tag{7}$$

$$a_2 = \frac{\sum_{i=1}^{10,000}\left(f_{er0i} - \overline{f_{er0}}\right)\left(f_{er1i} - \overline{f_{er1}}\right)}{\sum_{i=1}^{10,000}\left(f_{er0i} - \overline{f_{er0}}\right)^2}, b_2 = \overline{f_{er1}} - a_2\overline{f_{er0}},\ i = 1,2,3\ldots\ldots 10,000 \tag{8}$$

Through this method, the corresponding y_{2i} and x_{2i} can be obtained. Hence, the second revised stress/strain (f_{r2i}) and the different second revised stress/strain (f_{er2i}) can be derived:

$$f_{r2i} = f_{r1i} + y_{2i},\ i = 1,2,3\ldots\ldots 10,000 \tag{9}$$

$$f_{er2i} = f_{50i} - f_{r2i},\ i = 1,2,3\ldots\ldots 10,000 \tag{10}$$

f_{er2i} can be calculated by the difference in stress/strain between the 3D layer and the second revised stress/strain. Finally, based on the least square method, a general linear equation can be obtained in the series form of m-iteration. The difference in the m-2 revised stress/strain (x-axis) is considered with respect to the different m-1 (y-axis) revised stress/strain. This equation can be written as follows:

$$y_{mi} = a_m x_{mi} + b_m,\ i = 1,2,3\ldots\ldots 10,000, m = 2,3,4,\ldots \tag{11}$$

$$a_m = \frac{\sum_{i=1}^{10,000}\left(f_{er(m-2)i} - \overline{f_{er(m-2)}}\right)\left(f_{er(m-1)} - \overline{f_{er(m-1)}}\right)}{\sum_{i=1}^{10,000}\left(f_{er(m-2)i} - \overline{f_{er(m-2)}}\right)^2},\ b_m = \overline{f_{er(m-1)}} - a_m\overline{f_{er(m-2)}},\ i = 1,2,3\ldots\ldots 10,000, m = 2,3,4,\ldots \tag{12}$$

Through this method, the corresponding y_{mi} and x_{mi} can be obtained. Hence, the m revised stress/strain (f_{rmi}) and the different m revised stress/strain (f_{ermi}) can be derived as:

$$f_{rmi} = f_{r(m-1)i} + y_{mi},\ i = 1,2,3\ldots\ldots 10,000,\ m = 2,3,4,\ldots \tag{13}$$

$$f_{ermi} = f_{50i} - f_{rmi},\ i = 1,2,3\ldots\ldots 10,000,\ m = 2,3,4,\ldots \tag{14}$$

f_{ermi} can be calculated by the difference in stress/strain between the 3D layer and the m revised stress/strain.

4. Results

Although results for both the ferrite and martensite phase were processed and analyzed in this work, in this section, the results for the ferrite phase are presented. Similar numerical results (Tables A1 and A2) of the martensite phase are provided in Appendix A of this article for interested readers.

4.1. Local Stress/Strain Distribution in 2D and 3D

Initially, local stress/strain distributions for case 1 to case 9 are shown in Figure 3 to address how the local results vary between 2D and 3D separately sourced from 1- and 50-layers only for ferrite matrix. Note that case 1 to case 3 indicates different total grain numbers; case 4 to case 6 indicate different strain levels; case 7 to case 9 indicates different volume fractions. The detailed microstructure information is demonstrated in Table 1. To point out the difference between 2D and 3D-RVEs, the local stress and strains are arithmetically subtracted (3D−2D) as $f_{er0i} = f_{50i} - f_{1i}$, and the difference map with the true stress or strain scales −700 to 700 MPa −60 to 60%, respectively, is shown in Figure 3.

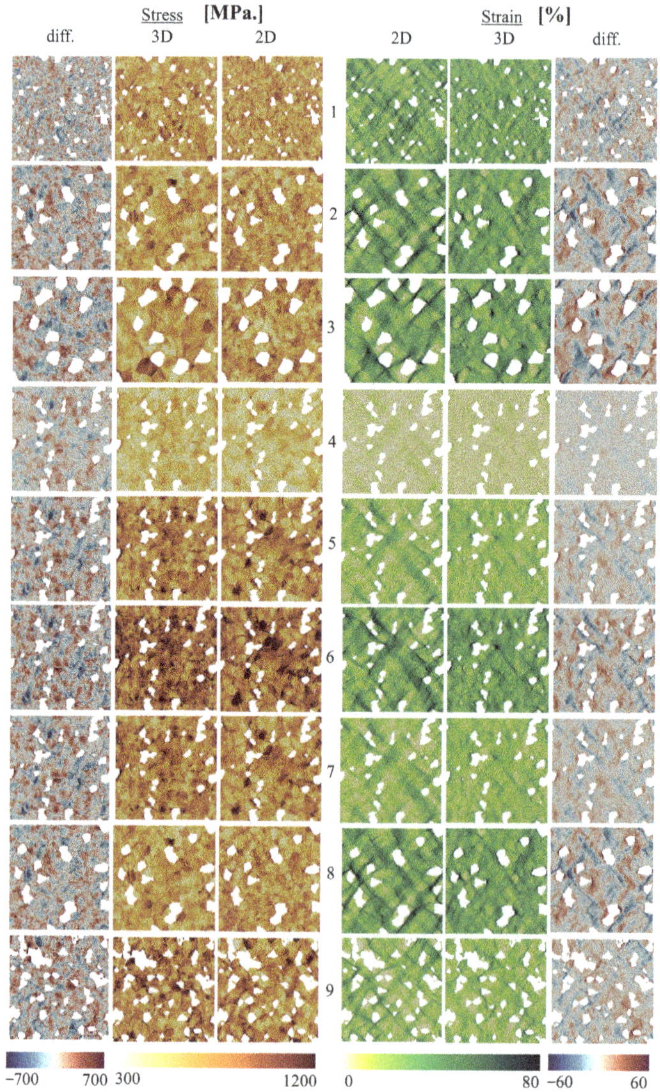

Figure 3. Local von Mises stress and strain distribution for ferrite phase from case 1 to case 9. Note

that case 1 to case 3 indicate different total grain numbers; case 4 to case 6 indicate different strain levels; case 7 to case 9 indicate different volume fractions. The extreme left and right column indicates the 2D and 3D local stress and strain distribution differences, respectively.

For 2D conditions, high contrast in the stress/strain concentration is observed. It can be noted that there is a significant difference between the 3D and 2D-RVEs local attribute distribution. In the 3D RVE case, the stress and strain are relatively homogeneously distributed and display lower contrast. Meanwhile, the higher stress/strain transfer to the martensite/ferrite interface. It can be concluded that the local stress/strain distribution is different from 3D and 2D-RVEs, regardless of the total grain numbers, strain levels, and volume fractions. Therefore, a straightforward transformation method can hardly be adapted to modify 2D results to 3D distribution for a specific element. The results from the proposed numerical statistical analysis are given in the next section.

4.2. Step by Step for Transformation from 2D to 3D

In Section 2, the proposed iterative method was derived for this specific problem. Figure 4 shows a scatter diagram of the ferrite phase with two extreme "total grain number" cases (case 1 (a, c) and case 3 (b, d)) via different iterative steps both for stress (a, b) and strain (c, d) value. Since all the cases are similar, this article only presents cases 1 and case 3 for discussion. Regarding the x-axis, the repaired stress/stain is indicative of the corresponding different revised stress/strain of 2D. It assumes 2D stress/strain (f_{1i}) for the zero iteration, but for the first iteration and onwards, it is derived to f_{er0i} and Equation (6) for the first and second iteration, respectively. For the y-axis, the different stress/strain comes from the general form in Equation (14), where the 2D revised stress/strain compared with 3D stress/strain is computed for each solution point. This term is now clearly used and mentioned in the explanation of f_{er0i} for the zero iteration. Then, Equations (6) and (10) indicate the first and second iteration, respectively.

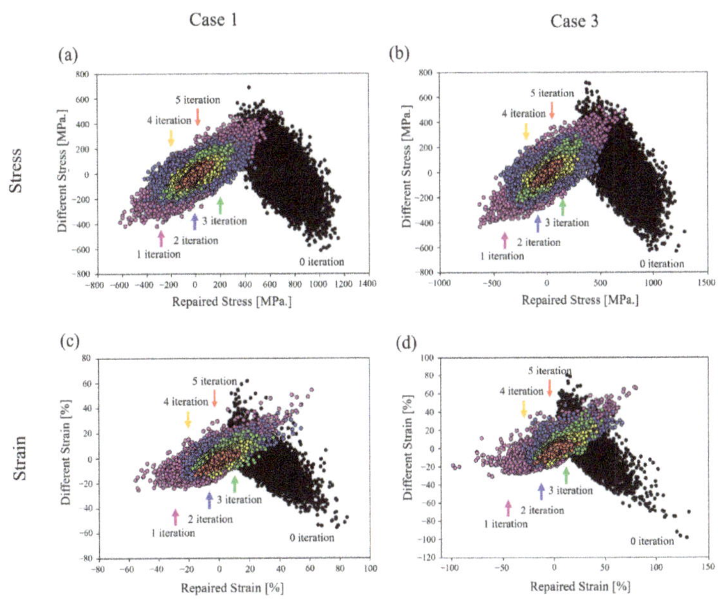

Figure 4. Scatter diagram showing the numerical tendency of stress in (**a**) case 1 (**b**) case 3 and strain in (**c**) case 1 (**d**) case 3 by the proposed iterative method for ferrite phase at 25% global true strain and 0.1 volume fraction.

For the same iterative step, the numerical distribution reveals a similar tendency regardless of different total grain numbers for stress cases, as shown in Figure 4a,b, or for strain cases, as shown in Figure 4c,d, respectively. For the initial condition, the stress/strain on the 2D layer result (f_{1i}) with respect to the difference of non-revised stress/strain between 2D and 3D layer results (f_{er0i}) are defined as black scatter, as shown in Figure 4. It can be observed that the stress/strain results in 2D have inverse behavior to the difference in stress/strain between the 3D and 2D layer results. Due to inverse behavior, it can be speculated that the larger value of stress/strain is overestimated, and the lower value of stress/strain is underestimated for the 2D layer result. Transferring the local stress/strain of a 2D layer to match the 3D layer better, a larger value of stress/strain should be forcibly applied to a negative value of stress/strain, and a positive value of stress/strain should be forcibly applied to a smaller one.

Based on the least square method, the linear equation for the original condition can be calculated in Equations (3) and (4). Tables 3 and 4 demonstrate the constants a_m,b_m and R_m^2 (m = 1, 2, 3, 4) of the linear equation by the proposed iterative method for stress/strain conditions during different iterative steps. Using a_1 and b_1, the value of stress/strain in 2D layer result will be forcibly modified by a corresponding value and become the first revised stress/strain in Equation (5). Meanwhile, the different first revised stress/strain can be calculated by Equation (6).

After applying the approach mentioned above, the first repaired stress/strain (f_{er0i}) concerning different first revised stress/strain (f_{er1i}) are shown in purple scatters in Figure 4. It can be observed that different first revised stress/strain results have proportional behavior to the result of the first repaired stress/strain.

Repeatedly, through the least square method, the linear equation for the second iteration condition can be calculated by Equations (7) and (8). Through a_2 and b_2, the first revised stress/strain result (f_{r1i}) will forcibly be modified with a corresponding value of a linear equation and become the second revised stress/strain result (f_{r2i}) in Equation (9). Additionally, the different second revised stress/strain (f_{er2i}) can be calculated by Equation (10). Followed by a similar iterative procedure, the second repaired stress strain (f_{er1i}) concerning the different second revised stress/strain (f_{er2i}) is defined as blue scatters, as shown in Figure 4. Similarly, third repaired stress/strain (f_{er2i}) concerning the different third revised stress/strain (f_{er3i}) is defined as green scatters.

Table 3. Constants a_m, b_m, and R_m^2 (m = 1, 2, 3, 4) for the ferrite phase of linear equation with stress condition by the proposed iterative method.

	Stress								
Analysis Parameter	**Total Grain**			**Stress/Strain Level, %**			**Volume Fraction, %**		
	8400	**1900**	**770**	**5**	**15**	**25**	**0.1**	**0.15**	**0.2**
Case	1	2	3	4	5	6	7	8	9
RVE									
Iterative	A	B	C	D	D	D	D	E	F
Const.									
a_1	−0.76	−0.81	−0.75	−0.63	−0.74	−0.78	−0.74	−0.66	−0.66
b_1	543.73	572.10	572.15	357.41	470.71	579.71	470.71	444.5	435.8
R_1^2	0.35	0.37	0.34	0.31	0.38	0.41	0.38	0.33	0.31
a_2	0.64	0.62	0.65	0.68	0.61	0.59	0.61	0.64	0.66
b_2	−8.62	−10.73	−5.70	−11.48	−9.46	−7.43	−9.46	−13.85	−12.25
R_2^2	0.64	0.64	0.68	0.69	0.62	0.60	0.62	0.69	0.72
a_3	0.35	0.36	0.32	0.31	0.37	0.40	0.37	0.30	0.27
b_3	−1.41	−2.97	−3.13	−1.38	-2.44	−3.03	−2.44	−4.95	−4.28
R_3^2	0.35	0.36	0.32	0.31	0.37	0.40	0.37	0.30	0.27
a_4	0.65	0.64	0.68	0.69	0.62	0.60	0.62	0.70	0.72
b_4	−0.18	−0.54	−1.03	−0.253	−0.36	−0.42	−0.36	−2.2	−2.4
R_4^2	0.65	0.64	0.68	0.69	0.62	0.59	0.62	0.70	0.72

Table 4. Constants a_m, b_m, and R^2_m (m = 1, 2, 3, 4) for the ferrite phase of linear equation with strain condition by the proposed iterative method.

Analysis Parameter	Total Grain			Strain Strain Level, %			Volume Fraction, %		
	8400	1900	770	5	15	25	0.1	0.15	0.2
Case	1	2	3	4	5	6	7	8	9
RVE									
Iterative	A	B	C	D	D	D	D	E	F
Const.									
a_1	−0.78	−0.85	−0.82	-0.72	−0.73	−0.75	−0.73	−0.74	−0.73
b_1	22.09	25.30	25.16	4.57	13.95	23.85	13.95	14.91	14.94
R_1^2	0.50	0.57	0.49	0.49	0.49	0.49	0.49	0.51	0.48
a_2	0.50	0.44	0.50	0.5	0.50	0.49	0.50	0.48	0.51
b_2	0.01	0.51	−0.31	−0.05	−1.87	−0.38	−1.87	0.21	0.16
R_2^2	0.50	0.44	0.51	0.50	0.50	0.49	0.50	0.49	0.53
a_3	0.50	0.56	0.49	0.49	0.49	0.50	0.49	0.50	0.47
b_3	−0.06	−0.11	−0.30	−0.02	−0.08	−0.16	−0.08	−0.21	−0.21
R_3^2	0.50	0.56	0.48	0.49	0.49	0.50	0.49	0.50	0.47
a_4	0.50	0.43	0.51	0.50	0.50	0.49	0.50	0.49	0.53
b_4	−0.005	−0.01	−0.04	−0.02	−0.007	−0.01	−0.007	−0.04	−0.04
R_4^2	0.50	0.43	0.51	0.50	0.50	0.49	0.50	0.49	0.52

4.3. Convergence and Statistical Analysis

A quantitative analysis method determines the convergent behavior and confirms the error analysis via different iterative steps. The function of average difference is defined as follows:

$$er_{\text{avg.}}(x) = \sum_{i=1}^{n} \frac{|f_{\text{ermi}}|}{n} \text{m} = 0, 1, 2 \ldots, \text{n} = 10{,}000 \quad (15)$$

where f_{ermi} is the difference in stress/strain between 3D and revised 2D layer results as derived in Equation (14), which can also be obtained by the magnitude of the y-axis in Figure 4, "m" is an iterative step from 0 to 5, and n is element number in 100 × 100 voxels RVEs. Figure 5 shows the average difference in stress/strain with different iterative steps. To avoid confusion, only case 1 to case 3 are shown in the convergence analysis. As can be seen, the convergent behavior with a small error is observed in the fourth iterative step. Therefore, it can be concluded that the fourth iterative step can reach the desired 3D local stress/strain distribution after almost 70% average difference in stress/strain conditions.

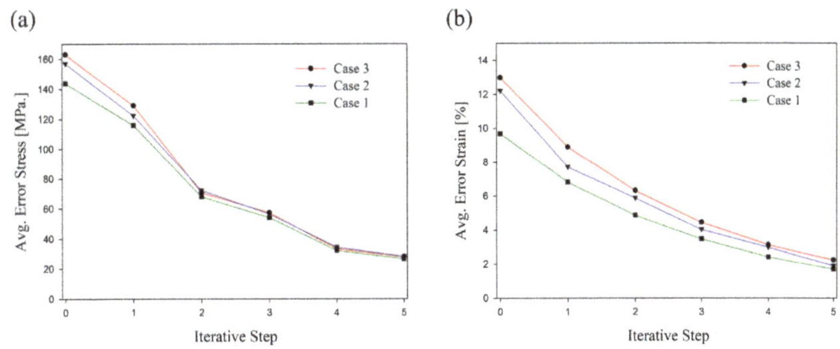

Figure 5. Convergence analysis for the ferrite phase at 25% global true strain. (**a**) Stress and (**b**) strain by the proposed iterative method with different iterative steps.

Previously presented qualitative comparisons of transformation are mainly based on individual visual perceptions. An appropriate statistical quantitative analysis is carried out here to compare the transformation results more accurately. For the statistical method, the probability distribution function is adopted. The probability distribution function can be derived in terms of μ and σ as follows:

$$p(x) = \frac{1}{\sigma\sqrt{2\pi}} \exp(-\frac{(x-\mu)^2}{2\sigma^2}), \quad -\infty < x < \infty \tag{16}$$

where μ is mean value of the probability distribution function σ is the standard deviation value. The standard normal distribution-based probability distribution function of the difference in stress/strain distribution for case 1 with different iterative steps is shown in Figure 6. Note that five different stress/strains from zero to the fourth iterative step have been considered in the probability distribution function. Again, the different stress/strain comes from the general form in Equation (14), where the 2D revised stress/strain compared with 3D stress/strain is computed for each solution point. It can be seen that the peak considerably protrudes with the increase in iterative steps in the probability diagram. Therefore, it can be concluded that the error is convergent as iterative steps increases, which represents the derived 3D stress and strain distribution, as effectively calculated by the proposed iterative method.

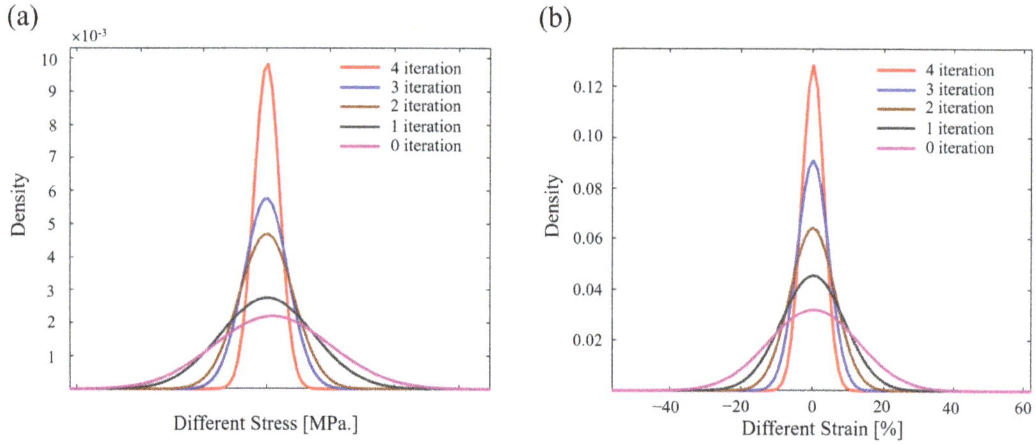

Figure 6. Probability distribution function of different (**a**) stress and (**b**) strain distributions between 3D result and iteratively revised 2D result for the ferrite phase of case 1 with different iterative steps.

4.4. Derivation of 3D Stress and Strain Distribution from the 2D Result

The graphic of derived 3D local stress and strain distribution with different iterative steps using the identified iterative constants will be demonstrated in this section. Due to the convergence and statistical analysis in Figures 5 and 6, the transformation is carried out up to the fourth iterative step. Local stress/strain distribution for the revised 2D layer result, 3D layer result, and differences between 2D and 3D layer results for the ferrite phase of case 1 with different iterative steps is as displayed in Figure 7. In the first iteration, the stress/strain distribution displays the virtually perfectly averaged arrangement when forcibly applying a negative value on a larger stress/strain value and a positive value on a smaller stress/strain value by modification from corresponding a_1 and b_1 in Tables 3 and 4. The high stress/strain zones disappear from the actual results. However, the difference in stress/strain between the first revised 2D and 3D layers can still be easily identified. The second revised stress/strain is obtained for the second iteration after modifying the first revised result by corresponding a_2 and b_2 in Tables 3 and 4. It can be observed that the stress/strain concentration is generated along with the ferrite/martensite interfaces. The second iteration result displays a dramatic difference compared with the first iteration result. Subsequently, the third and fourth revised stress/strain are obtained during the modification of the second and third results revised by the corresponding a_3 and b_3 as well as a_4 and b_4 in Tables 3 and 4. It can be seen that stress/strain concentration along the interface of the fourth revised result is more enhanced than the second revised result. It can be concluded that the derived 3D stress/strain distribution closely resembles the 3D result at the fourth iterative step.

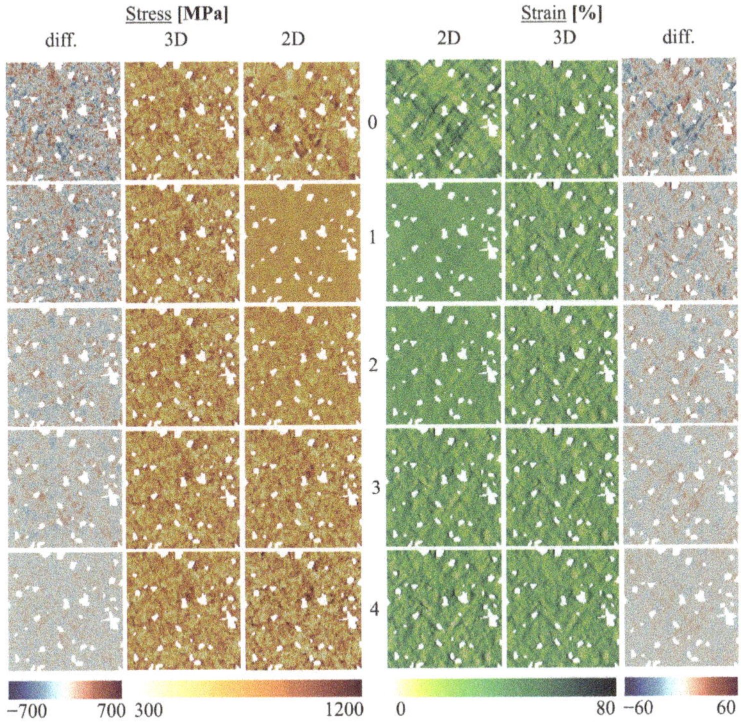

Figure 7. Local von Mises stress and strain distribution for the ferrite phase of case 1 with different iterative steps. The extreme left and right column indicate the 2D and 3D local stress and strain distribution differences, respectively.

5. Discussion

Due to the significantly higher stiffness of the martensite phase and strong "elastic mismatch" in the DP steels, higher interfacial stresses are induced due to applied mechanical or thermal loading [44,45]. Therefore, when comparing the stress and strain fields of the martensite and ferrite phases, extremely higher stress (≈2500 MPa) and lower strain are achieved in the martensite phase. On the contrary, smaller stress (≈800 MPa) and larger strain can be identified in the ferrite phase.

From the current and previous studies [28,37,38,46,47], in 2D crystal plasticity simulation, the obvious stress and strain concentrations exist in the ferrite phase due to the absence of the third dimension. As depth increases, the stress and strain are more distributed, and the concentration generally transfers to the matrix/particle interfaces and triple points of grains with a low Schmidt factor. As the thickness increases, the matrix's stress field is phenomenally influenced by the adjacent inclusion. Chao et al. [48] indicated that the distance between two circular inclusions strongly provoked the stress field of the matrix and then interfered with the stress intensity factor of the crack.

Additionally, Diehl et al. [28] investigated the interaction between stress/strain conditions, the phase distribution of neighboring grain, and their crystallographic orientation. It was concluded that there is a drastic difference in local stress/strain distribution between 2D and 3D-RVEs results, which means it is difficult to validate simulation results with similar experiments. If 3D geometry is measured by slicing the surface, there is no same specimen available for the in situ tests. On the contrary, if the specimen is tested by deformation loading, 2D EBSD only can be obtained, and the corresponding 2D simulation result differs from the 3D distributions [28].

In the database of parameters from the proposed iterative method, as displayed in Tables 3 and 4, it can be examined that the R_m^2 is practically similar in the same iterative step for ferrite phases regardless of total grain numbers, strain levels, and volume fractions. Additionally, the constants a_m and b_m nearly have a consistency tendency regardless of total grain numbers and volume fractions. There is a difference in the case of strain levels. That is, the constant a_m and R_m^2 display a similar value for different strain levels. Conversely, the constants b_m demonstrates larger value at larger strain levels. It is exhilarating that the 3D distribution can be straightforwardly responded to by 2D result by the same iterative constant owing to the similar a_m, b_m and statistical analysis, which means that the 3D distribution can be derived by the proposed iterative method regardless of total grain numbers and volume fractions.

The methodology of deriving 3D local distribution from 2D results has been thoroughly introduced. In the future, the constants a_m and b_m can provide a helpful suggestion to predict the 3D local stress/strain from 2D local stress/strain. Henceforward, the derived 3D local stress and strain distribution can be an excellent validation of the experiment and further recommends the optimal design of the multiple phase steel with different grain orientation, total grain numbers, composition, and loading conditions by the proposed iterative method.

6. Conclusions

In this article, the derived crystal plasticity simulation of 3D local stress and strain has been successfully obtained from the 2D simulation result. The difference in the local stress and strain distributions in 2D and 3D simulations of the same surface is significant. There are three different parameter series, including total grain numbers, strain levels, and volume fractions. After identifying the difference in the local distributions, the alternative error and least square methods are used to transform the 2D results to match with 3D observations. It is shown that using the presented proposed iterative technique, the 3D distribution can be quickly derived from the 2D result in a ferrite matrix of DP steel. It can be concluded that the desired transformation can be achieved in the fourth iterative step by using the identified iterative constants for each phase. For the iterative constants, almost similar values regardless of total grain numbers and volume fractions are displayed. The accuracy of the developed method is demonstrated by visual and statistical analysis, and the proposed numerical approach can be regarded as effective and efficient. In the future, this method can be applied to forecast 3D distributions of stress and strain in other multiphase materials, such as spheroidized steels and particle-based metal composites.

Author Contributions: Conceptualization, S.T. and F.Q.; methodology, S.T. and F.Q.; software, S.T. and F.Q.; validation, S.T. and F.Q.; formal analysis, S.T. and F.Q.; investigation, S.T. and F.Q.; resources, S.G.; data curation, S.T. and F.Q.; writing–original draft preparation, S.T. and F.Q.; writing–review and editing, S.T. and F.Q.; visualization, S.T., F.Q. and C.C.; supervision, S.G., C.C. and U.P.; project administration, U.P. All authors have read and agreed to the published version of the manuscript.

Funding: This research received no external funding.

Institutional Review Board Statement: Not applicable.

Informed Consent Statement: Not applicable.

Data Availability Statement: The simulation data are not publicly available but can be shared upon request.

Acknowledgments: The authors acknowledge the DAAD/MOST scholarship (Sandwich program) for doctoral candidates from Taiwan, 110-2927-I-011-505, for funding for sponsoring the stay of Shao-Chen Tseng at the IMF TU Freiberg. The authors acknowledge the DAAD Faculty Development for Candidates (Balochistan), 2016 (57245990)-HRDI-UESTP's/UET's funding scheme in cooperation with the Higher Education Commission of Pakistan (HEC) for sponsoring the stay of Faisal Qayyum at thr IMF TU Freiberg. The authors also acknowledge the support of Martin Diehl and Franz Roters (MPIE, Düsseldorf) for their help regarding the functionality of DAMASK.

Conflicts of Interest: The authors declare no conflict of interest.

Appendix A

Based on the proposed iterative method, the iterative constants for the martensite phase with stress and strain conditions are displayed in Tables A1 and A2, respectively.

Table A1. Constants a_m, b_m, and R_m^2 (m = 1, 2, 3, 4) for the martensite phase of linear equation with stress condition by the proposed iterative method.

	Stress								
Analysis	**Total Grain**			**Strain Level**			**Volume Fraction**		
Parameter	**8400**	**1900**	**770**	**5%**	**15%**	**25%**	**0.1**	**0.15**	**0.2**
Case	1	2	3	4	5	6	7	8	9
RVE	A	B	C	D	D	D	D	E	F
Iterative Const.									
a_1	−0.86	−0.78	−0.88	−0.7375	−0.7938	−0.82	−0.7938	−0.725	−0.766
b_1	1875	1614	1746	1526	2002	2222	2002	2092	1956
R_1^2	0.31	0.23	0.36	0.14	0.24	0.34	0.24	0.21	0.17
a_2	0.67	0.76	0.63	0.84	0.74	0.64	0.74	0.77	0.81
b_2	6.02	2.87	−1.003	28.96	14.86	6.64	14.86	20.6	−25.19
R_2^2	0.67	0.76	0.63	0.84	0.74	0.64	0.74	0.79	0.81
a_3	0.32	0.23	0.36	0.15	0.25	0.35	0.25	0.23	0.18
b_3	4.67	1.05	−1.00	5.69	6.49	6.37	6.49	6.55	−5.52
R_3^2	0.31	0.23	0.36	0.14	0.25	0.35	0.25	0.23	0.17
a_4	0.68	0.76	0.63	0.86	0.75	0.65	0.75	0.77	0.82
b_4	5.68	1.35	−1.00	8.39	8.43	7.14	8.43	6.24	−6.69
R_4^2	0.68	0.76	0.63	0.86	0.75	0.65	0.75	0.77	0.82

Table A2. Constants a_m, b_m, and R_m^2 (m = 1, 2, 3, 4) for the martensite phase of linear equation with strain condition by the proposed iterative method.

	Strain								
Analysis	**Total Grain**			**Strain Level**			**Volume Fraction**		
Parameter	**8400**	**1900**	**770**	**5%**	**15%**	**25%**	**0.1**	**0.15**	**0.2**
Case	1	2	3	4	5	6	7	8	9
RVE	A	B	C	D	D	D	D	E	F
Iterative Const.									
a_1	−0.31	0.09	0.5354	0.26	0.23	0.18	0.23	0.22	−0.03
b_1	2.19	−0.023	−1.3028	−0.30	−0.93	−1.38	-0.93	−0.3	0.60
R_1^2	0.005	0.0007	0.015	0.003	0.002	0.002	0.002	0.006	0.001
a_2	0.99	0.99	0.98	0.99	0.99	0.99	0.99	0.99	0.99
b_2	0.17	−0.12	−0.4026	0.028	0.104	0.16	0.104	−0.32	−0.53
R_2^2	0.99	0.99	0.98	0.99	0.99	0.99	0.99	0.99	0.99
a_3	0.005	0.0008	0.0164	0.003	0.003	0.002	0.003	0.006	0.001
b_3	0.0009	−0.00008	−0.0058	0	0	0	0	0	0

Table A2. *Cont.*

Analysis Parameter	Total Grain			Strain Strain Level			Volume Fraction		
	8400	1900	770	5%	15%	25%	0.1	0.15	0.2
R_3^2	0.005	0.0007	0.0156	0.003	0.003	0.002	0.003	0.006	0.001
a_4	0.99	0.99	0.9850	0.996	0.997	0.998	0.997	0.993	0.99
b_4	0.001	−0.00015	−0.0091	0	0	0	0	0	0
R_4^2	0.99	0.99	0.98	0.996	0.997	0.998	0.998	0.993	0.99

References

1. Tasan, C.C.; Diehl, M.; Yan, D.; Bechtold, M.; Roters, F.; Schemmann, L.; Zheng, C.; Peranio, N.; Ponge, D.; Koyama, M.; et al. An overview of dual-phase steels: Advances in microstructure-oriented processing and micromechanically guided design. *Annu. Rev. Mater. Res.* **2015**, *45*, 391–431. [CrossRef]
2. Hussein, T.; Umar, M.; Qayyum, F.; Guk, S.; Prahl, U. Micromechanical Effect of Martensite Attributes on Forming Limits of Dual-Phase Steels Investigated by Crystal Plasticity-Based Numerical Simulations. *Crystals* **2022**, *12*, 155. [CrossRef]
3. Biermann, H.; Martin, U.; Aneziris, C.G.; Kolbe, A.; Müller, A.; Schärfl, W.; Herrmann, M. Microstructure and compression strength of novel TRIP-steel/Mg-PSZ composites. *Adv. Eng. Mater.* **2009**, *11*, 1000–1006. [CrossRef]
4. Biermann, H.; Aneziris, C.G. *Austenitic TRIP/TWIP Steels and Steel-Zirconia Composites: Design of Tough, Transformation-Strengthened Composites and Structures*; Springer Nature: Berlin/Heidelberg, Germany, 2020; Volume 298.
5. Asghar, W.; Nasir, M.A.; Qayyum, F.; Shah, M.; Azeem, M.; Nauman, S.; Khushnood, S. Investigation of fatigue crack growth rate in CARALL, ARALL and GLARE. *Fatigue Fract. Eng. Mater. Struct.* **2017**, *40*, 1086–1100. [CrossRef]
6. Khan, F.; Qayyum, F.; Asghar, W.; Azeem, M.; Anjum, Z.; Nasir, A.; Shah, M. Effect of various surface preparation techniques on the delamination properties of vacuum infused Carbon fiber reinforced aluminum laminates (CARALL): Experimentation and numerical simulation. *J. Mech. Sci. Technol.* **2017**, *31*, 5265–5272. [CrossRef]
7. Qayyum, F.; Kamran, A.; Ali, A.; Shah, M. 3D numerical simulation of thermal fatigue damage in wedge specimen of AISI H13 tool steel. *Eng. Fract. Mech.* **2017**, *180*, 240–253. [CrossRef]
8. Mukhtar, F.; Qayyum, F.; Shah, M. Studying the effect of thermal fatigue on multiple cracks propagating in an SS316L thin flange on a shaft specimen using a multi-physics numerical simulation model. *J. Mech. Eng.* **2019**, *65*, 565–573. [CrossRef]
9. Hussain, N.; Qayyum, F.; Pasha, R.A.; Shah, M. Development of multi-physics numerical simulation model to investigate thermo-mechanical fatigue crack propagation in an autofrettaged gun barrel. *Def. Technol.* **2020**, *17*, 1579–1591. [CrossRef]
10. Ibrahim, I.; Mohamed, F.; Lavernia, E. Particulate reinforced metal matrix composites—A review. *J. Mater. Sci.* **1991**, *26*, 1137–1156. [CrossRef]
11. Kanouté, P.; Boso, D.P.; Chaboche, J.L.; Schrefler, B. Multiscale methods for composites: A review. *Arch. Comput. Methods Eng.* **2009**, *16*, 31–75. [CrossRef]
12. Hannard, F.; Simar, A.; Maire, E.; Pardoen, T. Quantitative assessment of the impact of second phase particle arrangement on damage and fracture anisotropy. *Acta Mater.* **2018**, *148*, 456–466. [CrossRef]
13. Zhang, J.F.; Zhang, X.X.; Wang, Q.Z.; Xiao, B.L.; Ma, Z.Y. Simulations of deformation and damage processes of SiCp/Al composites during tension. *J. Mater. Sci. Technol.* **2018**, *34*, 627–634. [CrossRef]
14. Lee, S.B.; Lebensohn, R.; Rollett, A.D. Modeling the viscoplastic micromechanical response of two-phase materials using Fast Fourier Transforms. *Int. J. Plast.* **2011**, *27*, 707–727. [CrossRef]
15. Lebensohn, R.A.; Kanjarla, A.K.; Eisenlohr, P. An elasto-viscoplastic formulation based on fast Fourier transforms for the prediction of micromechanical fields in polycrystalline materials. *Int. J. Plast.* **2012**, *32–33*, 59–69. [CrossRef]
16. Fujita, N.; Igi, S.; Diehl, M.; Roters, F.; Raabe, D. The through-process texture analysis of plate rolling by coupling finite element and fast Fourier transform crystal plasticity analysis. *Model. Simul. Mater. Sci. Eng.* **2019**, *27*, 085005. [CrossRef]
17. Kanit, T.; Forest, S.; Galliet, I.; Mounoury, V.; Jeulin, D. Determination of the size of the representative volume element for random composites: Statistical and numerical approach. *Int. J. Solids Struct.* **2003**, *40*, 3647–3679. [CrossRef]
18. Kadkhodapour, J.; Butz, A.; Ziaei-Rad, S.; Schmauder, S. A micro mechanical study on failure initiation of dual phase steels under tension using single crystal plasticity model. *Int. J. Plast.* **2011**, *27*, 1103–1125. [CrossRef]
19. Uthaisangsuk, V.; Prahl, U.; Bleck, W. Stretch-flangeability characterisation of multiphase steel using a microstructure based failure modelling. *Comput. Mater. Sci.* **2009**, *45*, 617–623. [CrossRef]
20. Bargmann, S.; Klusemann, B.; Markmann, J.; Schnabel, J.E.; Schneider, K.; Soyarslan, C.; Wilmers, J. Generation of 3D representative volume elements for heterogeneous materials: A review. *Prog. Mater. Sci.* **2018**, *96*, 322–384. [CrossRef]
21. Groeber, M.; Ghosh, S.; Uchic, M.D.; Dimiduk, D.M. A framework for automated analysis and simulation of 3D polycrystalline microstructures.: Part 1: Statistical characterization. *Acta Mater.* **2008**, *56*, 1257–1273. [CrossRef]
22. Groeber, M.; Ghosh, S.; Uchic, M.D.; Dimiduk, D.M. A framework for automated analysis and simulation of 3D polycrystalline microstructures. Part 2: Synthetic structure generation. *Acta Mater.* **2008**, *56*, 1274–1287. [CrossRef]

23. Diehl, M.; Groeber, M.; Haase, C.; Molodov, D.A.; Roters, F.; Raabe, D. Identifying structure–property relationships through DREAM. 3D representative volume elements and DAMASK crystal plasticity simulations: An integrated computational materials engineering approach. *Jom* **2017**, *69*, 848–855. [CrossRef]
24. Roters, F.; Diehl, M.; Shanthraj, P.; Eisenlohr, P.; Reuber, C.; Wong, S.L.; Maiti, T.; Ebrahimi, A.; Hochrainer, T.; Fabritius, H.O.; et al. DAMASK–The Düsseldorf Advanced Material Simulation Kit for modeling multi-physics crystal plasticity, thermal, and damage phenomena from the single crystal up to the component scale. *Comput. Mater. Sci.* **2019**, *158*, 420–478. [CrossRef]
25. Ramazani, A.; Mukherjee, K.; Quade, H.; Prahl, U.; Bleck, W. Correlation between 2D and 3D flow curve modelling of DP steels using a microstructure-based RVE approach. *Mater. Sci. Eng. A* **2013**, *560*, 129–139. [CrossRef]
26. Qayyum, F.; Guk, S.; Prüger, S.; Schmidtchen, M.; Saenko, I.; Kiefer, B.; Kawalla, R.; Prahl, U. Investigating the local deformation and transformation behavior of sintered X3CrMnNi16-7-6 TRIP steel using a calibrated crystal plasticity-based numerical simulation model. *Int. J. Mater. Res.* **2020**, *111*, 392–404. [CrossRef]
27. Qayyum, F.; Guk, S.; Schmidtchen, M.; Kawalla, R.; Prahl, U. Modeling the local deformation and transformation behavior of cast X8CrMnNi16-6-6 TRIP steel and 10% Mg-PSZ composite using a continuum mechanics-based crystal plasticity model. *Crystals* **2020**, *10*, 221. [CrossRef]
28. Diehl, M.; Shanthraj, P.; Eisenlohr, P.; Roters, F. Neighborhood influences on stress and strain partitioning in dual-phase microstructures. *Meccanica* **2016**, *51*, 429–441. [CrossRef]
29. Qayyum, F.; Guk, S.; Prahl, U. Studying the damage evolution and the micro-mechanical response of X8CrMnNi16-6-6 TRIP steel matrix and 10% zirconia particle composite using a calibrated physics and crystal-plasticity-based numerical simulation model. *Crystals* **2021**, *11*, 759. [CrossRef]
30. Qayyum, F.; Guk, S.; Kawalla, R.; Prahl, U. Experimental Investigations and Multiscale Modeling to Study the Effect of Sulfur Content on Formability of 16MnCr5 Alloy Steel. *Steel Res. Int.* **2018**, *90*, 1800369. [CrossRef]
31. Sedighiani, K.; Traka, K.; Roters, F.; Raabe, D.; Sietsma, J.; Diehl, M. Determination and analysis of the constitutive parameters of temperature-dependent dislocation-density-based crystal plasticity models. *Mech. Mater.* **2022**, *164*, 104117. [CrossRef]
32. Sedighiani, K.; Diehl, M.; Traka, K.; Roters, F.; Sietsma, J.; Raabe, D. An efficient and robust approach to determine material parameters of crystal plasticity constitutive laws from macro-scale stress–strain curves. *Int. J. Plast.* **2020**, *134*, 102779. [CrossRef]
33. Weidner, A.; Biermann, H. Review on Strain Localization Phenomena Studied by High-Resolution Digital Image Correlation. *Adv. Eng. Mater.* **2021**, *23*, 2001409. [CrossRef]
34. Weidner, A.; Biermann, H. Combination of Different In Situ Characterization Techniques and Scanning Electron Microscopy Investigations for a Comprehensive Description of the Tensile Deformation Behavior of a CrMnNi TRIP/TWIP Steel. *Jom* **2015**, *67*, 1729–1747. [CrossRef]
35. Weidner, A.; Berek, H.; Segel, C.; Aneziris, C.G.; Biermann, H. In situ tensile deformation of TRIP steel/Mg-PSZ composites. *Mater. Sci. Forum* **2013**, *738–739*, 77–81. [CrossRef]
36. Vinogradov, A.; Lazarev, A.; Linderov, M.; Weidner, A.; Biermann, H. Kinetics of deformation processes in high-alloyed cast transformation-induced plasticity/twinning-induced plasticity steels determined by acoustic emission and scanning electron microscopy: Influence of austenite stability on deformation mechanisms. *Acta Mater.* **2013**, *61*, 2434–2449. [CrossRef]
37. Qayyum, F.; Chaudhry, A.A.; Guk, S.; Schmidtchen, M.; Kawalla, R.; Prahl, U. Effect of 3D Representative Volume Element (RVE) Thickness on Stress and Strain Partitioning in Crystal Plasticity Simulations of Multi-Phase Materials. *Crystals* **2020**, *10*, 944. [CrossRef]
38. Qayyum, F.; Umar, M.; Guk, S.; Schmidtchen, M.; Kawalla, R.; Prahl, U. Effect of the 3rd Dimension within the Representative Volume Element (RVE) on Damage Initiation and Propagation during Full-Phase Numerical Simulations of Single and Multiphase Steels. *Materials* **2021**, *14*, 42. [CrossRef]
39. Umar, M.; Qayyum, F.; Farooq, M.U.; Guk, S.; Kirschner, M.; Korpala, G.; Prahl, U. Exploring the Structure–Property Relationship in Spheroidized C45EC Steel Using Full Phase Crystal Plasticity Numerical Simulations. *Steel Res. Int.* **2022**, *93*, 2100452. [CrossRef]
40. Qayyum, F.; Umar, M.; Elagin, V.; Kirschner, M.; Hoffmann, F.; Guk, S.; Prahl, U. Influence of Non-Metallic Inclusions on Local Deformation and Damage Behavior of Modified 16MnCrS5 Steel. *Crystals* **2022**, *12*, 281. [CrossRef]
41. Umar, M.; Qayyum, F.; Farooq, M.U.; Khan, L.A.; Guk, S.; Prahl, U. Analyzing the cementite particle size and distribution in heterogeneous microstructure of C45EC steel using crystal plasticity based DAMASK code. In Proceedings of the 2021 International Bhurban Conference on Applied Sciences and Technologies (IBCAST), Islamabad, Pakistan, 12–16 January 2021; pp. 15–20.
42. Jiang, Z.; Guan, Z.; Lian, J. Effects of microstructural variables on the deformation behaviour of dual-phase steel. *Mater. Sci. Eng. A* **1995**, *190*, 55–64. [CrossRef]
43. Eisenlohr, P.; Diehl, M.; Lebensohn, R.A.; Roters, F. A spectral method solution to crystal elasto-viscoplasticity at finite strains. *Int. J. Plast.* **2013**, *46*, 37–53. [CrossRef]
44. Tseng, S.C.; Chao, C.K.; Chen, F.M.; Chiu, W.C. Interfacial stresses of a coated polygonal hole subject to a point heat source. *J. Therm. Stresses.* **2020**, *43*, 1487–1512. [CrossRef]
45. Tseng, S.C.; Chao, C.K.; Chen, F.M. Interfacial stresses of a coated square hole induced by a remote uniform heat flow. *International J. Appl. Mech.* **2020**, *12*, 2050063. [CrossRef]

46. Diehl, M.; Naunheim, Y.; Yan, D.; Morsdorf, L.; An, D.; Tasan, C.C.; Zaefferer, S.; Roters, F.; Raabe, D. Coupled Experimental-Numerical Analysis of Strain Partitioning in Metallic Microstructures: The Importance of Considering the 3D Morphology. In Proceedings of the BSSM 12th International Conference on Advances in Experimental Mechanics, Sheffield, UK, 29–31 August 2017; pp. 1–2.
47. Diehl, M.; An, D.; Shanthraj, P.; Zaefferer, S.; Roters, F.; Raabe, D. Crystal plasticity study on stress and strain partitioning in a measured 3D dual phase steel microstructure. *Phys. Mesomech.* **2017**, *20*, 311–323. [CrossRef]
48. Chao, C.K.; Tseng, S.C.; Chen, F.M. Mode-III stress intensity factors for two circular inclusions subject to a remote uniform shear load. *J. Chin. Inst. Eng.* **2018**, *41*, 590–602. [CrossRef]

crystals

MDPI

Article

Description of Dynamic Recrystallization by Means of An Advanced Statistical Multilevel Model: Grain Structure Evolution Analysis

Peter Trusov *, Nikita Kondratev and Andrej Podsedertsev

Laboratory of Multilevel Structural and Functional Materials Modeling, Perm National Research Polytechnic University, 614990 Perm, Russia; kondratevns@gmail.com (N.K.); stevenmoore@yandex.ru (A.P.)
* Correspondence: tpv@matmod.pstu.ac.ru

Abstract: Physical multilevel models of inelastic deformation that take into account the material structure evolution hold promise for the development of functional materials. In this paper, we propose an advanced (modified via analyzing the mutual arrangement of crystallites) statistical multilevel model for studying thermomechanical processing of polycrystals that includes a description of the dynamic recrystallization process. The model is based on the consideration of homogeneous elements (grains, subgrains) aggregated into a representative volume (macropoint) under the Voigt hypothesis. In the framework of this statistical approach, there is no mandatory requirement for continuous filling of the computational domain with crystallites; however, the material grain structure cannot be created arbitrarily. Using the Laguerre polyhedra, we develop a method of grain structure simulation coupled with subsequent processing and transferring of the necessary data on the grain structure to the modified statistical model. Our research is of much current interest due to the fact that the mutual arrangement of crystallites, as well as the interfaces between them, has a significant impact on the properties of polycrystals, which are particularly important for physical mechanisms that provide and accompany the processes of inelastic deformation (recrystallization, grain boundary hardening, grain boundary sliding, etc.). The results of the simulations of the high-temperature deformation of a copper polycrystal, including the description of the recrystallization process, are presented.

Keywords: multilevel models; dynamic recrystallization; grain shape and grain size; defect and grain structure evolution

Citation: Trusov, P.; Kondratev, N.; Podsedertsev, A. Description of Dynamic Recrystallization by Means of An Advanced Statistical Multilevel Model: Grain Structure Evolution Analysis. *Crystals* **2022**, *12*, 653. https://doi.org/10.3390/cryst12050653

Academic Editor: Zheng Mingyi

Received: 31 March 2022
Accepted: 29 April 2022
Published: 2 May 2022

1. Introduction

Methods for processing polycrystalline materials, especially hard-to-deform alloys, via severe plastic deformation (stamping, drawing, extrusion, forging, etc.) usually include high homological temperatures [1,2]. An actual problem that manufacturers face during this processing is to obtain structures and products with optimal mechanical properties (high strength, high yield strength, low weight, etc.) [3,4]. Plastic deformation in these processes is usually accompanied by the appearance and evolution of structural defects at various structural-scale levels, which in turn causes inhomogeneity of the physical and mechanical properties [5,6]. The high temperature of the plastic deformation process and subsequent heating reduce the number of defects and promote their redistribution, increasing the uniformity of the structure. This is possible due to the implementation of a combination of two high-temperature processes, namely dynamic recovery (DRV) and dynamic recrystallization (DRX) [5,7–10]. DRV is followed by reduction and redistribution of defects (primarily, dislocations) and densities, which usually results in the formation of subgrain, block–cellular, and cellular structures [5,7,8,11]. DRX refers to the formation and further development of new, less-defective grains by absorbing neighboring grains [5,7,8]. If this process is implemented in the material in a heterogeneous way, where the stages of formation of DRX nuclei and their subsequent growth are clearly separated, then it is called discontinuous dynamic recrystallization (DDRX) [5] or classical DRX [7]. However, if the DRX process occurs homogeneously in such a way that the stages of formation and growth of new grains are not clearly distinguished, then it is called continuous dynamic recrystallization (CDRX) [5,7]. The main material characteristic of the development of a grain and defect structure during DRX and its type is the stacking fault energy (SFE). The increased SFE values promote the formation of a subgrain structure under active recovery effects, and accordingly the implementation of CDRX, while the lower SFE values are responsible for the occurrence of DDRX [5,7].

At present, the multilevel models of inelastic deformation with internal variables are widely used to simulate changes in the structure and responses of the materials [6,12–16]. There are three classes of multilevel models: statistical, direct, and self-consistent. Statistical models consider a representative macrovolume of a material ("macropoint") as a set of individual grains combined into a polycrystalline aggregate, as hypothesized by Voigt, Reiss, and Kroener [17–19]. Self-consistent (mean-field) models are based on solving the boundary single inclusion problem of a crystallite (grain, subgrain, fragment) embedded in a matrix by imaging the averaged (effective) characteristics of a polycrystal [20–22]. Direct models rely on solving boundary value problems and on determining the field values of the model (stresses, strains, and internal variables) for each crystallite included in the representative macrovolume under consideration [13,23,24]. Statistical models have been recognized by many authors as the least accurate among all three classes of models, yet their modifications make it possible to explicitly consider the current state of neighboring grains by highlighting their contact interactions along the conjugate flat sections of boundaries (facets) of grains [25,26]. Additionally, these models have high computational efficiency [27]. One of the main drawbacks of the self-consistent models is that they are unable to take into account real physical interactions between the neighboring crystallites (e.g., exchange of dislocation flows, formation of orientation misfit dislocations) and require the replacement of the environment by a continuum with averaged properties, which do not reflect the real state of the grain and defect structures. Direct models are known to be the most accurate models, yet they remain the most resource-intensive [28]. The implementation of direct models of crystallites with varying compositions and moving boundaries involves great computational difficulties, especially when studying a wider range of physical processes, including solid-state phase transitions and recrystallization processes.

Numerous approaches and methods have been developed to model DRX [5,29,30]. Phenomenological models are the most simplified models, which include an empirical or semi-empirical description of a recrystallized material fraction. The Johnson–Mehl–Avrami–Kolmogorov (JMAK) kinetic equations form the basis for describing the evolution of the recrystallized volume fraction of polycrystals in these models [31–33]. The development of this approach is associated with creating models able to describe such characteristics of the grain structure, as the grain size, grain shape, and grain boundary orientation [34–36]. The JMAK equations often turn out to be insufficient for studying the grain structure evolution because they do not take into account the heterogeneous nature of the DRX process and are only applicable within a narrow range of variation in the initial structure, temperature, strain, and strain rate [37,38]. Recent developments in the computer technology provide an opportunity to develop physical approaches that permit an explicit description of the grain structure evolution [5,6]. In the framework of this class of models, internal variables and parameters are introduced to characterize the material structure and to assess factors responsible for the formation of DRX nuclei and for the migration of grain boundaries. Due to this, these models are applied to describe the formation of the required material structure, and they are well suited to technological regimes in the metallurgical and manufacturing industries. Within this physical approach, there are several types of models that include descriptions of dynamic recrystallization, in particular self-consistent [8,21,39,40] and direct [41–43] models. One of the main problems encountered in modeling DRX processes is the description of the grain structure evolution that occurs due to the formation of new grains and due to the migration of their high-angle boundaries. To describe the grain structure evolution by applying the models of the above type, various approaches are typically used, including the Monte Carlo method [44], cellular automaton method [38,45,46], phase-field method [47], level set method [43,48], graph theory based methods [49,50], and some others.

Despite the great number of approaches and methods, the problem of modeling the grain structure evolution during DRX still remains an urgent research issue and requires further research. Recent papers [8,38,40,43] are indicative of a current interest in this subject. The topicality of the problem is also related to the presence of a relationship between the current state of the grain structure and the macro-properties of polycrystals. This fact has been repeatedly confirmed by the results of many independent experiments [5,51,52]. The most well-known manifestation of the influence of the state of the grain structure on the properties of the material is the dependence of the yield strength on the average grain size (Hall–Petch equation) [53–55]. Severe hot plastic deformation produces significant changes (usually refinement) in the grain structure, which are mainly associated with DRX [5,7]. The grain structure determined by the shape and size of crystallites is of decisive importance in providing effective mechanical properties (plasticity, creep, hardness, fatigue, etc.) of a polycrystalline material [56,57]. As an example, the formation of equiaxed grains and their refinement during recrystallization facilitate the implementation of grain boundary sliding, and consequently the initiation of deformation in the structural superplasticity mode under appropriate temperature and velocity conditions [26,58]. This causes abnormally large deformations to appear at almost constant and relatively low values of plastic flow stress.

In view of the foregoing, the problem of developing an advanced constitutive physical model of a polycrystalline material relying on an explicit physical description of the DRX process needed to analyze the evolution of grain and defect structures is of particular importance. In this paper, we propose an original method of statistical modeling to form and reconstruct the grain structure by applying Laguerre polyhedra [59]. The developed geometric model for describing the grain shape and grain size allows one to overcome the above-mentioned difficulties inherent in using self-consistent and direct models. It has high computational efficiency and is able to describe the peculiarities of physical mechanisms, where the decisive factor is the interaction between contacting grains.

Section 2 presents the structure, scale levels, and relationships of an advanced statistical multilevel model for describing inelastic deformation with allowance for the DDRX and the mechanism governing the formation of nuclei and their development into individual low-defect grains. Section 3 discusses a method for grain structure formation and reconstruction during high-temperature deformation using the Laguerre polyhedra, as well as subsequent processing and transfer of the obtained data on the grain structure to the advanced statistical model. Section 4 gives a description and analysis of the obtained results.

2. The Advanced Statistical Multilevel Model for Describing Inelastic Deformation during Discontinuous Dynamic Recrystallization

This study investigates DDRX, which is a common phenomenon encountered in plastic formation of a wide class of polycrystalline materials with medium and low SFE values, including copper, nickel, austenitic steels, and some aluminum alloys [5,7]. At the initial stage of hot plastic deformation, DDRX is characterized by the formation of low-defect nuclei, which absorb the neighboring more-defective grains due to boundary migration. Thus, two processes are responsible for the occurrence of DDRX: (1) the formation of nuclei and their further development into new grains; (2) the migration of the high-angle boundaries of DRX grains [5,7]. DRX nuclei have a characteristic size of the order of a micrometer and are not formed due to random fluctuations, as in the case of phase transitions [5]. They appear in the course of directed rearrangement of defect and subgrain structures. For most metals, the commonly accepted mechanism of nuclei formation during DDRX involves the movement ("bending", "swelling") of the grain boundary regions associated with the original subgrains of a polycrystal [60–63]. This mechanism, first investigated for static recrystallization, was described in [64]. The presence of this phenomenon during DDRX was confirmed by the theoretical and experimental results of many researchers [60,61,63,65]. The development of the DDRX process governed by this mechanism is presented schematically in Figure 1. Initially, the first nuclei layer is formed at the original grain boundaries, and it has orientation coherency with the parent grain (Figure 1b). In a similar way, new grains (their average size decreases gradually) are formed at the boundaries of previously recrystallized grains (Figure 1c,d). These grains occur due to the addition of atoms to a subgrain (a DRX nucleus) along the grain boundaries and due to their subsequent penetration deep into the defective grain. In Figure 1, the following designations are used: ε_c is the critical deformation up to which no DRX occurs; ε_m is the deformation corresponding to the maximum value of stresses σ_m in the load diagram; ε_s is the deformation corresponding to almost complete recrystallization of the material; ε_x is the deformation corresponding to one recrystallization cycle ($\varepsilon_s = \varepsilon_c + \varepsilon_x$).

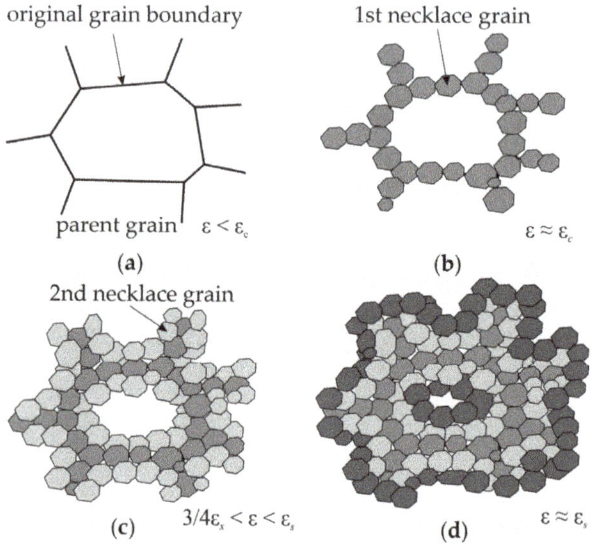

Figure 1. Scheme for the formation of a fine-grained structure during DDRX at various degrees of deformation ε: $\varepsilon < \varepsilon_c$ (**a**), $\varepsilon \approx \varepsilon_c$ (**b**), $3/4\varepsilon_x < \varepsilon < \varepsilon_s$ (**c**), $\varepsilon \approx \varepsilon_s$ (**d**) (it was developed based on [61]).

Previously, the authors developed an advanced statistical multilevel model intended for describing static [25] and dynamic recrystallization [66,67]. In these papers, the processes of preliminary plastic deformation and the subsequent annealing of the material or its deformation at a given temperature were investigated to evaluate the initial stage of recrystallization. The present article provides a description of a significantly modified model in regard to refining the simulation of the polycrystal grain structure evolution. The model is designed to study the processes of severe plastic deformation of polycrystalline materials at elevated temperatures of inelastic deformation.

In solving the formulated problem, we distinguish three structural levels in the structure of a statistical multilevel model. They are the macrolevel, mesolevel-I, and mesolevel-II. At the macrolevel, the statistical description of a representative number of grains with uniform distribution of orientations of crystallographic axes in the space of the reference configuration is carried out. At mesolevel-I, the behavior of a separate grain consisting of mesolevel-II elements (homogeneous subgrains) is investigated. The grains are separated by high-angle boundaries, whose misorientation is greater than 10–15 degrees; the subgrain misorientation is about several degrees, and as a rule it increases with increasing inelastic deformation [5,68]. A scheme illustrating the relationship between the scale-based submodels is shown in Figure 2.

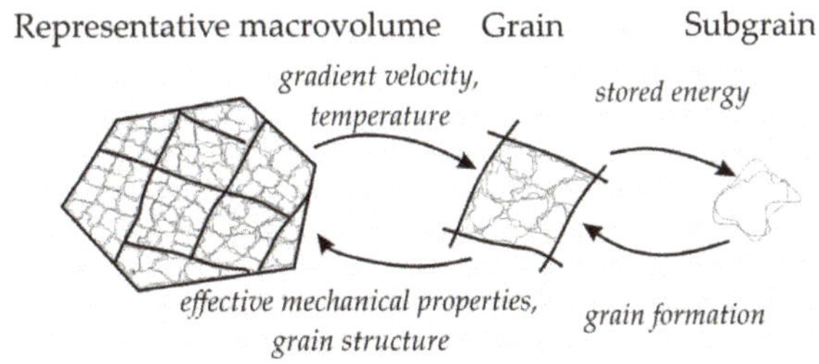

Figure 2. Scale levels and the connection between the structural elements of a multilevel model (thin lines correspond to subgrain boundaries, bold lines show grains boundaries).

At the macrolevel, the effective properties of the material are determined and the impacts are transferred from this level to the underlying scale levels. For each moment of time t, the mechanical and thermal effects are prescribed and the velocity gradient $\hat{\nabla}\mathbf{V}(t)$ and temperature $\Theta(t)$ are assumed to be known at the macrolevel. Therefore, for each grain, the following equality is true [6]:

$$\hat{\nabla}\mathbf{v}(t) = \hat{\nabla}\mathbf{V}(t), \theta(t) = \Theta(t), \tag{1}$$

where $\hat{\nabla}\mathbf{v}(t)$ is the velocity gradient of the grain's material particle; $\theta(t)$ is the grain temperature; $\hat{\nabla}$ is the Hamilton operator in the current Lagrangian coordinate system; $\mathbf{V}$, $\mathbf{v}$ are the velocity vectors obtained at macro and mesolevels. The indexes denoting the grain (subgrain) number are omitted hereinafter, unless otherwise specified.

At mesolevel-I, the problem of determining the stress–strain state of a grain and the corresponding internal variables of the model is solved using an advanced statistical model. The mathematical formulation of a mesolevel-I model contains the following relations [66,67]:

$$\boldsymbol{\kappa}^{cr} = \mathbf{\Pi} : \left(\hat{\nabla}\mathbf{v}^{\mathrm{T}} - \boldsymbol{\omega} - \mathbf{z}^{in}\right), \tag{2}$$

$$\mathbf{z}^{in} = \sum_{k=1}^{K} \dot{\gamma}^{(k)}\mathbf{b}^{(k)}\mathbf{n}^{(k)}, \tag{3}$$

$$\tau^{(k)} = \mathbf{b}^{(k)}\mathbf{n}^{(k)} : \boldsymbol{\kappa}, \tag{4}$$

$$\dot{\gamma}^{(k)} = \dot{\gamma}_0 \left(\frac{\tau^{(k)}}{\tau_c^{(k)}}\right)^m H\left(\tau^{(k)} - \tau_c^{(k)}\right), \tag{5}$$

$$\dot{\tau}_c^{(k)} = \dot{\tau}_{cs}^{(k)} + \dot{\tau}_{cgb}^{(k)}, \tag{6}$$

$$\dot{\tau}_{cs}^{(k)} = \sum_{j=1}^{N_s} h^{(kj)}\dot{\gamma}^{(j)}, \tag{7}$$

$$h^{(kl)} = \left[q_{lat} + (1 - q_{lat})\delta^{(kl)}\right]h^{(l)},\ h^{(l)} = h_0\left|1 - \tau_c^{(l)}/\tau_{sat}\right|^a, \tag{8}$$

$$\dot{\tau}_{cgb}^{(k)} = \eta \sum_{j=1}^{N_f} \frac{S_j}{S} \sum_{s=1}^{N_s} \xi^{(s,j)}\dot{\gamma}^{(s)}\tau_{bs}^{(k,s)}, \tag{9}$$

$$\boldsymbol{\omega} = \dot{\mathbf{o}} \cdot \mathbf{o}^{\mathrm{T}} = \mathbf{I} \times (\mathbf{k}_3\mathbf{k}_1\mathbf{k}_2 - \mathbf{k}_2\mathbf{k}_1\mathbf{k}_3 + \mathbf{k}_1\mathbf{k}_2\mathbf{k}_3) : \mathbf{l}^e, \tag{10}$$

$$\mathbf{z}^e = \mathbf{z} - \mathbf{z}^{in}, \tag{11}$$

$$e_{st} = \frac{\alpha}{G}\left(\tau_c^2 - \tau_{c0}^2\right). \tag{12}$$

The corresponding initial conditions are written as follows [6]:

$$\boldsymbol{\kappa}\big|_{t=0} = \boldsymbol{\kappa}_0,\ \mathbf{o}\big|_{t=0} = \mathbf{o}_0,\ \tau_{cs}^{(k)}\Big|_{t=0} = \tau_{cs0}^{(k)},\ \tau_{ctw}^{(k)}\Big|_{t=0} = \tau_{ctw0}^{(k)},\ \gamma_s^{(k)}\Big|_{t=0} = \gamma_{s0}^{(k)}. \quad (13)$$

Here, $\boldsymbol{\kappa} = \frac{\rho}{\hat{\rho}}\boldsymbol{\sigma}$ is the mesolevel weighted Kirchoff stress tensor; $\boldsymbol{\kappa}^{cr} = d\boldsymbol{\kappa}/dt + \boldsymbol{\kappa}\cdot\boldsymbol{\omega} - \boldsymbol{\omega}\cdot\boldsymbol{\kappa}$ is its corotational derivative; $\boldsymbol{\sigma}$ is the mesolevel Cauchy stress tensor; $\mathbf{z} = \hat{\nabla}\mathbf{v}^{\mathrm{T}} - \boldsymbol{\omega}$ is the strain measure satisfying the objectivity requirement; $\mathbf{z}^e$, $\mathbf{z}^{in}$ are the mesolevel elastic and inelastic components of the strain rate measure; $\mathbf{l}^e$ is the elastic component of the transposed velocity gradient; ρ,$\hat{\rho}$ denote the material density in the reference and current configurations, respectively; $\dot{\gamma}^{(k)}$ is the shear rate on the k-th intragranular slip system; $\dot{\gamma}_0$ is the shear rate on the slip system when the tangential stress reaches the critical shear stress; m is the strain rate sensitivity exponent of the material; $\mathbf{b}^{(k)}, \mathbf{n}^{(k)}$ are the unit vectors of the slip direction and the normal to the k-th slip system, respectively; H is the Heaviside function; $\tau^{(k)}, \tau_c^{(k)}$ are the shear and critical stresses of the k-th system, respectively; $\tau_{cs}^{(k)}, \tau_{cgb}^{(k)}$ are the components of critical stresses of the k-th system, which occur due to the interactions of dislocations with the dislocation structure and grain boundaries, respectively; $\tau_c^2 = \sum\limits_{k=1}^{N_s}\left(\tau_c^{(k)}\right)^2, \tau_{c0}^2 = \sum\limits_{k=1}^{N_s}\left(\tau_{c0}^{(k)}\right)^2$ are the total squared critical stresses in the grain under study at the current and initial instants of deformation, respectively; $h^{(kl)}$ is the matrix describing the crystal hardening caused by the interactions between dislocations and forest dislocations; q_{lat} is the latent hardening parameter; τ_{sat} is the saturation stress; h_0, a are the parameters describing material hardening; N_s, N_f denote the number of slip systems and the number of facets of the considered grain, respectively; S, S_j are the grain area and the j-th grain facet, respectively; $\xi^{(s,j)}$ is the parameter characterizing the interactions between the dislocations of the s-th slip system of the considered grain and those of the neighboring grain along the j-the facet, respectively (the definition of this parameter is given in [69]); $\tau_{bs}^{(k,s)}$ is the component of the shear barrier stresses acting on the k-th slip on the side of orientation misfit dislocations that appear in the facets because of the dislocation motion on the s-the system; η is the dimensionless parameter defined during the model identification procedure; π is the elastic tensor of the crystallite, the components of which π^{ijmn} were defined and turned out to constant in the actual basis $\mathbf{k}_i$ (in the reference configuration, the basis vectors are denoted by $\mathbf{k}_0^i$) of the rigid moving coordinate system (MCS) rotating with spin $\boldsymbol{\omega}$. The rigid orthonormal coordinate system is related to one crystallographic direction and the crystallographic plane containing this direction. In the reference configuration, the basis of the MCS coincides with the basis of the crystallographic coordinate system (CCS). The basis of the CCS is distorted during deformation, while the basis of the MCS remains rigid. The motion relative to the CCS is a deformation motion. The definition of MCS and spin is given in [70]; $\mathbf{o} = \mathbf{k}_i\mathbf{k}_0^i$ is the tensor that combines the MCS with the laboratory coordinate system (LCS); α is the dimensionless experimentally determined correction parameter characterizing the fraction of energy stored in dislocations; G is the shear modulus; $\delta^{(kl)}$ is the Kronecker delta; the dot above the corresponding mesolevel variables denotes the material time derivative t; the subscript T denotes the transpose of the value of the second rank tensor. The transition to macrolevel variables is implemented via volume averaging of the corresponding mesolevel variables. To describe hardening by analyzing the interaction of mobile dislocations with forest dislocations (relation for $\dot{\tau}_{cs}^{(k)}$), we apply the known hardening law [71,72] and the grain boundary hardening law (relation for $\dot{\tau}_{cgb}^{(k)}$) described in [69].

At the macrolevel, we calculate the average values of stresses $\boldsymbol{\Sigma}$, elastic properties $\mathbf{II}$, and an inelastic component of the velocity gradient $\hat{\nabla}\mathbf{V}^{in}$ [6]:

$$\boldsymbol{\Sigma} = \langle\boldsymbol{\sigma}\rangle,\ \mathbf{II} = \langle\pi\rangle,\ \hat{\nabla}\mathbf{V}^{in} = \left\langle\hat{\nabla}\mathbf{v}^{in}\right\rangle, \quad (14)$$

where $\langle\ \rangle$ is the volume averaging operator.

At the mesolevel-II (the level of individual subgrains), which is an auxiliary level, there is an opportunity to accurately evaluate the above DDRX recrystallization mechanism when modeling the formation of the nuclei located near the grain boundary. When the Bailey–Hirsch criterion is satisfied, the DRX nuclei start to grow, and as a result new grains are formed. According to this criterion, a decrease in local volume energy due to defect elimination must be greater or equal to an increase in grain boundary energy that is caused by an increase in the boundary area during the formation of a recrystallized grain [73]:

$$f^{(i,j)} = e_{dst}^{(i,j)} - e_{gb}^{(i,j)} \Delta s / \Delta \mathrm{v} \geq 0, \tag{15}$$

where Δs is the increase in the boundary area when a subgrain volume (a DRX nucleus) changes by $\Delta \mathrm{v}$, $e_{dst}^{(i,j)}$ is the difference between the specific energies of adjacent subgrains stored per unit volume i and j, and $e_{gb}^{(i,j)}$ is the energy per area unit for the boundary between the i-th and j-th subgrains.

In order to apply criterion (15) and calculate the values of Δs, $\Delta \mathrm{v}$, we introduce an additional mesolevel-II variable (a characteristic subgrain size) r. We speak of nuclei that are spherical in shape and assume that the initial subgrain size distribution obeys the Rayleigh law [62,63]. If the necessary experimental data are available, then we can use other distribution laws. It is assumed that the size of subgrains does not change during deformation. The recrystallization nuclei are located near the boundary (a set of facets). Each facet of the grain under consideration is assigned a sample of subgrain sizes (recrystallization nuclei) r according to the Rayleigh law. The validity of the recrystallization criterion (15) is verified for every subgrain separated by the high-angle boundary from the neighboring ones.

We do not consider here the mechanism governing the formation of new grains from subgrains during their rotation until the high-angle misorientations with the parent grain occur. We suppose that new grains are formed only during the DRX process. In the volume of one grain, the subgrains are separated by the low-angle boundaries and are slightly misoriented relative to each other. Thus, it is reasonable to believe that the stored energy e_{st}, determined by relation (15), is approximately the same in all subgrains within one grain. The data on the stored energy e_{st}^{I} calculated at mesolevel-I and needed to verify the validity of the recrystallization criterion (15) are transmitted to mesolevel-II. Thus, we have:

$$e_{st}^{II} = e_{st}^{I} \stackrel{def}{=} e_{st}, \tag{16}$$

where e_{st}^{II} is the stored energy at mesolevel-II. The difference in stored energy between the subgrains belonging to different grains but having a common high-angle boundary may be significant.

In fulfilling criterion (15), we assume that DRX nuclei are active and that they are separated into individual grains. This separation proceeds in two stages. At the first stage, when condition (15) is satisfied, the active nuclei are added to the static model; at the second stage, the polyhedral grain structure is rearranged by considering new recrystallized grains. The description of this method is given in Section 3. The new DRX grains are considered to be low-defective; thus, all internal variables of a recrystallized grain in the annealed material correspond to the reference configuration, with the exception of the new grain orientation determined by the tensor $\mathbf{o}$ and the geometric characteristics. A refined definition of these characteristics is also given in Section 3. The volume of the new DRX grain is subtracted from the volume of the parent grain. Experimental data [5,61] indicated that for the mechanism of DRX nuclei formation considered here, the new (recrystallized) grains have a consistent orientation coherency with the parent grain. It was shown that the new grain is separated from the parent grain by a high angle boundary, and it has initial orientation with respect to the parent grain within the range from 10 to 15° with a random axis of rotation [5,61]. The DRX grain orientation with respect to the neighboring (absorbed) grain is arbitrary (Figure 1b). At this stage of modeling, in order to evaluate the influence of grain boundary hardening, we assume that the new grain is surrounded by six identical facets, and the neighboring grains are determined in a random way.

The migration of the high-angle boundary of a new grain occurs at temperatures sufficient to ensure its mobility. The Arrhenius law is used to study the effect of temperature on the high-angle mobility [5]:

$$m = m_0 \exp\left(-\frac{Q}{R\theta}\right), \tag{17}$$

where θ is the absolute temperature; R is the universal gas constant; Q is the activation energy of the grain boundary migration; m_0 is the experimentally determined pre-exponent. The high angle boundary migration velocity v_m is determined by the product of the driving force f (the indexes denoting the neighboring crystallites are omitted) and the mobility m [5]:

$$v_m = fm. \tag{18}$$

At this stage of model development, we assume for simplicity that the shape of the recrystallization nuclei is spherical (the refine geometry of recrystallized grains is described in Section 3) and that each new recrystallized grain penetrates deep into the neighboring grain. In this case, v_m is equal to the rate of change of the ball radius r, which describes the shape of the recrystallized grain, i.e., $\dot{r} = v_m$ (the dot above r shows the material derivative). The absorbed grain volume is reduced by the volume of recrystallized grains. The volume fraction of the recrystallized material X in a polycrystal is determined through the following relation:

$$X = \frac{V_r}{V_0}, V_r = \sum_{i=1}^{N_r} \mathrm{v}_r^{(i)}, \tag{19}$$

where V_0 is the initial value of the representative volume of a polycrystal under consideration; V_r is the DRX material volume; N_r is the number of DRX grains; $\mathrm{v}_r^{(i)} = \frac{4}{3}\pi r^{(i)3}$ is the i-th DRX grain volume.

3. Rearrangement of the Grain Structure Formed during High-Temperature Deformation with Recrystallization

During recrystallization, the grain structure varies continuously; changes are related to changes in the number, shape, and size of the grains under study. We propose a method for the formation of the grain structure and its rearrangement during deformation, which allows consideration of the formation of new DRX grains. The developed method is applied for a statistical model, one of the main tasks of which is to solve the problem of grain structure rearrangement during recrystallization and to establish a correspondence between the geometric image of a real grain structure described by polyhedra (Figure 3a) and its model representation in the framework of the modified statistical model (Figure 3b). In this model, the grain structure is determined by such internal variables as the grain volume v_g, the facets with the normal $\mathbf{n}_j$ and the area s_j (where j denotes the number of a facet), and through assigning neighboring grains to the considered grain. First, the initial polyhedral grain structure is formed in the software package Neper [59]. Then, the obtained data are transferred to the calculation module of the statistical model for the implementation of the inelastic deformation model described above.

The modeling of active recrystallization nuclei proceeds in two stages. Information on the first stage of the simplified description is given in the previous section of this article. During this stage, subgrains (DRX nuclei) are assigned to each facet of the initial grains. As the recrystallization criterion (15) is met, these subgrains are considered as independent grains added, without taking into account their geometric shape, to sample data of the statistical model. When the volume fraction of active recrystallization nuclei f_{cr} reaches its critical value of 20% (found in computational experiments), the second stage at which the new grains are introduced into the geometric model of a representative volume begins. To this end, we again start the procedure for generating a polyhedral grain structure in the Neper program [59]. This allows us to take into account new grains and to find whether the statistical model is consistent with the calculation scheme. With the software package Neper, we construct a polyhedral structure on the basis of the laws of distribution through the sphericity parameters ψ_g (the ratio of the surface area of the sphere, whose volume is equal to the volume of a considered grain to the grain surface area) and the normalized grain size d_{eq} ($d_{eq} = \frac{d}{\langle d \rangle}$, where d is the grain size defined as the radius of a sphere of equivalent volume and $\langle d \rangle$ is the average grain size). To this end, the Neper involves the numerical implementation of optimization methods [59].

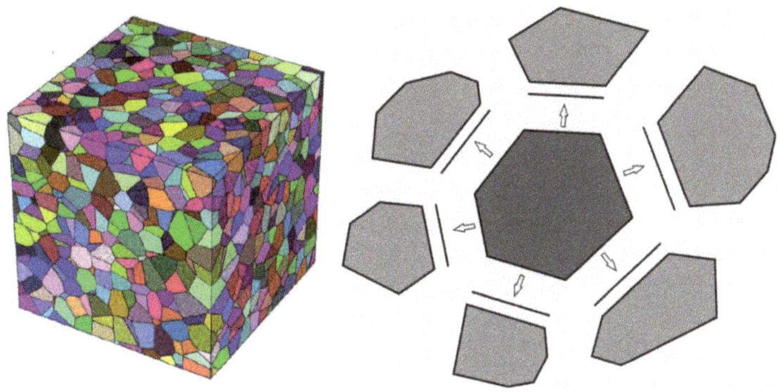

(**a**) (**b**)

Figure 3. Polyhedral representation of polycrystal grains (**a**) and scheme of the grain structure in the statistical model (**b**).

After the polyhedral grain structure is formed in Neper, the obtained data (grain volumes, normals, and facet areas with indication of neighboring grains) are transferred to the statistical model, where the grain structure is redefined. Note that new crystallites are introduced into the geometric structure of the considered region through DRX simulations; previously, the statistical model does not take into account the geometry of this region. To compare the initial and newly deformed structures, we consider two types of grains: original defective grains and new DRX grains. The main geometric characteristic of grains is their size, which is determined by the volume of an equivalent sphere. The defect structure is primarily characterized by the stored energy e_{st}. At critical deformation ε_c, large non-recrystallized grains usually coexist with fine recrystallized grains (Figure 1b), and a bimodal grain size distribution is generally observed. In this case, we distinguish a cluster of grains, namely a large "consumed" defective grain surrounded by fine DRX grains for the initial grain structure (Figure 4). Initially, we compare all grains according to their characteristic sizes. For this purpose, the square of difference in grain sizes is assumed to be minimal for original and new grain structures. At later recrystallization stages where the grain size distribution is close to unimodal (Figure 1d), the grains are also compared vs. their sizes. In both cases, the variant of the grain structure, determined with regard to the best fit of grain sizes, is taken as the initial approximation. Then, the genetic algorithm method [74] is applied to perform a purposeful selection of grains. This allows us to reach a minimum in distribution of the difference between the energies stored in the grains of original and new structures (comparison is made by the mathematical expectation of the stored energy distribution and grain sizes). The conditions for the migration of grain boundaries before and after the geometric structure rearrangement will be statistically equivalent, which ensures "continuity" in the simulation of the recrystallization process.

Next, we describe the grain structure evolution in terms of a statistical model. The grain structure is subjected to thermomechanical loading change for the following two reasons. The first is due to mechanical influences, mainly due to the movement of dislocations during inelastic deformation, whereby the areas and normals of grain facets vary according to the known formulas of continuum mechanics [75]:

$$\hat{s}_i = s_i \det(\mathbf{f}) \left(\mathbf{n}_i \cdot \mathbf{f}^{-1} \cdot \mathbf{f}^{-\mathrm{T}} \cdot \mathbf{n}_i\right)^{-1/2}, \hat{\mathbf{n}}_i = \left(\mathbf{n}_i \cdot \mathbf{f}^{-1} \cdot \mathbf{f}^{-\mathrm{T}} \cdot \mathbf{n}_i\right)^{-1/2} \mathbf{f}^{\mathrm{T}} \cdot \mathbf{n}_i \tag{20}$$

where $s_i, \hat{s}_i$ are the areas of facets in the reference and actual configurations with corresponding normal $\mathbf{n}_i, \hat{\mathbf{n}}_i$; det( ) is the operation of a determinant; is the facet number. The changes in grain volume observed during elastoviscoplastic deformation are neglected. The statistical model of inelasticity is based on the Voigt hypothesis, and a deformation gradient $\mathbf{f}$ is known at every instant of deformation.

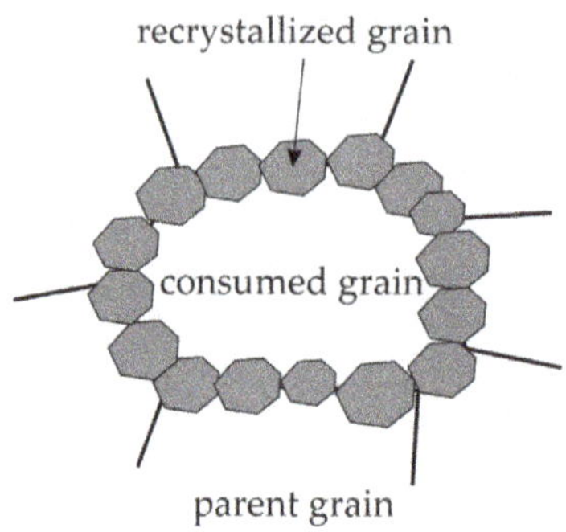

Figure 4. Scheme used for modeling a cluster of grains (a "consumed" defective grain surrounded by DRX grains).

The second mechanism governing the grain structure changes is diffusion, which is realized when the high-angle boundaries migrate deep into the defective grains in the direction of the outer normal. Due to this process, both the numbers of grains and their neighbors (adjacent grains) change, and hence there arises a need for reconstructing a polyhedral grain structure in the software package Neper. In DRX simulation, the known value is the boundary migration velocity v_m, which is determined by Equation (18) for each facet at every instant of thermomechanical process. The driving force f is determined by relation (15), where Δs, Δv mean that the area and volume, respectively, of recrystallized grains increase. As the grain grows, the second term in (15) does not contribute to f, meaning this value is almost completely determined by the difference in stored energy between neighboring grains. It is assumed that the migration velocity of the boundary facet is directed along the outer normal $\hat{\mathbf{n}}$ of the boundary facet, i.e., the following ratio is true:

$$\mathbf{v}_m = v_m \hat{\mathbf{n}}. \tag{21}$$

Due to its significant non-linearity, the problem is solved by a step-by-step method, where the step size Δt is determined from the analysis of the results of a series of numerical experiments. It is assumed that in the time interval Δt, v_m is constant. The velocity vector u_m resulting from the boundary migration is calculated for Δt by integrating (21) as:

$$\mathbf{u}_m = \int_{\tau}^{\tau+\Delta t} v_m \hat{\mathbf{n}} dt = v_m \hat{\mathbf{n}} \Delta t. \tag{22}$$

Figure 5 presents a scheme illustrating how the grain shape changes during the migration of the facet with the normal $\hat{\mathbf{n}}$. Further, we will consider only the migrating facet and the adjacent facets. The elements of the migrating facet are denoted by the index m and the adjacent facets by the index f. In this model, the grain structure is represented by flat sections, meaning the "bending" of the boundary cannot be described. However, there is a possibility of considering the rotation of the adjacent facets under the assumption that an increase in volume is provoked by this effect (see Figure 5). In this case, it is assumed that the polyhedron remains convex. In the statistical model, the change in the geometry of the polyhedra is not actually simulated. The characteristic size and sphericity of grains needed for rearrangement of the polyhedral grain structure in Neper are determined in a simplified way after a certain number of time steps.

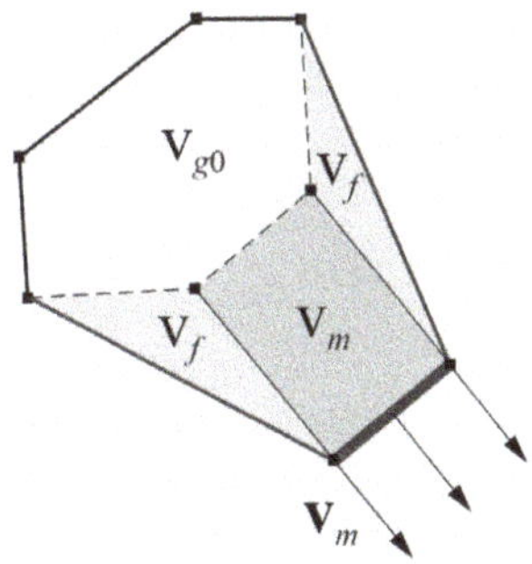

Figure 5. Scheme of facet migration.

According to the scheme given in Figure 5, the volume of the recrystallized part of the grain occurred due to the facet migration v_r is calculated as the sum of two terms:

$$v_r = v_m + v_{\sum f} = \beta v_m, \tag{23}$$

where v_m is the volume swept by the mobile facet; $v_{\sum f}$ is the total volume made up of the "rotation" of the facets; $\beta > 1$ is the parameter describing the change in the volume of a DRX grain due to the migration of the facets. The relation for v_m is determined assuming that the swept volume is the volume of a rectangular parallelepiped:

$$v_m = u_m s_m, \tag{24}$$

where s_m is the area of the migrating facet. During the facet migration and assuming that the rectangular shape of the facets remains unchanged, the area of each of the adjacent facets increases by the value s_r:

$$s_r = u_m \sqrt{s_f}, \tag{25}$$

where s_f is the initial area of the facet adjacent to the migrating facet. For the consumed grain adjacent to the migrating facet, the volume decreases by the value v_r calculated by relation (23), and the areas of the neighboring grain facets adjacent to the migrating facet decrease by the values determined by (25). If the volume of the neighboring grain reduces to zero (to an accuracy of computational error), then it is replaced by the grain from a representative volume of the statistical model. This grain is determined from the facet of the closest orientation, which is assumed to be conjugated with the migrating facet.

Finally, the characteristic size r_g (to do this, the sphere of equal volume is associated with a grain) and sphericity ψ_g are obtained for each grain:

$$r_g = \left(\frac{3}{4\pi} v_g\right)^{1/3}, \psi_g = \frac{4\pi r_g{}^2}{s_g}, \tag{26}$$

where $v_g = v_{g0} + v_r$, $s_g = s_{g0} + s_r$ is the volume and area of the grain determined by the above-mentioned relations. The distributions of r_g and ψ_g obtained for the considered representative volume are transferred to Neper to construct a new polyhedral structure, and then the grain structure data are introduced into the statistical model; this procedure is repeated many times.

4. Simulation Results and Their Analysis

This paper investigates the inelastic deformation of a copper polycrystal sample. The multilevel model parameters are given in Table 1.

Table 1. The parameters of the multilevel model of inelastic deformation.

Parameter	Value	Literature Source
π_{1111}	140 GPa	[76]
π_{1122}	104 GPa	[76]
π_{1313}	63 GPa	[76]
G	39 GPa	[77]
τ_{c0}	10 MPa	Identification procedure
τ_{sat}	27 MPa	Identification procedure
h_0	200 MPa	Identification procedure
a	1.4	Identification procedure
q_{lat}	1.4	[72]
$\dot{\gamma}_0$	0.001 s^{-1}	[72]
m	83	[72]
η	1.0	Identification procedure
α	0.0012	Identification procedure
e_{gb}	0.625 J/m^2	[5]
r_0	0.25 μm	[62]
Q	121 kJ/mol	[5]
m_0	1.5×10^{-6} $s \cdot m^2/kg$	[5]

The identification of material parameters for the inelastic deformation model (Table 1) was carried out on the basis of the experimental data [78]. This table includes the results of experimental studies of the stress–strain state of cylindrical samples measuring 8 mm in diameter and 12 mm in height. The uniaxial compression tests were carried out at a temperature of 775 K and at a strain rate of 2×10^{-3} s^{-1} for a polycrystal of commercial pure copper, which was pre-treated by annealing at 973 K for two hours [78]. The values of the anisotropic elastic moduli of the grain π_{ijkl} [76] and the shear modulus of the polycrystal G [77] correspond to the absolute temperature 775 K adopted for the numerical experiment. In order to perform specific calculations by applying an advanced statistical model, the grain structure of a polycrystal should be given in the reference configuration (the procedure for grain structure formation is described in [79]). Based on the available experimental data, statistic laws were derived for the distribution of normalized grain size d_{eq} and sphericity ψ_g of the copper polycrystal grains. The average polycrystalline copper grain size d_0 was 78 μm [78]. To approximate the grain size distribution, the lognormal law was used [80,81]. According to the data collected from the microsectional analysis [78], there is a scattering in grain sizes, which is characteristic of annealed materials. The following parameters are needed to specify a log-normal distribution: $\mu = 3.86$, $\sigma = 1$ such that the mathematical expectation $\exp(\mu + \sigma^2/2)$ be equal to the average grain size $d_0 = 78$ μm. If the representative experimental data on grain size distribution exist, then there is no difficulty in solving the optimization problem of determining the statistical distribution law and identifying its parameters, as was done in [82] for subgrain size distributions. We suppose that in the initial state, the copper sample after annealing has predominantly equiaxed grains with an average sphericity $\langle \psi_g \rangle = 0.90$; the data on sphericity are given in [83]. Figure 6 presents the grain size distribution d_{eq} and grain sphericity ψ_g histograms obtained based on existing experimental data [78,83].

The applied statistical two-level mathematical model is based on Voigt's hypothesis. The kinematic influence is prescribed and the velocity gradient is known. Plastic deformations corresponding to uniaxial deformation are considered [6]:

$$\hat{\nabla}\mathbf{V} = \dot{\gamma}\mathbf{k}_{01}\mathbf{k}_{01} - \frac{\dot{\gamma}}{2}\mathbf{k}_{02}\mathbf{k}_{02} - \frac{\dot{\gamma}}{2}\mathbf{k}_{03}\mathbf{k}_{03}, \quad (27)$$

where $\mathbf{k}_{0i} = \mathbf{k}_0^i$ is the orthonormal basis of the laboratory coordinate system; $\dot{\gamma}$ is the prescribed deformation velocity.

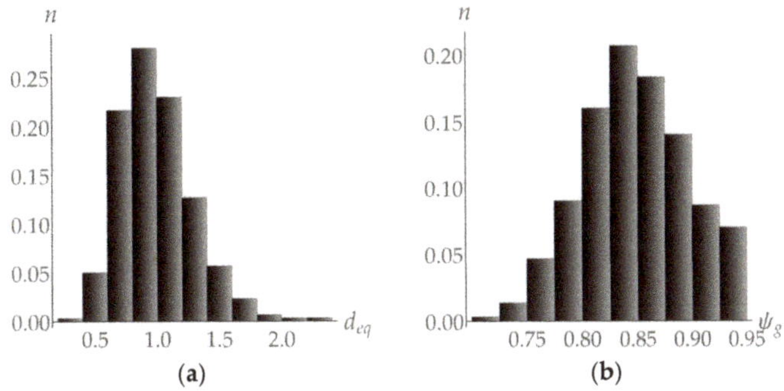

Figure 6. Normalized grain size distribution d_{eq} (non-dimensional) (**a**) and grain sphericity ψ_g (non-dimensional) (**b**) histograms.

Further, we present the results of the computational experiments carried out at deformation velocity $\dot{\gamma} = 10^{-3}\ \text{s}^{-1}$ and temperature $\theta = 775$ K. The loading diagram (with and without the recrystallization process) is given in Figure 7a, and the corresponding deformation textures formed after the end of plastic deformation are shown in Figure 7b. We emphasize a significant contribution of recrystallization to the deformation texture "blurring", which is attributed to the appearance of new DRX grains with an orientation consistent with the parent grain—numerous points for DRX grains are seen on the pole figure near the point reflecting the orientation of the parent grain.

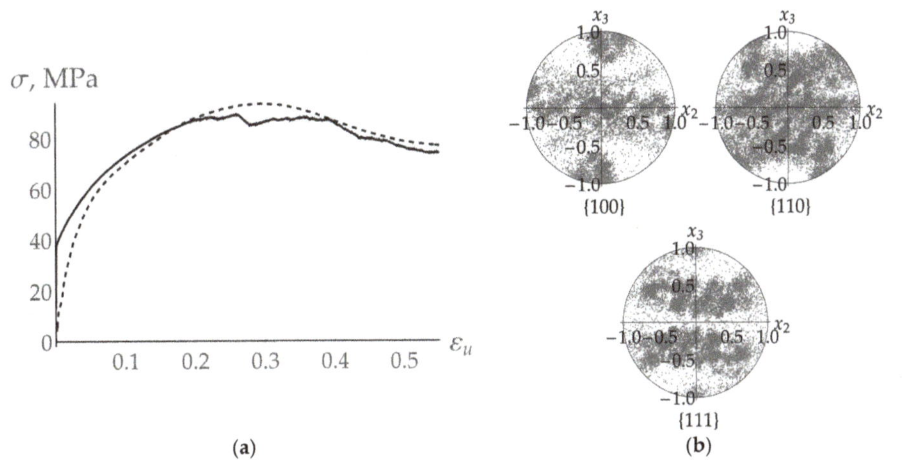

Figure 7. Loading diagram for the copper polycrystal at deformation velocity $\dot{\gamma} = 10^{-3}\ \text{s}^{-1}$ and temperature $\theta = 775$ K (dashed curve—experiment [78], solid curve—calculations) (**a**). Calculated pole figures {100}, {110}, and {111} at the end of deformation (**b**).

Figure 8a illustrates the evolution of the average grain size $\langle r_g \rangle$ during plastic deformation. Figure 8b shows a change in the average value of the facet area $\langle s_g \rangle$. Figure 8c demonstrates the evolution of the ratio of the average grain volume to the facet area $\phi_g = \langle s_g \rangle / \langle v_g \rangle$. Notice that the recrystallization process refines the initially coarse polycrystal grains. At the end of deformation, the average grain size reaches a value of 14 μm. The fraction of high-angle boundaries also increases due to the appearance of many recrystallized grains (Figure 8c). Initially, the representative volume of the polycrystal contains

300 grains, and at the end of deformation it consists of 2017 grains. In the reference configuration, the edge of the representative volume in the form of a cube is about 421 μm. At the end of deformation, the size of the edge along which tension occurred is 767 μm, and the remaining two edges are 312 μm in size. The violations of the curves correspond to the instants of time at which the polyhedral grain structure rearranges in Neper.

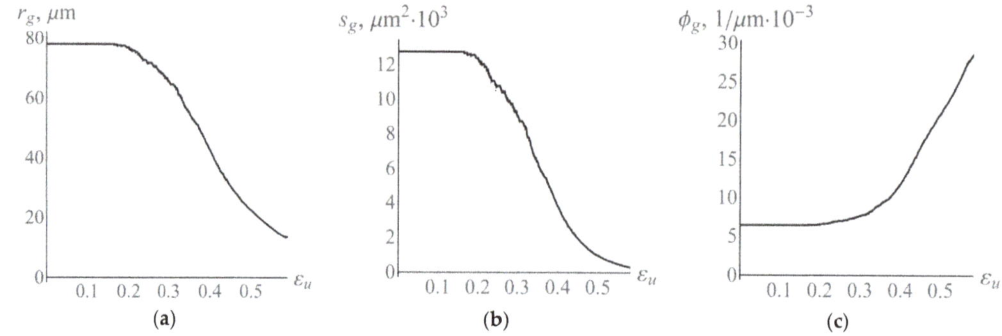

Figure 8. Evolution of average grain size (**a**) and average grain facet area (**b**) and the ratio of the average grain volume to the facet area (**c**) during plastic deformation.

Figure 9 presents the polyhedral structures formed in Neper. These structures correspond to the instant of deformation $\varepsilon_m = 0.25$ at which the stress tensor σ_m reaches its maximum (Figure 9a) and to the end of deformation $\varepsilon_f = 0.53$ (Figure 9b). It is interesting that at the instant of deformation ε_m, both large initial grains and new grains with a small size appear. At the instant of deformation ε_f, practically no large grains are kept.

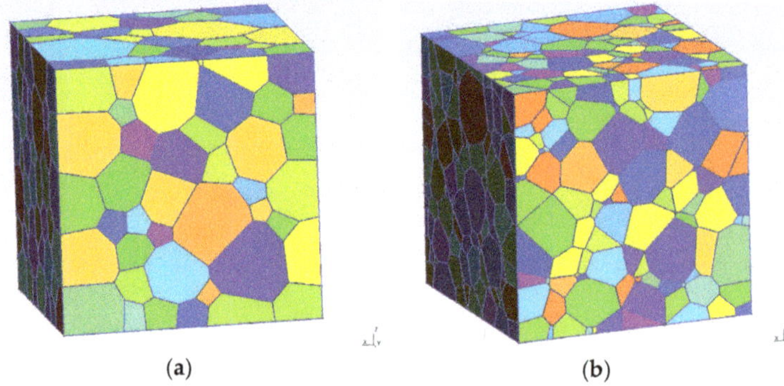

Figure 9. Polyhedral grain structure obtained using Neper at the instants of deformation ε_m (**a**) and ε_f (**b**).

This is confirmed by the grain size distribution r_g (Figure 10a,b) and sphericity ψ_g (Figure 10c,d) histograms, plotted for two different instants of deformation ε_m, ε_f. For ε_m, the distribution of grain sizes r_g and sphericity ψ_g is a bimodal distribution with two characteristic peaks that reflect the initial stage of DRX implementation. At the end of deformation ε_f, the structure becomes fine-grained (Figure 10b) and equiaxed (Figure 10d) with an average grain size of about 14 μm. The grain size distribution (Figure 10b) is characterized by the presence of a fraction (about 9%) of non-recrystallized grains with sizes that range from 30 μm to 150 μm. The sizes of non-recrystallized grains significantly exceed those of recrystallized grains, as evidenced by the data from Figure 10a,b. The level of energy stored on defects in non-recrystallized grains is much higher as compared to recrystallized grains. Therefore, the current critical stresses caused by the interaction of mobile dislocations with forest dislocations (relation (8)) exhibit high values. On the other hand, in large non-recrystallized grains, the contribution of the grain size to strengthening due to the interaction of mobile dislocations with grain boundaries is smaller (relation (9)). During the recrystallization process, large initial grains are gradually replaced by less defective recrystallized grains of a smaller size, which causes the current critical stress values to reduce. Due to the competing processes of hardening and softening, the resistance to deformation becomes stationary. The original non-recrystallized grain boundaries are non-equilibrium (new recrystallized grains are formed on these boundaries), which promotes the subsequent absorption of non-recrystallized grains. Under further deformation and at increasing stored energy level in recrystallized grains, these grains may become absorbed because of the appearance of less defective grains, which is an indication of the cyclic nature of the process of intermittent dynamic recrystallization. Thus, the amount of recrystallized material should be determined for each cycle separately.

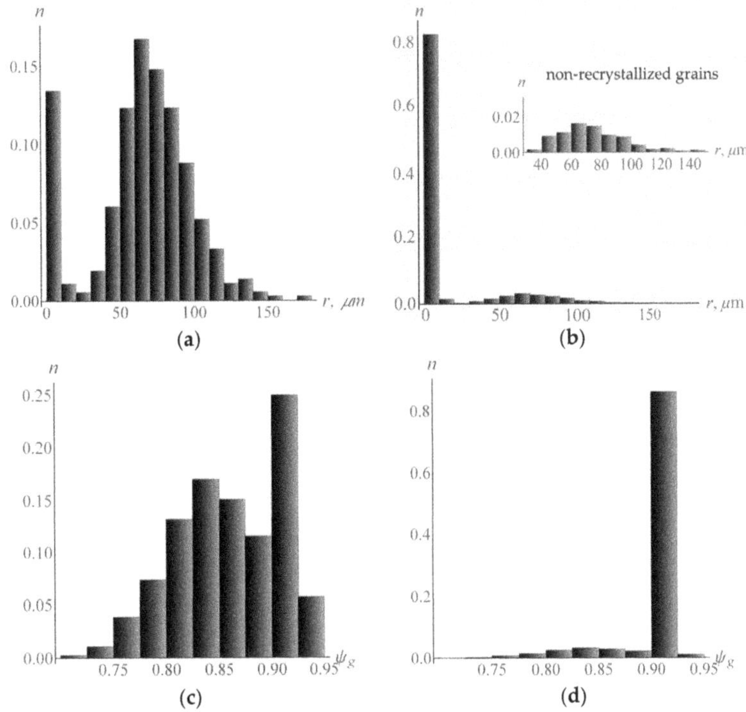

Figure 10. Grain size distribution r_g (**a**,**b**) and grain sphericity ψ_g (**c**,**d**) histograms at different instants of deformation ε_m and ε_f, respectively.

5. Discussion and Conclusions

The construction of models that are suitable for designing functional materials is a complex task that requires a careful study of many aspects of their behavior. The polycrystalline materials used in industry for the manufacture of real products are alloys. As a rule, these materials must meet the yield stress and strength requirements, meaning they are difficult to deform and their processing takes place at elevated deformation temperatures. During material processing, processes accompanying this deformation occur, among which the most significant with regard to the material structure evolution is recrystallization. This is why the description of recrystallization should be included in the models under study. In contrast to pure metals, recrystallization that develops in alloys depends on the behavior of impurity atoms and particles of secondary phases, which have a retarding effect on the motion of recrystallized grain boundaries. Moreover, the particles of secondary phases are also the sites for the predominant formation of recrystallization nuclei. The above-mentioned structural metals exhibit the ability to undergo solid-state phase transformations, meaning they have two or more phases in original billets, or experience solid-state phase transformations during thermomechanical processing. The occurrence of different phases and the increasing complexity of the model parameter identification process raise the non-trivial question of modeling the interaction between them. Hence, it follows that the development of a universal model capable of describing a variety of processes that occur during the thermomechanical processing of real materials is a challenging but feasible task. By analyzing the obtained results, we can conclude that the multilevel modeling approach based on the introduction of internal variables offers the possibility of developing additional blocks of a mathematical constitutive model needed to describe different physical mechanisms observed during thermo-mechanical treatment and for their further integration into in the full model to describe metal processing. The statistical model proposed in this work was primarily used to study the grain structure evolution in a copper polycrystal during its recrystallization and the coupled response of the stress–strain state of a half-crystal representative volume. Furthermore, this model has potential for determining thermomechanical factors that govern the formation of the necessary (given by statistical laws) grain structure, which ensures the required material properties.

We modified the advanced statistical model of inelastic deformation with the intent of introducing a process of dynamic recrystallization. For this purpose, we developed a method for describing the grain structure evolution in the framework of a geometrical approach based on Laguerre polyhedra. The obtained data were analyzed and transferred to the modified statistical model. The developed model has a number of advantages compared to existing models, namely it has higher computational efficiency and is able to take into account the physical interactions between neighboring grains. The results of the computational experiments obtained with this model are in good agreement with the experimental data. The model makes it possible to evaluate the effects associated with the refinement of the grain structure and its transformation to an equiaxed shape and to analyze the tendency of the average grain size to decrease. We also proposed a method to introduce a description of new recrystallized grains into the model and a method to establish a correspondence between the geometric image of a real grain structure, described by a polyhedral, and the grain structure of a statistical model, determined by a set of volumes, normals, and grain facet areas, indicating the neighboring grains in the recrystallization process.

Author Contributions: Conceptualization, P.T. and N.K.; methodology, P.T. and N.K.; software, A.P.; validation, A.P., N.K. and P.T.; formal analysis, N.K. and A.P.; investigation, N.K., P.T. and A.P.; data curation, N.K. and A.P.; writing—original draft preparation, N.K.; writing—review and editing, P.T., N.K. and A.P.; visualization, N.K. and A.P.; supervision, P.T.; funding acquisition, N.K. All authors have read and agreed to the published version of the manuscript.

Funding: The study was carried out with a financial support from the Ministry of Education and Science of the Russian Federation as part of the implementation of the national project "Science and Universities" (the state task fulfillment in the laboratory of multilevel structural and functional materials modeling, project no. FSNM-2021-0012).

Conflicts of Interest: The authors declare no conflict of interest.

References

1. Hsu, E.; Carsley, J.E.; Verma, R. Development of Forming Limit Diagrams of Aluminum and Magnesium Sheet Alloys at Elevated Temperatures. *J. Mater. Eng. Perform.* **2008**, *17*, 288–296. [CrossRef]
2. Zhang, R.; Shao, Z.; Lin, J. A Review on Modelling Techniques for Formability Prediction of Sheet Metal Forming. *Int. J. Lightweight Mater. Manuf.* **2018**, *1*, 115–125. [CrossRef]
3. Zheng, G.; Pang, T.; Sun, G.; Wu, S.; Li, Q. Theoretical, Numerical, and Experimental Study on Laterally Variable Thickness (LVT) Multi-Cell Tubes for Crashworthiness. *Int. J. Mech. Sci.* **2016**, *118*, 283–297. [CrossRef]
4. Sun, G.; Zhang, H.; Wang, R.; Lv, X.; Li, Q. Multiobjective Reliability-Based Optimization for Crashworthy Structures Coupled with Metal Forming Process. *Struct. Multidisc. Optim.* **2017**, *56*, 1571–1587. [CrossRef]

5. Humphreys, F.J.; Hatherly, M. *Recrystallization and Related Annealing Phenomena*; Elsevier: Amsterdam, The Netherlands, 2012; ISBN 978-0-08-098388-2.
6. Trusov, P.V.; Shveykin, A.I. *Multilevel Models of Mono- and Polycrystalline Materials: Theory, Algorithms, Application Examples*; SB RAS: Novosibirsk, Russia, 2019. (In Russian)
7. Montheillet, F. Moving Grain Boundaries During Hot Deformation of Metals: Dynamic Recrystallization. In *Moving Interfaces in Crystalline Solids*; Fischer, F.D., Ed.; CISM International Centre for Mechanical Sciences; Springer: Vienna, Austria, 2005; pp. 203–256. ISBN 978-3-211-27404-0.
8. Zhou, G.; Li, Z.; Li, D.; Peng, Y.; Zurob, H.S.; Wu, P. A Polycrystal Plasticity Based Discontinuous Dynamic Recrystallization Simulation Method and Its Application to Copper. *Int. J. Plast.* **2017**, *91*, 48–76. [CrossRef]
9. Chen, S.F.; Li, D.Y.; Zhang, S.H.; Han, H.N.; Lee, H.W.; Lee, M.G. Modelling Continuous Dynamic Recrystallization of Aluminum Alloys Based on the Polycrystal Plasticity Approach. *Int. J. Plast.* **2020**, *131*, 102710. [CrossRef]
10. Liu, F.; Fa, T.; Chen, P.H.; Wang, J.T. Steady-State Characteristics of Fcc Pure Metals Processed by Severe Plastic Deformation: Experiments and Modelling. *Philos. Mag.* **2020**, *100*, 62–83. [CrossRef]
11. Hansen, N. Cold Deformation Microstructures. *Mater. Sci. Technol.* **1990**, *6*, 1039–1047. [CrossRef]
12. Asaro, R.J. Micromechanics of Crystals and Polycrystals. *Adv. Appl. Mech.* **1983**, *23*, 1–115. [CrossRef]
13. Roters, F.; Eisenlohr, P.; Hantcherli, L.; Tjahjanto, D.D.; Bieler, T.R.; Raabe, D. Overview of Constitutive Laws, Kinematics, Homogenization and Multiscale Methods in Crystal Plasticity Finite-Element Modeling: Theory, Experiments, Applications. *Acta Mater.* **2010**, *58*, 1152–1211. [CrossRef]
14. McDowell, D.L. A Perspective on Trends in Multiscale Plasticity. *Int. J. Plast.* **2010**, *9*, 1280–1309. [CrossRef]
15. Acar, P.; Ramazani, A.; Sundararaghavan, V. Crystal Plasticity Modeling and Experimental Validation with an Orientation Distribution Function for Ti-7Al Alloy. *Metals* **2017**, *7*, 459. [CrossRef]
16. Galán-López, J.; Hidalgo, J. Use of the Correlation between Grain Size and Crystallographic Orientation in Crystal Plasticity Simulations: Application to AISI 420 Stainless Steel. *Crystals* **2020**, *10*, 819. [CrossRef]
17. Taylor, G.I. Plastic Strain in Metals. *J. Inst. Met.* **1938**, *62*, 307–324.
18. Kocks, F.; Mecking, H. Physics and Phenomenology of Strain Hardening: The FCC Case. *Prog. Mater. Sci.* **2003**, *48*, 171–273. [CrossRef]
19. Zhang, K.; Holmedal, B.; Hopperstad, O.S.; Dumoulin, S.; Gawad, J.; Van Bael, A.; Van Houtte, P. Multi-Level Modelling of Mechanical Anisotropy of Commercial Pure Aluminium Plate: Crystal Plasticity Models, Advanced Yield Functions and Parameter Identification. *Int. J. Plast.* **2015**, *66*, 3–30. [CrossRef]
20. Lebensohn, R.A.; Liu, Y.; Ponte Castañeda, P. On the Accuracy of the Self-Consistent Approximation for Polycrystals: Comparison with Full-Field Numerical Simulations. *Acta Mater.* **2004**, *52*, 5347–5361. [CrossRef]
21. Lebensohn, R.A.; Tomé, C.N.; CastaÑeda, P.P. Self-Consistent Modelling of the Mechanical Behaviour of Viscoplastic Polycrystals Incorporating Intragranular Field Fluctuations. *Philos. Mag.* **2007**, *87*, 4287–4322. [CrossRef]
22. Beyerlein, I.J.; Knezevic, M. Review of Microstructure and Micromechanism-Based Constitutive Modeling of Polycrystals with a Low-Symmetry Crystal Structure. *J. Mater. Res.* **2018**, *33*, 3711–3738. [CrossRef]
23. Yaghoobi, M.; Ganesan, S.; Sundar, S.; Lakshmanan, A.; Rudraraju, S.; Allison, J.E.; Sundararaghavan, V. PRISMS-Plasticity: An Open-Source Crystal Plasticity Finite Element Software. *Comput. Mater. Sci.* **2019**, *169*, 109078. [CrossRef]
24. Feather, W.G.; Lim, H.; Knezevic, M. A Numerical Study into Element Type and Mesh Resolution for Crystal Plasticity Finite Element Modeling of Explicit Grain Structures. *Comput. Mech.* **2021**, *67*, 33–55. [CrossRef]
25. Trusov, P.V.; Kondratev, N.S.; Yanz, A.Y. A Model for Static Recrystallization through Strain-InducedBoundary Migration. *Phys. Mesomech.* **2020**, *23*, 97–108. [CrossRef]
26. Shveykin, A.; Trusov, P.; Sharifullina, E. Statistical Crystal Plasticity Model Advanced for Grain Boundary Sliding Description. *Crystals* **2020**, *10*, 822. [CrossRef]
27. Trusov, P.V.; Shveykin, A.I. Multilevel Crystal Plasticity Models of Single- and Polycrystals. Statistical Models. *Phys. Mesomech.* **2013**, *16*, 23–33. [CrossRef]
28. Trusov, P.V.; Shveykin, A.I. Multilevel Crystal Plasticity Models of Single- and Polycrystals. Direct Models. *Phys. Mesomech.* **2013**, *16*, 99–124. [CrossRef]
29. Rollett, A.D. Overview of Modeling and Simulation of Recrystallization. *Prog. Mater. Sci.* **1997**, *42*, 79–99. [CrossRef]
30. Huang, K.; Logé, R.E. A Review of Dynamic Recrystallization Phenomena in Metallic Materials. *Mater. Des.* **2016**, *111*, 548–574. [CrossRef]
31. Tan, K.; Li, J.; Guan, Z.; Yang, J.; Shu, J. The Identification of Dynamic Recrystallization and Constitutive Modeling during Hot Deformation of Ti55511 Titanium Alloy. *Mater. Des.* **2015**, *84*, 204–211. [CrossRef]
32. Irani, M.; Joun, M. Determination of JMAK Dynamic Recrystallization Parameters through FEM Optimization Techniques. *Comput. Mater. Sci.* **2018**, *142*, 178–184. [CrossRef]
33. Puchi-Cabrera, E.S.; Guérin, J.D.; La Barbera-Sosa, J.G.; Dubar, M.; Dubar, L. Plausible Extension of Anand's Model to Metals Exhibiting Dynamic Recrystallization and Its Experimental Validation. *Int. J. Plast.* **2018**, *108*, 70–87. [CrossRef]
34. Vandermeer, R.A.; Rath, B.B. Microstructural Modeling of Recrystallization in Deformed Iron Single Crystals. *Metall. Mater. Trans. A* **1989**, *20*, 1933–1942. [CrossRef]

35. Momeni, A.; Dehghani, K. Prediction of Dynamic Recrystallization Kinetics and Grain Size for 410 Martensitic Stainless Steel during Hot Deformation. *Met. Mater. Int.* **2010**, *16*, 843–849. [CrossRef]
36. Lin, F.; Zhang, Y.; Tao, N.; Pantleon, W.; Juul Jensen, D. Effects of Heterogeneity on Recrystallization Kinetics of Nanocrystalline Copper Prepared by Dynamic Plastic Deformation. *Acta Mater.* **2014**, *72*, 252–261. [CrossRef]
37. Matsumoto, H.; Velay, V. Mesoscale Modeling of Dynamic Recrystallization Behavior, Grain Size Evolution, Dislocation Density, Processing Map Characteristic, and Room Temperature Strength of Ti-6Al-4V Alloy Forged in the (A+β) Region. *J. Alloy. Compd.* **2017**, *708*, 404–413. [CrossRef]
38. Arun Babu, K.; Prithiv, T.S.; Gupta, A.; Mandal, S. Modeling and Simulation of Dynamic Recrystallization in Super Austenitic Stainless Steel Employing Combined Cellular Automaton, Artificial Neural Network and Finite Element Method. *Comput. Mater. Sci.* **2021**, *195*, 110482. [CrossRef]
39. Beltran, O.; Huang, K.; Logé, R.E. A Mean Field Model of Dynamic and Post-Dynamic Recrystallization Predicting Kinetics, Grain Size and Flow Stress. *Comput. Mater. Sci.* **2015**, *102*, 293–303. [CrossRef]
40. Zecevic, M.; Lebensohn, R.A.; McCabe, R.J.; Knezevic, M. Modelling Recrystallization Textures Driven by Intragranular Fluctuations Implemented in the Viscoplastic Self-Consistent Formulation. *Acta Mater.* **2019**, *164*, 530–546. [CrossRef]
41. Chuan, W.; He, Y.; Wei, L.H. Modeling of Discontinuous Dynamic Recrystallization of a Near-α Titanium Alloy IMI834 during Isothermal Hot Compression by Combining a Cellular Automaton Model with a Crystal Plasticity Finite Element Method. *Comput. Mater. Sci.* **2013**, *79*, 944–959. [CrossRef]
42. Zhao, P.; Wang, Y.; Niezgoda, S.R. Microstructural and Micromechanical Evolution during Dynamic Recrystallization. *Int. J. Plast.* **2018**, *100*, 52–68. [CrossRef]
43. Ruiz Sarrazola, D.A.; Pino Muñoz, D.; Bernacki, M. A New Numerical Framework for the Full Field Modeling of Dynamic Recrystallization in a CPFEM Context. *Comput. Mater. Sci.* **2020**, *179*, 109645. [CrossRef]
44. Yu, P.; Wu, C.; Shi, L. Analysis and Characterization of Dynamic Recrystallization and Grain Structure Evolution in Friction Stir Welding of Aluminum Plates. *Acta Mater.* **2021**, *207*, 116692. [CrossRef]
45. Raabe, D. Introduction of a Scalable Three-Dimensional Cellular Automaton with a Probabilistic Switching Rule for the Discrete Mesoscale Simulation of Recrystallization Phenomena. *Philos. Mag. A* **1999**, *79*, 2339–2358. [CrossRef]
46. Zhu, H.; Chen, F.; Zhang, H.; Cui, Z. Review on Modeling and Simulation of Microstructure Evolution during Dynamic Recrystallization Using Cellular Automaton Method. *Sci. China Technol. Sci.* **2020**, *63*, 357–396. [CrossRef]
47. Chang, K.; Chen, L.-Q.; Krill, C.E.; Moelans, N. Effect of Strong Nonuniformity in Grain Boundary Energy on 3-D Grain Growth Behavior: A Phase-Field Simulation Study. *Comput. Mater. Sci.* **2017**, *127*, 67–77. [CrossRef]
48. Maire, L.; Scholtes, B.; Moussa, C.; Bozzolo, N.; Muñoz, D.P.; Settefrati, A.; Bernacki, M. Modeling of Dynamic and Post-Dynamic Recrystallization by Coupling a Full Field Approach to Phenomenological Laws. *Mater. Des.* **2017**, *133*, 498–519. [CrossRef]
49. Mora, L.A.B.; Gottstein, G.; Shvindlerman, L.S. Three-Dimensional Grain Growth: Analytical Approaches and Computer Simulations. *Acta Mater.* **2008**, *56*, 5915–5926. [CrossRef]
50. Mellbin, Y.; Hallberg, H.; Ristinmaa, M. A Combined Crystal Plasticity and Graph-Based Vertex Model of Dynamic Recrystallization at Large Deformations. *Model. Simul. Mater. Sci. Eng.* **2015**, *23*, 045011. [CrossRef]
51. Glezer, A.M.; Kozlov, E.V.; Koneva, N.A.; Popova, N.A.; Kurzina, I.A. *Plastic Deformation of Nanostructured Materials*; CRC Press: Boca Raton, FL, USA, 2017; ISBN 978-1-315-11196-4.
52. Dyakonov, G.S.; Mironov, S.; Semenova, I.P.; Valiev, R.Z.; Semiatin, S.L. EBSD Analysis of Grain-Refinement Mechanisms Operating during Equal-Channel Angular Pressing of Commercial-Purity Titanium. *Acta Mater.* **2019**, *173*, 174–183. [CrossRef]
53. Hall, E.O. The Deformation and Ageing of Mild Steel: III Discussion of Results. *Proc. Phys. Soc. B* **1951**, *64*, 747–753. [CrossRef]
54. Petch, N.J. The Cleavage Strength of Polycrystals. *J. Iron Steel Inst.* **1953**, *174*, 25–28.
55. Armstrong, R.W. The Influence of Polycrystal Grain Size on Several Mechanical Properties of Materials. *Metall. Trans.* **1970**, *1*, 1169–1176. [CrossRef]
56. Knezevic, M.; Beyerlein, I.J. Multiscale Modeling of Microstructure-Property Relationships of Polycrystalline Metals during Thermo-Mechanical Deformation. *Adv. Eng. Mater.* **2018**, *20*, 1700956. [CrossRef]
57. Zhou, X.; Feng, Z.; Zhu, L.; Xu, J.; Miyagi, L.; Dong, H.; Sheng, H.; Wang, Y.; Li, Q.; Ma, Y.; et al. High-Pressure Strengthening in Ultrafine-Grained Metals. *Nature* **2020**, *579*, 67–72. [CrossRef] [PubMed]
58. Tan, J.C.; Tan, M.J. Superplasticity and Grain Boundary Sliding Characteristics in Two Stage Deformation of Mg–3Al–1Zn Alloy Sheet. *Mater. Sci. Eng. A* **2003**, *339*, 81–89. [CrossRef]
59. Quey, R.; Renversade, L. Optimal Polyhedral Description of 3D Polycrystals: Method and Application to Statistical and Synchrotron X-Ray Diffraction Data. *Comput. Methods Appl. Mech. Eng.* **2018**, *330*, 308–333. [CrossRef]
60. Roberts, W.; Boden, H.; Ahlblom, B. Dynamic Recrystallization Kinetics. *Metal. Sci.* **1979**, *13*, 195–205. [CrossRef]
61. Ponge, D.; Gottstein, G. Necklace Formation during Dynamic Recrystallization: Mechanisms and Impact on Flow Behavior. *Acta Mater.* **1998**, *46*, 69–80. [CrossRef]
62. Zurob, H.S.; Bréchet, Y.; Dunlop, J. Quantitative Criterion for Recrystallization Nucleation in Single-Phase Alloys: Prediction of Critical Strains and Incubation Times. *Acta Mater.* **2006**, *54*, 3983–3990. [CrossRef]
63. Cram, D.G.; Zurob, H.S.; Brechet, Y.J.M.; Hutchinson, C.R. Modelling Discontinuous Dynamic Recrystallization Using a Physically Based Model for Nucleation. *Acta Mater.* **2009**, *57*, 5218–5228. [CrossRef]

64. Beck, P.A.; Sperry, P.R. Strain Induced Grain Boundary Migration in High Purity Aluminum. *J. Appl. Phys.* **1950**, *21*, 150–152. [CrossRef]
65. Hallberg, H.; Wallin, M.; Ristinmaa, M. Simulation of Discontinuous Dynamic Recrystallization in Pure Cu Using a Probabilistic Cellular Automaton. *Comput. Mater. Sci.* **2010**, *49*, 25–34. [CrossRef]
66. Kondratev, N.S.; Trusov, P.V.; Podsedertsev, A.N. Multilevel model of polycrystals: Application to assessing the effect of texture and grains misorientation on the critical deformation of the dynamic recrystallization initiation. *PNRPU Mech. Bull.* **2021**, *4*, 83–97. (In Russian) [CrossRef]
67. Kondratev, N.S.; Trusov, P.V.; Podsedertsev, A.N. Multilevel physical-oriented model: Applicationto the description of the initial stageof dynamic recrystallization of polycrystals. *Probl. Strength Plast.* **2021**, *83*, 451–461. [CrossRef]
68. Lebensohn, R.A.; Brenner, R.; Castelnau, O.; Rollett, A.D. Orientation Image-Based Micromechanical Modelling of Subgrain Texture Evolution in Polycrystalline Copper. *Acta Mater.* **2008**, *56*, 3914–3926. [CrossRef]
69. Kondratev, N.S.; Trusov, P.V. Description of Hardening Slip Systems Due to the Boundaries of the Crystallines in a Polycrystalline Aggregate. *PNRPU Mech. Bull.* **2012**, *3*, 78–97. (In Russian)
70. Trusov, P.V.; Shveykin, A.I. On Motion Decomposition and Constitutive Relations in Geometrically Nonlinear Elastoviscoplasticity of Crystallites. *Phys. Mesomech.* **2017**, *20*, 377–391. [CrossRef]
71. Anand, L. Single-Crystal Elasto-Viscoplasticity: Application to Texture Evolution in Polycrystalline Metals at Large Strains. *Comput. Methods Appl. Mech. Eng.* **2004**, *193*, 5359–5383. [CrossRef]
72. Bronkhorst, C.A.; Kalidindi, S.R.; Anand, L. Polycrystalline Plasticity and the Evolution of Crystallographic Texture in FCC Metals. *Phil. Trans. R. Soc. Lond. A* **1992**, *341*, 443–477. [CrossRef]
73. Bailey, J.E.; Hirsch, P.B. The Recrystallization Process in Some Polycrystalline Metals. In *Proceedings of the Royal Society of London. Series, A. Mathematical and Physical Sciences*; The Royal Society: London, UK, 1962; Volume 267, pp. 11–30. [CrossRef]
74. Mitchell, M. *An Introduction to Genetic Algorithms*; MIT Press: Massachusetts, USA, 1998; ISBN 978-0-262-63185-3.
75. Pozdeev, A.A.; Trusov, P.V.; Nyashin, Y.I. *Large Elastoplastic Deformations: Theory, Algorithms, Applications*; Nauka: Moscow, Russia, 1986. (In Russian)
76. Chang, Y.A.; Himmel, L. Temperature Dependence of the Elastic Constants of Cu, Ag, and Au above Room Temperature. *J. Appl. Phys.* **1966**, *37*, 3567–3572. [CrossRef]
77. Ledbetter, H.M.; Naimon, E.R. Elastic Properties of Metals and Alloys. II. Copper. *J. Phys. Chem. Ref. Data* **1974**, *3*, 897–935. [CrossRef]
78. Blaz, L.; Sakai, T.; Jonas, J.J. Effect of Initial Grain Size on Dynamic Recrystallization of Copper. *Metal Sci.* **1983**, *17*, 609–616. [CrossRef]
79. Kondratev, N.S.; Trusov, P.V.; Podsedertsev, A.N. The polycrystals grain structure formation for modified two-level crystal plasticity statistical models. *Procedia Struct. Integr.* **2022**, *in press*.
80. Vaz, M.F.; Fortes, M.A. Grain Size Distribution: The Lognormal and the Gamma Distribution Functions. *Scr. Metall.* **1988**, *22*, 35–40. [CrossRef]
81. Raeisinia, B.; Sinclair, C.W. A Representative Grain Size for the Mechanical Response of Polycrystals. *Mater. Sci. Eng. A* **2009**, *525*, 78–82. [CrossRef]
82. Kondratev, N.S.; Trusov, P.V. To Determination a Distribution Law of Subgrain Sizes Formed in the Cold Plastic Deformation Process. *AIP Conf. Proc.* **2020**, *2216*, 040010. [CrossRef]
83. Suresh, K.S.; Rollett, A.D.; Suwas, S. Evolution of Microstructure and Texture During Deformation and Recrystallization of Heavily Rolled Cu-Cu Multilayer. *Metall. Mater. Trans. A* **2013**, *44*, 3866–3881. [CrossRef]

crystals

MDPI

Article

Study of the Effect of Grain-Boundary Misorientation on Slip Transfer in Magnesium Alloy Using a Misorientation Distribution Map

Jie Sun [1,2,*], Jianhua Liu [1,2], Qingqiang Chen [1,2], Laixiao Lu [1,2] and Yanhua Zhao [1]

1 School of Mechanical and Electronic Engineering, Shandong Jianzhu University, Jinan 250101, China; ljh@sdjzu.edu.cn (J.L.); 13770@sdjzu.edu.cn (Q.C.); lulaixiao@sdjzu.edu.cn (L.L.); zyh@sdjzu.edu.cn (Y.Z.)
2 Research Center of High Performance Materials Forming, Shandong Jianzhu University, Jinan 250101, China
* Correspondence: sunjie20@sdjzu.edu.cn

Abstract: The microstructure evolution of a Mg–Gd–Y alloy was studied using uniaxial tension combined with an electron backscatter diffraction technique. The results show that large amounts of slip transfer phenomena can be observed around the grain–boundary area after tension, and the activation of these slips depends largely on the misorientation of grain boundaries. The Mg–Gd–Y alloy shows almost randomized grain–boundary misorientation, but transferred slip traces were preferred at boundaries with misorientation around the [0001] axis between 0–30°. Theoretically, materials with a higher fraction of slip transfer at the grain–boundary area would improve the ductility. Upon comparing the two groups of magnesium alloy with different grain–boundary misorientation distributions, the one with more grain boundaries favored for slip transfer achieved higher elongation during a tension test. Therefore, in addition to weakening the texture, adjusting the misorientation of the grain boundaries appears to be a new method to improve the ductility of magnesium alloys.

Keywords: magnesium alloy; slip transfer; crystallographic misorientation; ductility

Citation: Sun, J.; Liu, J.; Chen, Q.; Lu, L.; Zhao, Y. Study of the Effect of Grain-Boundary Misorientation on Slip Transfer in Magnesium Alloy Using a Misorientation Distribution Map. *Crystals* **2022**, *12*, 388. https://doi.org/10.3390/cryst12030388

Academic Editors: Dmitry Medvedev and Mingyi Zheng

Received: 31 January 2022
Accepted: 10 March 2022
Published: 13 March 2022

1. Introduction

Magnesium is the lightest structural metal, and has received substantial attention as a potential material in the transportation industry [1,2]. However, the application of wrought magnesium alloys is restricted by its poor formability at room temperature, caused by the strong basal texture developed during processing and the lack of available deformation modes [3].

It is well known that deformation behavior of metals depends largely on the compatibility of neighboring grains. Recently, the intergranular deformation behavior of magnesium alloys has attracted increasing attention [4–8]. Jonas [9] and Barnett [10] suggested that the strain accommodation required by the neighboring grains might affect the variant selection of twins in magnesium alloys. The formation of a twin involves shearing the matrix. When twin nucleation occurs at a grain boundary, the strain must be shared by the neighboring grain to accommodate the shape change. The amount and type of deformation modes required for strain accommodation in the neighboring grains would affect the variant selection process. In our previous study [11], the activation of $\{10\bar{1}2\}$ $<10\bar{1}1>$ twinning was influenced by slip–induced twinning behavior and strain compatibility among surrounding grains, except for the Schmid factor related to grain orientation. In addition to twinning, slips are also affected by neighboring grains. Martin [6] found that the activation of basal slip adjacent to grains that were deformed predominantly by non–basal slip would accommodate high local strain.

Since the development of a mathematical treatment for orientation distribution function, texture analysis has been proven to be a very useful method to understand the properties of metallic materials. Combined with studies on deformation behavior of single–crystal and texture analysis, the mechanical properties of most polycrystalline materials could be predicted.

Additionally, polycrystalline magnesium alloys develop internal strain/stress characteristics of three length–scales, i.e., macro–, inter– and intragranular. At the intergranular scale, strain mismatch between grains with different orientations due to elastic and plastic anisotropy at the grain level will result in stress relaxation in individual grains, and then the activation of various intragranular deformation modes. Thus, the deformation behavior of Mg alloys depends largely on the compatibility of neighboring grains, and then the interactions between neighboring grains will significantly affect the mechanical properties.

In our previous works, with the combination of electron backscatter diffraction (EBSD) and micro–scale digital–image correlation (DIC), the relationship between local strain and grain orientation was fully studied. The average intergranular strains were found to be higher than the average intragranular strains. Additionally, a grain boundary with a high m' value and high Schmid factor for certain slip systems in adjacent grains could accommodate enough local strains and improve the deformation compatibility [12]. Finally, a higher local strain at the grain–boundary area was found to be related to higher elongation during a tension test [13].

The interaction behavior of neighboring grains depends largely on the grain–boundary misorientation. However, achieving a clear relationship between grain–boundary misorientation and mechanical properties used to be difficult, until the development of a mathematical tool to construct continuous distribution functions using axis–angle parameters in a series of hyperspherical harmonics [14,15]. By using this method, we believe a better understanding of the relationship between microstructure and mechanical properties can be achieved.

Here, we focus on studying intergranular deformation behavior in magnesium alloys. In order to achieve that, experiments were performed on an extruded Mg–Gd–Y alloy, which shows both discrete grain orientation distribution and grain–boundary misorientation distribution. Interactions between grains during deformation were carefully analyzed using continuous misorientation distribution functions (MDF) to develop a statistical relationship between grain–boundary misorientation and deformation behavior at a micro–scale.

2. Experiment Methods

The alloy used in this paper, with a nominal composition of Mg–8.0Gd–3.0Y–0.5Zr (wt%) (GW83), was prepared by semi–continuous casting. The billet was solution–treated at 520 °C for eight hours, then extruded at 450 °C with an extrusion ratio of 25:1 and an extrusion speed of 6 mm/s.

Tension tests for stress–strain curves were conducted using a ZWICK/Roell 20KN mechanical testing machine at room temperature. Tension specimens had a rectangular cross section of 3 mm × 1.4 mm with a 10 mm gauge length. Deformation microstructure was observed at a tensile strain of 0.1. The observation planes of tension samples were polished before the tensile test using a sequence of ethanol–based diamond suspensions of 6, 3, and 1 μm, respectively. This was followed by fine–polishing using colloidal silica suspension (OPS), and a final 2–4 s etching using a solution of 5% HNO_3, 15% acetic acid, 20% H_2O and 60% ethanol before SEM and EBSD observations.

The EBSD analysis was carried using a Quanta 250 SEM equipped with a TSL^{TM} EBSD camera and an OIM software package. SEM pictures were also collected in the Quanta 250 SEM. The EBSD scanning process was conducted at a step size of 1 μm with a voltage of 20 kV and a current of 107 nA.

Misorientation distribution function (MDF) [14,15], calculated using the symmetrized hyperspherical harmonic formulation, was applied to quantify misorientation statistics for the magnesium samples in this paper.

Grain orientation on each side of a boundary was represented by three Euler angles (ϕ_1, Φ, ϕ_2). Then, misorientation, represented by axis–angle parameter (ω, $\boldsymbol{n}$), was calculated.

A minimum misorientation angle and an axis of rotation lying in a pre-defined standard region were selected, and the subset of the rotation space—commonly referred to as the fundamental zone—was obtained. Rotation axis was then written as a function of polar coordinates (θ, ϕ). A generic MDF, which is a real-valued probability density function, can be expanded with real coefficients $f_{L,\,M}^{NC}$ and $f_{L,\,M}^{NS}$, as:

$$f(\omega,\,\theta,\,\phi) = \sum_{N=0}^{\infty}\sum_{L=0}^{N}\left[f_{L,\,0}^{NC}Z_{L,\,0}^{NC} + \sum_{M=1}^{L}\left(f_{L,\,M}^{NC}Z_{L,\,M}^{NC} + f_{L,\,M}^{NS}Z_{L,\,M}^{NS}\right)\right] \tag{1}$$

Coefficients $f_{L,\,M}^{NC}$ and $f_{L,\,M}^{NS}$ were calculated using ω, θ, ϕ as variables.

$$f_{LM}^{NC} = \int_0^{2\pi}\int_0^{\pi}\int_0^{\pi} Z_{L,\,M}^{NC} f(\omega,\,\theta,\,\phi)(sin\alpha)^2 d\omega sin\theta d\theta d\phi \tag{2}$$

$$f_{LM}^{NS} = \int_0^{2\pi}\int_0^{\pi}\int_0^{\pi} Z_{L,\,M}^{NS} f(\omega,\,\theta,\,\phi)(sin\alpha)^2 d\omega sin\theta d\theta d\phi$$

$$Z_{L,\,M}^{NC} = (-1)^{L+M} K \times \cos(M\phi)$$

$$Z_{L,\,M}^{NS} = (-1)^{L+M} K \times sin(M\phi)$$

$$K = \frac{2^L L!}{\pi}\left[(2L+1)\frac{(L-M)!(N+1)(N-L)!}{(L+M)!(N+L+1)!}\right]^{\frac{1}{2}} \times \left[\sin\left(\frac{\omega}{2}\right)\right]^L \times C_{N-L}^{L+1}\left(\cos\left(\frac{\omega}{2}\right)\right) \times P_L^M(\cos(\theta))$$

with integer indices $0 \leq N$, $0 \leq L \leq N$ and $-L \leq M \leq L$. C_{N-L}^{L+1} is a Gegenbauer polynomial and P_L^M is an associate Legendre function.

Then, sections of constant misorientation angles (10–90°) were selected to visualize the misorientation space. In each section, 36 misorientation relationships were calculated (same misorientation angle, but different axis). The detailed derivation of the above equations and the meaning of each parameter can be found in the literature [14,15]. The maximum N was selected as 16 in this work. The MDF was then normalized to the random distribution of misorientation and are plotted in this paper.

3. Results and Discussion

3.1. Initial Microstructure

Figure 1 shows the initial microstructure information. The material used in this paper is the same as in our previous study [12]. The figure is rotated 90° clock-wise, in order to show the frame of reference used in Figures 2–6. A fully recrystallized microstructure with an average grain size of 25 μm can be found (Figure 1a). The initial texture of this Mg alloy, shown in Figure 1b, is relatively weak, and the highest texture intensity is only 2.752 which is lower than most of the current wrought Mg alloys.

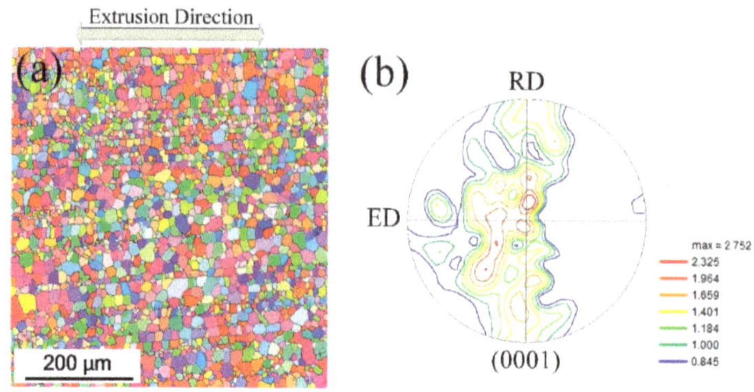

Figure 1. Initial Mg–8.0Gd–3.0Y–0.5 Zr alloy microstructure: (**a**) IPF map; (**b**) PF map.

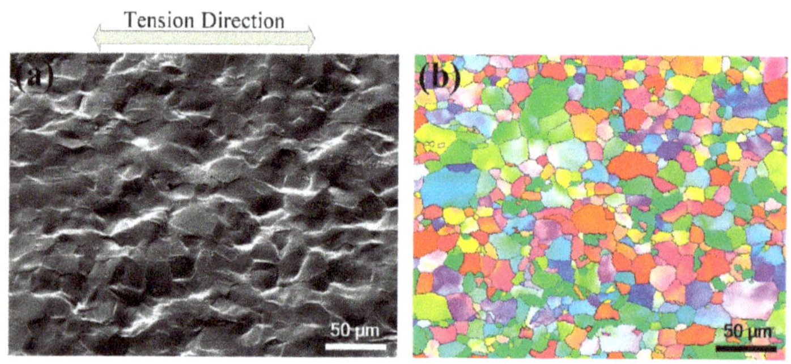

Figure 2. SEM map (**a**) and the corresponding IPF (**b**) after 10% tensile strain in Mg–8.0Gd–3.0Y–0.5 Zr alloy.

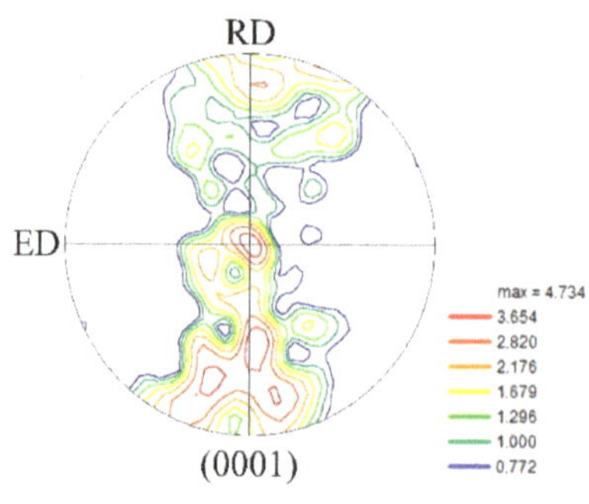

Figure 3. PF map after 10% tensile strain in Mg–8.0Gd–3.0Y–0.5 Zr alloy.

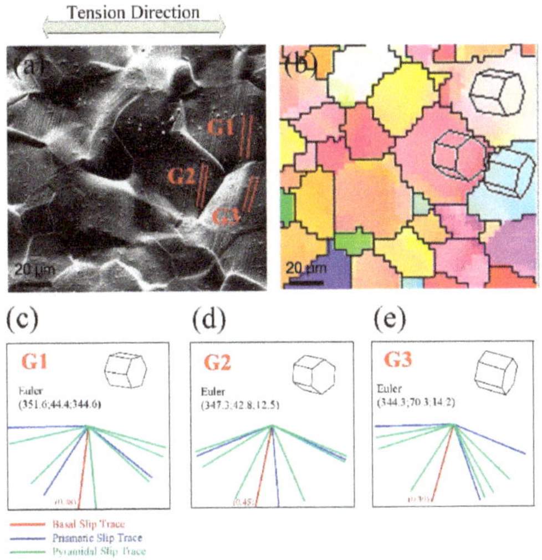

Figure 4. (**a**) SEM micrograph of Mg–8.0Gd–3.0Y–0.5 Zr alloy after tension test at strain of 0.1; (**b**) the corresponding IPF map; (**c**) possible slip trace directions in Grain 1; (**d**) possible slip trace directions in Grain 2; (**e**) possible slip trace directions in Grain 3.

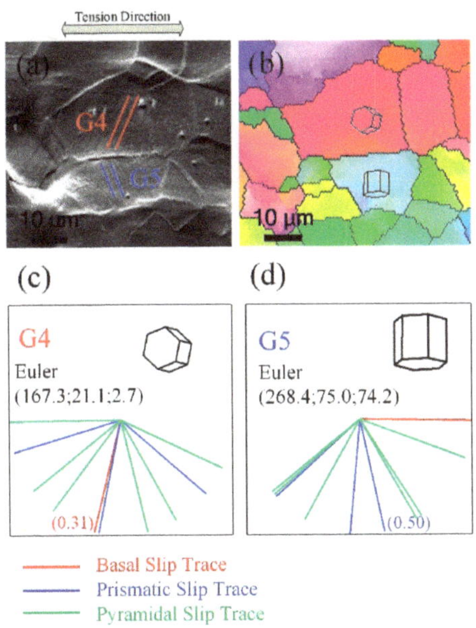

Figure 5. (**a**) SEM micrograph of Mg–8.0Gd–3.0Y–0.5 Zr alloy after tension test at strain of 0.1; (**b**) the corresponding IPF map; (**c**) possible slip trace directions in Grain 4; (**d**) possible slip trace directions in Grain 5.

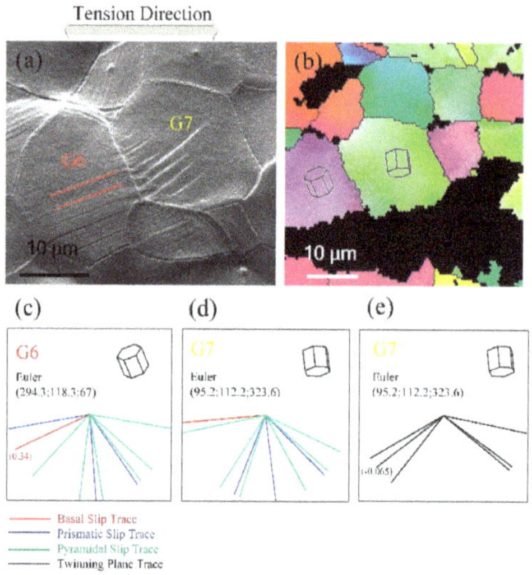

Figure 6. (**a**) SEM micrograph of Mg–8.0Gd–3.0Y–0.5 Zr alloy after tension test at strain of 0.1; (**b**) the corresponding IPF map; (**c**) possible slip trace directions in Grain 6; (**d**) possible slip trace directions in Grain 7; (**e**) possible twinning boundary trace directions in Grain 7.

3.2. Slip Transfer at Grain–Boundary Area

Figure 2 shows the deformed microstructure after the tension test (ε = 0.1), and Figure 3 shows the pole figure (PF) after tension. Obvious slip traces in the grains and fluctuation at the grain boundaries can be seen. Using slip trace analysis, basal slip and first-order prismatic slip were found to be the dominant slip deformation modes and only a few $\{10\bar{1}2\}$ $<10\bar{1}1>$ instances of twinning can be found. In addition, many slip traces were found to be transferred through grain boundaries. SEM photographs of different kinds of slip traces with high magnification are shown in Figures 4–6. Several kinds of slip and twinning in pairs can be found around grain boundaries. Basal slip traces were found to come in pairs with basal slip (named B–B, Figure 4), prismatic slip (named B–P, Figure 5), and $\{10\bar{1}2\}$ $<10\bar{1}1>$ twinning (named B–T, Figure 6). The statistical data in Figure 7a show that the number fraction of transferred slips across grain boundaries has a high proportion of above 80%. Additionally, the vast majority of these transferred slips were identified as basal slips (Figure 7b).

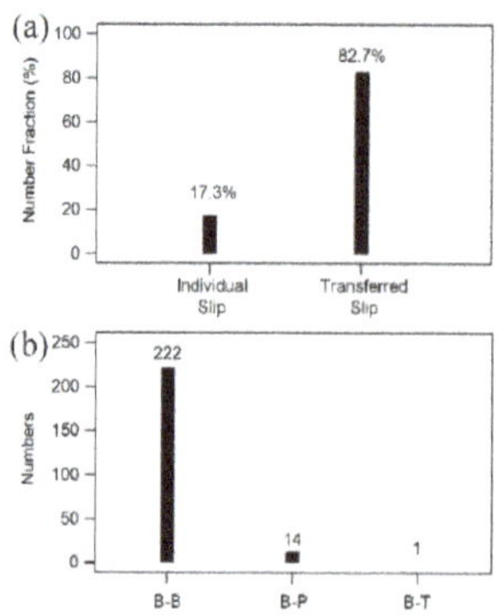

Figure 7. (**a**) The number fractions of grains with slip traces observed, and (**b**) the number of grains with different transferred slip modes observed (94% are B–B).

In Figure 4a, slip traces can be found in Grain 1, Grain 2 and Grain 3. Using slip trace analysis, as shown in Figure 4c–e, these slip traces were identified as basal slip traces with high Schmid factors of 0.48, 0.45 and 0.39, respectively.

A geometric compatibility factor m', first defined by Luster and Morris [16], was employed.

$$m' = cos\psi \times cos\kappa \tag{3}$$

where ψ and κ are the angles between the active normal slip plane and slip directions, respectively.

Using this parameter, the geometric compatibility factor between the slip systems in adjacent grains may vary between 0 and 1. For m' = 1, complete compatibility exists between the slip systems in the neighboring grains, and in this case, both the slip directions and the slip planes in each grain will be parallel. In contrast, m' = 0 indicates that the slip systems are completely incompatible, such that either the slip directions or slip planes are orthogonal. In general, m' will assume intermediate values between 0 and 1.

The geometric compatibility factors between the two slip systems in Grain 1 and Grain 2 are listed in Table 1. The highest value of geometric compatibility factor is 0.91 which indicates good compatibility for the activation of basal slip systems in the adjacent grains.

Table 1. m' value between basal slip systems in Grain 1 and Grain 2.

G2 \ G1	Basal [11-20] (0.27)	Basal [1-210] (0.48)	Basal [-2110] (0.23)
Basal [11-20] (0.41)	0.91	0.09	0.82
Basal [1-210] (0.45)	0.82	0.91	0.09
Basal [-2110] (0.04)	0.09	0.81	0.91

Schmid factor of basal slip in grains wherein the occurrence of transferred basal slip was calculated; the result is shown in Figure 8, with the comparison with the Schmid factor distribution in all grains (1231 grains were collected for the analysisi of Schmid factor). It can be seen that the relationship between the Schmid factor of the basal slip system and the activation of slip transfer at the grain–boundary area is weak. Therefore, it seems that the criterion for the activation of transferred basal slip cannot depend only on the value of the Schmid factor.

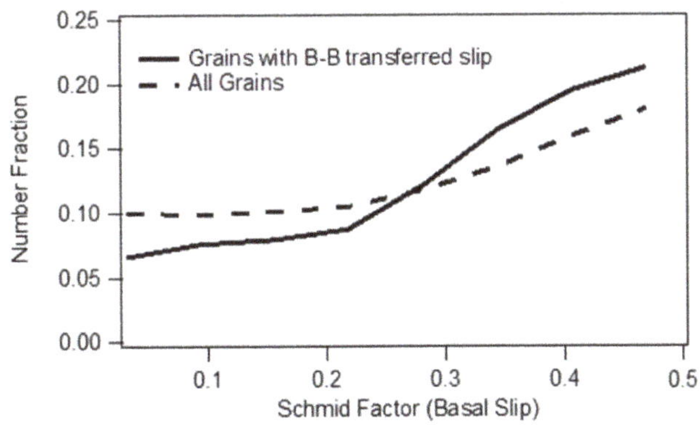

Figure 8. Relationship between Schmid factor of basal slip and number fraction of B–B transferred slip.

3.3. *Effect of Grain–Boundary Misorientation Distribution on Slip Transfer*

As the above results show, deformation behavior at grain level does not depend only on grain orientations, and the interaction between grains around grain-boundary areas appears to be significant during the deformation process. In this case, whether certain types of grain boundaries might favor transferred slips should be considered.

Figure 9 shows the misorientation angle distribution of boundaries with transferred basal slips across and all boundaries in this material (1632 grain boundaries were collected for the analysis of grain–boundary misorientation distribution). A significant difference can be found between these two distributions, and low–angle boundaries appear to favor transferred basal slips.

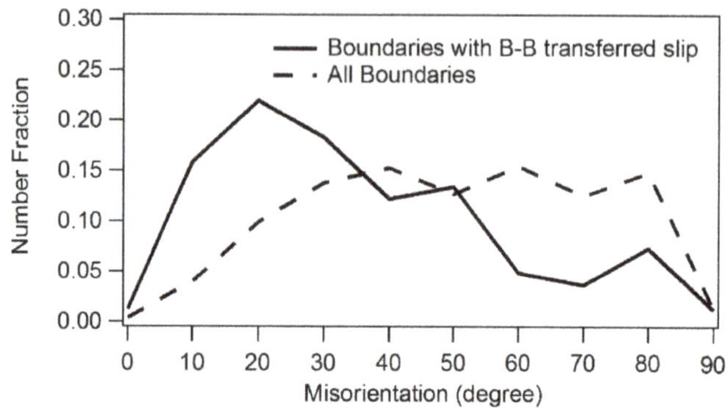

Figure 9. Relationship between misorientation angle and number fraction of B–B transferred slip.

In order to fully understand the effect of grain boundary on the slip transfer phenomenon, MDF maps using the axis–angle parameters were draw to fully understand the geometric characteristics of these boundaries favored for transferred basal slip. Figure 10 shows the MDF of all boundaries in the Mg–Gd–Y alloy. The misorientation distribution is quite randomized, and the highest intensity is only 1.76. Figure 11 shows the MDF of the boundaries with transferred basal slips across them. Obvious high intensification can be found at low misorientation angles and especially around the axis around [0001]. By using the MDF method, it can easily be seen that the activation of the transferred basal slip system is preferred at boundaries with misorientation at the axis around [0001] and an angle of 0–30°.

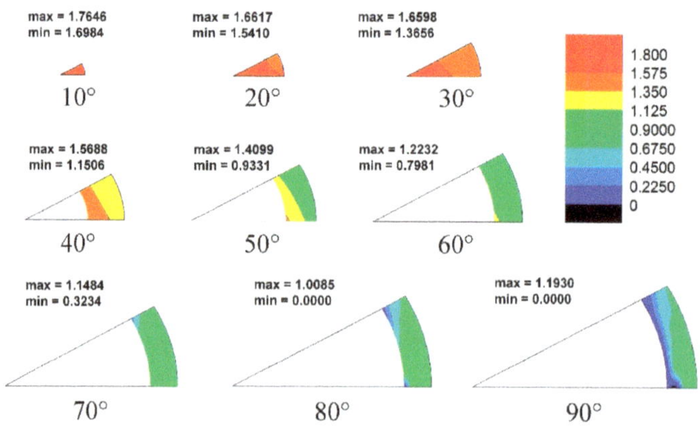

Figure 10. Misorientation distribution map of Mg–8.0Gd–3.0Y–0.5 Zr alloy.

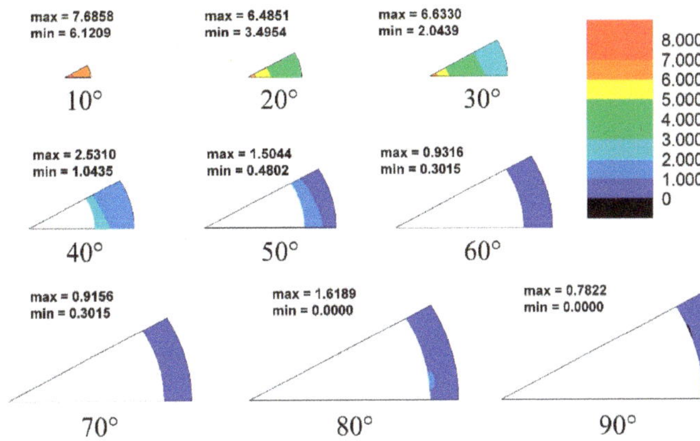

Figure 11. Misorientation distribution map of B–B slip transferred boundaries.

3.4. Effect of Grain–Boundary Misorientation Distribution on Ductility

The grain–boundary area preferred for more slip transfer would accommodate higher strain. If one polycrystalline magnesium alloy has more boundaries of the sort, the ductility of this material, theoretically, should be higher.

In order to support the arguments made in the present manuscript, data from two groups of Mg–3Al–1Zn (AZ31) alloy have been included. Figure 12 exhibits two groups of data, all from the Mg–3Al–1Zn (AZ31) alloy, which is the most common of the magnesium alloys. The manufacturing process for each is different; one involves rolling and the other involves extrusion. These two kinds of material show a similar Schmid factor distribution during tension along their prior working (rolling or extrusion) direction. However, the rolled AZ31 alloy shows much higher elongation than the extruded one (Figure 12b, data collected from the literature [17–29]). This phenomenon could not be explained based on the theory that a higher ductility is achieved by grain orientations favored for basal slip.

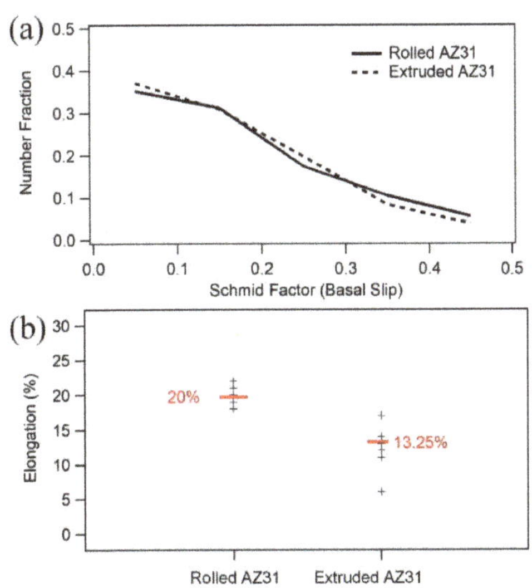

Figure 12. (**a**) Schmid factor distribution, and (**b**) elongation during tension tests of typical rolled AZ31 alloys and extruded alloys (data collected from the literature).

By analyzing the MDF of these two AZ31 alloys, in Figures 13 and 14, a higher fraction of low-angle misorientation distribution was found in rolled AZ31 (600 boundaries were collected for the analysis of the misorientation distribution map for each of the AZ31 alloys). Therefore, with a higher fraction of boundaries favored for transferred slip, a higher elongation can be obtained in the rolled AZ31 alloy.

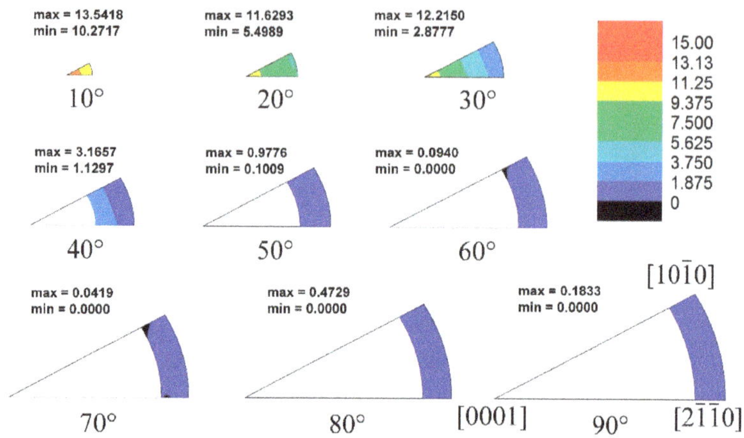

Figure 13. Misorientation distribution map of rolled AZ31 alloy.

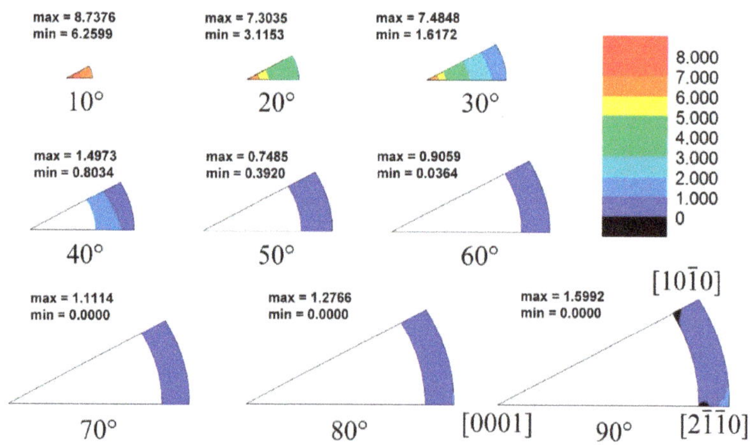

Figure 14. Misorientation distribution map of extruded AZ31 alloy.

In past decades, texture-weakening has been proven to be a good method for improving the ductility of magnesium alloys [20,30]. With more grains favored for basal slips, the elongation of magnesium alloys during a tension test could be increased dramatically. The Mg–Gd–Y alloy shown in this paper has a weakened texture, and the maximum texture intensity is only 2.752. The results shown in this paper already prove that with a better deformation compatibility, the grain-boundary area can accommodate a higher strain and the magnesium materials can achieve higher ductility. Since it has been reported that misorientation distribution can be altered after heat treatment [31,32], the deformation compatibility of magnesium alloys could also be improved. Furthermore, high deformation compatibility, when combined with a randomized texture, would further improve the ductility of magnesium alloys.

4. Conclusions

(1) The activation of slip crossed the grain boundary during tension deformation, and the basal slip system was the dominate type. Not a lot of twinning was observed on the surface after tension.
(2) The activation of dislocation slip pairs was affected by the misorientation of grain boundaries. Grain boundaries with a misorientation around the [0001] axis and at a 10–30° angle promoted the activation of basal slip pairs.
(3) The extruded AZ31 alloy showed higher elongation, because the extruded AZ31 alloy exhibited a higher fraction of grain-boundary misorientation around the [0001] axis at an angle of 0–30°.

Author Contributions: Formal analysis, J.S.; Methodology, J.S., J.L., Q.C. and L.L.; Project administration, J.S.; Writing—original draft, J.S.; Writing—review & editing, Y.Z. All authors have read and agreed to the published version of the manuscript.

Funding: This work was supported by the National Natural Science Foundation of China (Grant No. 52001188).

Data Availability Statement: Not applicable.

Conflicts of Interest: The authors declare no conflict of interest.

References

1. Mordike, B.L.; Ebert, T. Magnesium: Properties—Applications—Potential, Mater. *Sci. Eng. A* **2001**, *302*, 37–45. [CrossRef]
2. Luo, A.A. Magnesium: Current and potential automotive applications. *JOM* **2002**, *54*, 42–48. [CrossRef]
3. Yoo, M.H. Slip, twinning, and fracture in hexagonal close-packed metals. *Metall. Trans. A* **1981**, *12*, 409–418. [CrossRef]
4. Wang, F.; Sandlöbes, S.; Diehl, M.; Sharma, L.; Roters, F.; Raabe, D. In situ observation of collective grain-scale mechanics in Mg and Mg–rare earth alloys. *Acta Mater.* **2014**, *80*, 77–93. [CrossRef]
5. Martin, G.; Sinclair, C.W.; Schmitt, J.-H. Plastic strain heterogeneities in an Mg–1Zn–0.5Nd alloy. *Scr. Mater.* **2013**, *68*, 695–698. [CrossRef]
6. Martin, G.; Sinclair, C.W.; Lebensohn, R.A. Microscale plastic strain heterogeneity in slip dominated deformation of magnesium alloy containing rare earth. *Mater. Sci. Eng. A* **2014**, *603*, 37–51. [CrossRef]
7. Stanford, N.; Sotoudeh, K.; Bate, P. Deformation mechanisms and plastic anisotropy in magnesium alloy AZ31. *Acta Mater.* **2011**, *59*, 4866–4874. [CrossRef]
8. Panicker, R.; Chokshi, A.; Mishra, R.; Verma, R.; Krajewski, P. Microstructural evolution and grain boundary sliding in a superplastic magnesium AZ31 alloy. *Acta Mater.* **2009**, *57*, 3683–3693. [CrossRef]
9. Jonas, J.J.; Mu, S.; Al-Samman, T.; Gottstein, G.; Jiang, L.; Martin, É. The role of strain accommodation during the variant selection of primary twins in magnesium. *Acta Mater.* **2011**, *59*, 2046–2056. [CrossRef]
10. Barnett, M.R.; Keshavarz, Z.; Beer, A.G.; Ma, X. Non-Schmid behaviour during secondary twinning in a polycrystalline magnesium alloy. *Acta Mater.* **2008**, *56*, 5–15. [CrossRef]
11. Jin, L.; Dong, J.; Sun, J.; Luo, A.A. In-situ investigation on the microstructure evolution and plasticity of two magnesium alloys during three-point bending. *Int. J. Plast.* **2015**, *72*, 218–232. [CrossRef]
12. Sun, J.; Jin, L.; Dong, J.; Ding, W.; Luo, A.A. Microscopic deformation compatibility during monotonic loading in a Mg-Gd-Y alloy. *Mater. Charact.* **2016**, *119*, 195–199. [CrossRef]
13. Sun, J.; Jin, L.; Dong, J.; Wang, F.; Dong, S.; Ding, W.; Luo, A.A. Towards high ductility in magnesium alloys—The role of intergranular deformation. *Int. J. Plast.* **2019**, *123*, 121–132. [CrossRef]
14. Patala, S.; Mason, J.K.; Schuh, C.A. Improved representations of misorientation information for grain boundary science and engineering. *Prog. Mater. Sci.* **2012**, *57*, 1383–1425. [CrossRef]
15. Mason, J.; Schuh, C. Hyperspherical harmonics for the representation of crystallographic texture. *Acta Mater.* **2008**, *56*, 6141–6155. [CrossRef]
16. Luster, J.; Morris, M.A. Compatibility of deformation in two-phase Ti-Al alloys: Dependence on microstructure and orientation relationships. *Metall. Mater. Trans. A* **1995**, *26*, 1745–1756. [CrossRef]
17. Keshavarz, Z.; Barnett, M.R. EBSD analysis of deformation modes in Mg–3Al–1Zn. *Scr. Mater.* **2006**, *55*, 915–918. [CrossRef]
18. Yi, S.-B.; Davies, C.; Brokmeier, H.-G.; Bolmaro, R.; Kainer, K.; Homeyer, J. Deformation and texture evolution in AZ31 magnesium alloy during uniaxial loading. *Acta Mater.* **2006**, *54*, 549–562. [CrossRef]
19. Wang, F.; Feng, M.; Jiang, Y.; Dong, J.; Zhang, Z. Cyclic deformation and fatigue of extruded Mg–Gd–Y magnesium alloy. *J. Mater. Sci. Technol.* **2013**, *561*, 403–410. [CrossRef]
20. Hirsch, J.; Al-Samman, T. Superior light metals by texture engineering: Optimized aluminum and magnesium alloys for automotive applications. *Acta Mater.* **2013**, *61*, 818–843. [CrossRef]

21. Xiong, Y.; Yu, Q.; Jiang, Y. Multiaxial fatigue of extruded AZ31B magnesium alloy. *Mater. Sci. Eng. A* **2012**, *546*, 119–128. [CrossRef]
22. Yi, S.; Bohlen, J.; Heinemann, F.; Letzig, D. Mechanical anisotropy and deep drawing behaviour of AZ31 and ZE10 magnesium alloy sheets. *Acta Mater.* **2010**, *58*, 592–605. [CrossRef]
23. Wu, L.; Agnew, S.; Ren, Y.; Brown, D.; Clausen, B.; Stoica, G.; Wenk, H.; Liaw, P. The effects of texture and extension twinning on the low-cycle fatigue behavior of a rolled magnesium alloy, AZ31B. *Mater. Sci. Eng. A* **2010**, *527*, 7057–7067. [CrossRef]
24. Begum, S.; Chen, D.; Xu, S.; Luo, A.A. Low cycle fatigue properties of an extruded AZ31 magnesium alloy. *Int. J. Fatigue* **2009**, *31*, 726–735. [CrossRef]
25. Stanford, N.; Barnett, M. Effect of composition on the texture and deformation behaviour of wrought Mg alloys. *Scr. Mater.* **2008**, *58*, 179–182. [CrossRef]
26. Chino, Y.; Kimura, K.; Mabuchi, M. Twinning behavior and deformation mechanisms of extruded AZ31 Mg alloy. *Mater. Sci. Eng. A* **2008**, *486*, 481–488. [CrossRef]
27. Chino, Y.; Kimura, K.; Hakamada, M.; Mabuchi, M. Mechanical anisotropy due to twinning in an extruded AZ31 Mg alloy. *Mater. Sci. Eng. A* **2008**, *485*, 311–317. [CrossRef]
28. Agnew, S.R.; Duygulu, Ö. Plastic anisotropy and the role of non-basal slip in magnesium alloy AZ31B. *Int. J. Plast.* **2005**, *21*, 1161–1193. [CrossRef]
29. Barnett, M.; Keshavarz, Z.; Beer, A.; Atwell, D. Influence of grain size on the compressive deformation of wrought Mg–3Al–1Zn. *Acta Mater.* **2004**, *52*, 5093–5103. [CrossRef]
30. Luo, A.; Mishra, R.; Sachdev, A. High-ductility magnesium–zinc–cerium extrusion alloys. *Scr. Mater.* **2011**, *64*, 410–413. [CrossRef]
31. Yi, S.; Brokmeier, H.-G.; Letzig, D. Microstructural evolution during the annealing of an extruded AZ31 magnesium alloy. *J. Alloy. Compd.* **2010**, *506*, 364–371. [CrossRef]
32. Wu, W.; Jin, L.; Zhang, Z.; Ding, W.; Dong, J. Grain growth and texture evolution during annealing in an indirect-extruded Mg–1Gd alloy. *J. Alloy. Compd.* **2014**, *585*, 111–119. [CrossRef]

crystals

MDPI

Article

Development of the Concurrent Multiscale Discrete-Continuum Model and Its Application in Plasticity Size Effect

Zhenting Zhang, Zhen Tong * and Xiangqian Jiang

EPSRC Future Metrology Hub, Centre for Precision Technologies, University of Huddersfield, Huddersfield HD1 3DH, UK; zhenting.zhang@hud.ac.uk (Z.Z.); x.jiang@hud.ac.uk (X.J.)
* Correspondence: z.tong@hud.ac.uk

Abstract: A concurrent multiscale model coupling discrete dislocation dynamics to the finite element method is developed to investigate the plastic mechanism of materials at micron/submicron length scales. In this model, the plastic strain is computed in discrete dislocation dynamics (DDD) and transferred to the finite element method (FEM) to participate in the constitutive law calculation, while the FEM solves the complex boundary problem for DDD simulation. The implementation of the whole coupling scheme takes advantage of user subroutines in the software ABAQUS. The data structures used for information transferring are introduced in detail. Moreover, a FE mesh-based regularization method is proposed to localize the discrete plastic strain to continuum material points. Uniaxial compression tests of single crystal micropillars are performed to validate the developed model. The results indicate the apparent dependence of yield stress on sample size, and its underlying mechanisms are also analyzed.

Keywords: discrete dislocation dynamics; finite element method; multiscale model; size effects

Citation: Zhang, Z.; Tong, Z.; Jiang, X. Development of the Concurrent Multiscale Discrete-Continuum Model and Its Application in Plasticity Size Effect. *Crystals* **2022**, *12*, 329. https://doi.org/10.3390/cryst12030329

Academic Editors: Shujun Zhang and Pavel Lukáč

Received: 28 January 2022
Accepted: 25 February 2022
Published: 26 February 2022

1. Introduction

The effect of sample size or shape on material deformation behavior [1–3] has received attention for decades. For instance, the sample size of concrete influences its kinetics of external sulfate attack [4]. At the microscale, it has been reported that the deformation mechanism of crystal materials is different from that of bulk materials [5–7]. It is recognized that the strength of micro/submicron crystals depends not only on the material itself but also the sample/grain size. Many efforts have been made to study the size effects of crystal materials at the micro/submicron scale in order to improve material performance for high-quality products [8–11]. The size effect is induced by many physical mechanisms, such as dislocation motion, nucleation, and starvation [12,13]. It remains difficult to capture the full picture of the deformation behavior of crystal material at the micro/nano-scale only by experimental tests. Therefore, it is necessary to employ a suitable simulation method to reveal the underlying deformation mechanisms of crystal materials at the micro/nano-scale.

Computational simulation methodologies have been continuously developed to predict material deformation behaviors. According to the difference in simulation scale, the main simulation methods can be broadly classified into the finite element method (FEM) [14,15], crystal plasticity finite element methods (CPFEMs) [16,17], discrete dislocation dynamics (DDD) [8,18,19], and molecular dynamics (MD) [20]. Based on empirical constitutive equations, the FEM shows a great advantage in dealing with various complex boundary problems at the macroscale. Compared to the classic FEM, CPFEMs consider the material microscopic plastic shear deformation by employing critical parameters that represent the microstructural state of materials beneath the materials' macroscopic deformation. Typical examples are the slip systems-based phenomenological CPFEMs and the dislocation density-based CPFEM. Moreover, by considering the density of geometrically necessary dislocations (GNDs) in the constitutive equations, it can be used to predict the size effect of crystal materials because GNDs are related to the strain gradient and has been considered in the constitutive equation [21–24]. However, CPFEMs cannot provide an explicit microstructure evolution during the deformation process, because of the empirical constitutive equation used. To obtain insight into the physical deformation mechanism, a microscale simulation model that considers the intrinsic microstructure characteristics of crystal materials is needed. MD simulation directly depicts the interaction of atoms. In recent decades, it has been widely employed to investigate the mechanical properties of single crystals and nanopolycrystalline materials (e.g., nano-polycrystal silicon [25] and copper [26]) by means of nanoindentation or many other tests [11]. However, MD simulation is only suitable for nanoscale specimens due to the extremely heavy computational load. Indeed, the plastic deformation of crystals results from its dislocation activities. To overcome those limitations, DDD has been proposed by many researchers to capture the size effect and uncover its underlying mechanisms. It omits the direct atomic interactions and takes dislocation as the intrinsic carrier of plastic strain. DDD modeling is often employed together with the FEM to deal with complex boundary conditions. The mostly used coupling schemes are the discrete continuous model (DCM) [27] and superposition method (SPM) [28].

To solve the boundary value problem of a finite crystal, van der Giessen and Needleman [28] first proposed the concept of the SPM. In the SPM, the total stress field in a finite crystal is the sum of two stress fields. One is the analytical stress field due to the existence of dislocations in an infinite unbounded crystal and the other is the complementary elastic stress field calculated by finite element analysis in a finite crystal. The method is able to capture the short-range dislocation stress field accurately and has been improved by incorporating a 3D dislocation mechanism [13,29] and dislocation climb [30]. Nevertheless, the calculation of the analytical stress field between all existing dislocations is time-consuming. In addition, there is no explicit introduction of the plastic strain that arises from dislocation activities. The DCM is another hybrid approach coupling DDD to the FEM. The numerical calculation in the DCM is straightforward and has clear physical sense. The moving dislocation generates plastic strain along its gliding path, and then the plastic strain is localized to the Gaussian integration points of the FEM. By making use of the plastic strain, the equilibrium stress is computed in the FEM module and is used to drive the movement of discrete dislocations.

Since Lemarchand [27] first proposed the DCM in 2001, it has been widely used to study the plastic behavior of materials at micron/submicron length scales, for example, the effect of hard coating [31], low-amplitude cyclic loading [32], temperature [33], and shock loading [34] on the plastic deformation of micropillars. The dislocation mechanisms causing the channel size effects of superalloys [12], incipient plastic instability of nanoindentation [35], and strain bursts of micropillars [10] have been investigated using the DCM. Recently, Cui et al. [31] studied the confined plasticity in coated micropillars using the DCM and found that the operation of a single arm source plays a key role in the plastic deformation of coated pillars. Santos-Güemes et al. [36] studied the interaction mechanism of dislocation and precipitates using a developed multi-scale DCM. The intrinsic hardening mechanism of a nickel-based superalloy and its dependence on the interfacial dislocation network and lattice mismatch were reported [37]. By introducing a dislocation climb model into the DCM, Liu et al. [38] investigated the high-temperature anneal hardening behavior of an Au submicron-pillar.

Most studies are focused on the application of the DCM in plastic mechanisms at micron/submicron scales; however, less attention has been paid to its numerical algorithms for improved computational efficiency. Zbib et al. [39] proposed a simple method of regularizing the plastic strain induced by dislocations to the material points of their localized elements. Liu et al. [40] introduced the Burgers vector density function to apportion the discrete plastic strain to the corresponding material points. Vattré [41] and Cui [42] also discussed the detailed numerical algorithm of localizing the discrete strain to continuum material points in a 3D DCM simulation. In addition, the strain regularization method in a 2D DCM was developed by Huang [30] by considering the interaction of dislocation with its neighboring inter-phase boundaries and other dislocations. However, the computational load remains a challenge in computational simulation. There is always a balance of numerical accuracy and computational efficiency.

In this paper, a concurrent multiscale model coupling DDD to the FEM is developed to investigate the plastic mechanism of materials at micron/submicron length scales. A FE mesh-based regularization method is proposed to improve the computational efficiency. The implementation details of how to apply the DCM principle to couple DDD with FEM calculation modules during the deformation process are introduced. The optimized data structure and the achievement of the simultaneous coupling scheme by subroutines are presented for the first time. The description of the 2D-DCM is provided in Section 2. Section 3 discusses transfer methods of plastic strain and stress among DDD and FEM modules. The user subroutines of Abaqus used in the multiscale framework are described in detail in Section 4. In Section 5, a compression test of a single crystal is performed using the developed method, and it is followed with some main conclusions drawn in Section 6.

2. Description of DCM Coupling Framework

As schematically illustrated in Figure 1, the multiscale framework consists of two modules, i.e., DDD and FEM module. The key point of the 2D-DCM is that the plastic strain calculated by DDD directly participates in the constitutive law of the FEM. The stress tensor at simulation step n, σ_n, is expressed as,

$$\sigma_n = [C^e] : \left(\varepsilon_n^t - \varepsilon_{n-1}^p\right) \quad (1)$$

where ε_{n-1}^p is the plastic strain at the last step by DDD, ε_n^t is the total strain at simulation step n, and C^e is the elastic moduli tensor. The explicit computation of movement, evolution, and interaction between dislocations and other defects (e.g., obstacles, dislocation, and sources) is performed in DDD. In addition, the calculation of the equilibrium stress field in the FEM module and the coupling algorithm of both modules are discussed.

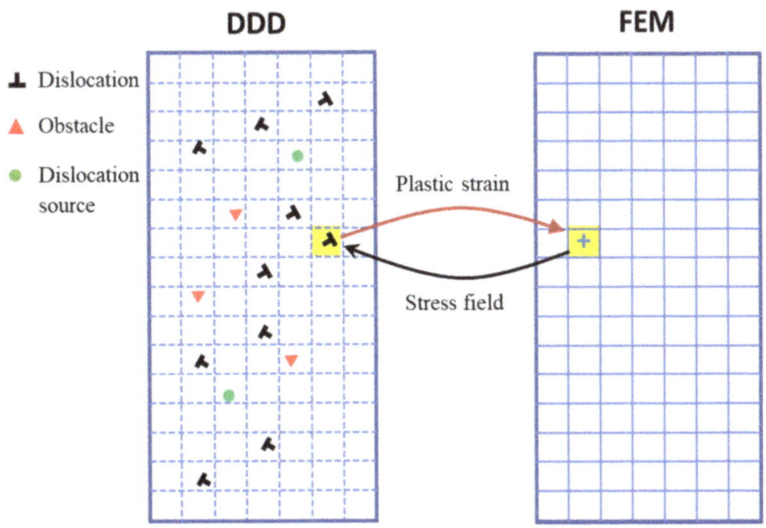

Figure 1. Schematic of the coupling scheme of DCM (Yellow block represents the single element in FEM module.).

2.1. Two-Dimensional Dislocation Dynamics

For the 2D plane strain problem discussed in this paper, only edge dislocations are considered. The effect of dislocation climb is also negligible compared with that of dislocation glide. The driving force applied on dislocations can be expressed as

$$f_i = n_i \cdot \sigma_i \cdot b_i, \tag{2}$$

where n_i is a unit normal to the slip plane, b_i is the Burgers vector, and σ_i is the stress tensor at the position of dislocation i, which is evaluated in the FEM.

Assuming that the edge dislocations can reach a stable velocity instantly, the glide velocity of dislocation i is

$$v^i = \frac{f^i}{B}, \tag{3}$$

where B is the viscous drag coefficient.

Static point sources are distributed on the slip plane randomly to mimic the operation of Frank–Read sources. A dislocation dipole is generated once the shear stress at the point source exceeds its critical strength, τ_{nuc}, during a time period. The initial distance L_{nuc} between two newly generated dislocations is

$$L_{nuc} = \frac{\mu}{2\pi(1-\nu)} \frac{b}{\tau_{nuc}}, \tag{4}$$

where μ is the shear modulus and ν is the Poisson ratio of crystal material.

Two dislocations with opposite Burgers vectors annihilate each other when their distance is less than the critical annihilation distance L_e, which is taken as 6b [43,44].

The motion of dislocations can be impeded by various obstacles, such as small precipitates and forest dislocation. The point obstacles are modeled in 2D-DDD to consider the effect of the obstacles existing in real crystals [43,45]. If one obstacle lies on the gliding path of a dislocation, the dislocation is pinned by the obstacle when its resolved shear stress is less than the strength of the obstacle. Dislocations can be de-pinned or bypass the obstacle if its shear stress exceeds the obstacle strength.

2.2. Calculation in Finite Element Module

The standard FEM is employed to solve the boundary value problem of a finite crystal as follows:

$$[K]\{U\} = \{f^a\} + \{f^p\}, \tag{5}$$

$$\{f^p\} = \int_V [C^e]\varepsilon^p[B]dV, \tag{6}$$

where $[K]$ is stiffness matrix, $\{U\}$ is the nodal displacement vector, $\{f^a\}$ is the applied force vector, $\{f^p\}$ is the force vector from plastic strain, $[B]$ = grad[N], and [N] is the shape function vector. The DDD module contributes the plastic strain ε^p to the unified continuum mechanics framework.

3. Plastic Strain–Stress Transfer in Coupling Scheme

3.1. Plastic Strain and Stress Transfer Scheme

In this computational simulation model, the plastic strain is explicitly calculated in the DDD module and is then transferred to Gaussian integration points of the FEM module for the constitutive calculation. The plastic strain is induced by the glide of dislocation, which can be calculated as:

$$\varepsilon_i^p = \frac{A_i}{2V}(n_i \otimes b_i + b_i \otimes n_i), \tag{7}$$

where A_i is the swept area of the dislocation i and V is the representative volume.

The plastic strain ε_i^p obtained from the DDD module should be located at the corresponding integration points as the eigen-strains. Different regularization methods have been proposed to solve this problem. In the early work by Lemarchand [27], it was assumed that a slab with finite thickness h surrounding the glide plane represents the elementary slip events. The elementary volume associated with each Gauss integration point is taken as the representative volume. The plastic strain distributed at each integration point depends on the intersection volume between the slab and elementary volume. Based on the work of Lemarchand [27], Liu [40] proposed a weight function (i.e., the Burgers vector density function) to distribute the total plastic strain to each integration point of finite element. Vattré [41] and Cui [42] also made efforts to improve the regularization method of plastic strain. For the 2D computational model, Huang [30] presented a detailed regularization procedure, which is similar to that of Liu [40]. A local domain along the slip plane of dislocation was defined and half of its width was set as a, which is analogous to the dislocation core spreading radius. The integration points in the local domain would be allocated with plastic strain, and the allocation proportion of each integration point was decided by its distance to the slip plane of dislocation. This regularization method can relieve a high gradient of plastic strain among neighboring elements. However, it causes a heavy computational burden, especially when numerous dislocations exist in the simulation model, due to the distances between each integration point, and the slip plane should be calculated.

An efficient regularization method is proposed in this paper by considering the characteristics of meshes of the FEM model. Elements in the FEM model are divided into elementary domains (i.e., integration point domain), which are associated with each integration point in the elements. As shown in Figure 2, a dislocation glides along its slip plane from D to D′. These elementary domains intersecting with the gliding path are regarded as 'core domains,' which are shown as the red domains in Figure 2. If the distance from the integration point of an elementary domain to a core domain is less than the given value a, the elementary domain k is defined as a 'secondary domain.' The hatched and meshed areas represent the secondary domains in Figure 2. The parameter a has the same meaning and evaluation as the one used in Huang's method. The core and secondary domains together compose the domain Ω_i. The localized plastic strain of integration point j, ε_{ij}^{p}, in the domain Ω_i is given according to a weight function,

$$\varepsilon_{ij}^{p} = \frac{\omega_j}{\sum_j \omega_j}\frac{A_i}{2V}(n_i \otimes b_i + b_i \otimes n_i), \tag{8}$$

where ω_j is the distribution coefficient for plastic strain caused by dislocation i on integration point j. The value of ω_j, which is selected as the Burgers vector spreading function developed by Cai [46], can be calculated as

$$\omega_j = \frac{1}{\left(r_j^2 + a^2\right)^{3.5}}, \tag{9}$$

where r_j is the distance between the integration points of the j-th elementary domain and its nearest core domain. For the integration point O shown in Figure 2, the distance is equal to the length of OB, and it replaces the length of OA used by Huang [47]. The simplification of r_j makes it easier to obtain the distribution coefficient. Instead of solving the formula for the distance from point to line, it only needs the distance between the integration points of the core and secondary domain. Once the core domains are obtained, the secondary domains around them can be quickly captured according to the numbering order of integration points and elements. Figure 3 shows the numbering order of integration points in different elements. The new method omits the determination of the gliding path equations of each dislocation by replacing the glide path with the core domain. Thus, with the present method, it is computationally efficient to capture the influenced domain, Ω_i, of each moving dislocation and calculate the distribution coefficient through directly solving the distance formula of two points.

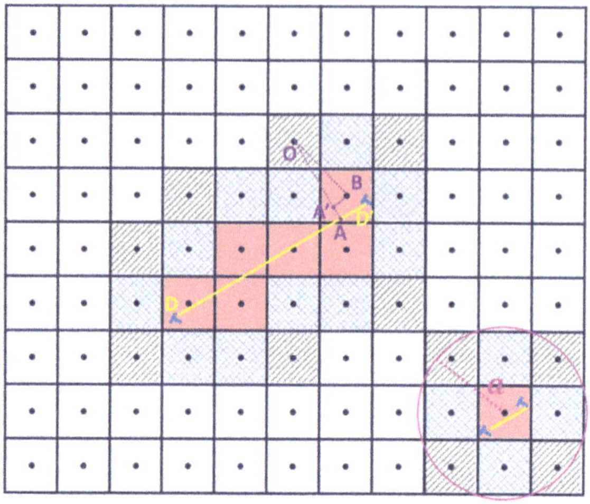

Figure 2. Schematic of distributing plastic strain to integration points of the FEM. The black points represent integration points. The red areas and the shaded areas are the core domains and secondary domains, respectively.

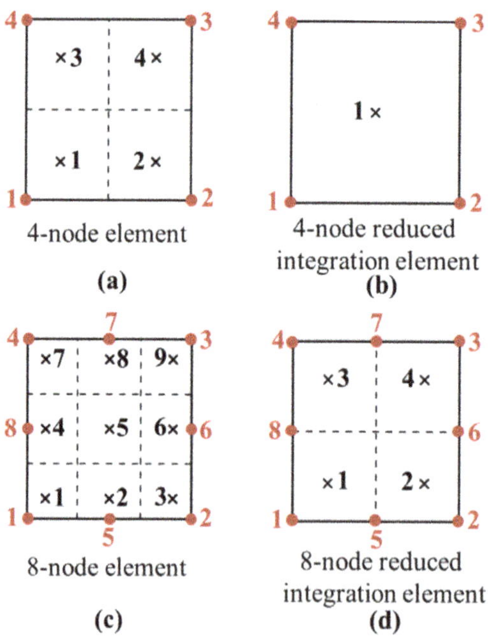

Figure 3. The numbering of integration points and nodes in 2D quadrilateral parent elements: (**a**) 4-node element, (**b**) 4-node reduced integration element, (**c**) 8-node element, and (**d**) 8-node reduced integration element. The dashed lines divide the element into integration point domains. The red dot and black cross represent the node and integration point, respectively.

After FEM calculation, there are two steps to transfer the stress field from the FEM module to the DDD module, i.e., (i) stress extrapolation from integration points to nodes by the inverse shape function, and (ii) stress interpolation from nodes to dislocations by the shape function of the element [41]. In finite element analysis, the numerical integration is carried out at the Gaussian points and the stress value at the integration point is the most accurate. Therefore, in this work, the stress of dislocations is determined by the integration point domain where dislocations reside. The stress tensor of a dislocation is the same as its nearest Gaussian integration point. The division of various elements into integration point domains is illustrated in Figure 3. Each domain contains one integration point and any point in the domain has the shortest distance to the contained integration point compared to other integration points in the parent element. The number of integration points in the parent element is determined by the degree of the integrated polynomial terms in the element's stiffness matrix and the desired accuracy of results. According to the number of integration points required, their locations in the parent element can be obtained by Table 1.

Table 1. Gauss numerical integration constants.

Number of Integration Points (n)	Integration Point Location x_i
1	$x_1 = 0$
2	$x_1 = 1/\sqrt{3},\ x_2 = -1/\sqrt{3}$
3	$x_1 = \sqrt{0.6},\ x_2 = 0,\ x_3 = -\sqrt{0.6}$

3.2. Coordinate System Conversion Involved in Coupling Scheme

In order to complete the data exchange between DDD and the FEM module, the transformation of the coordinate system should be taken into account.

The calculation in the DDD module, such as glide force, dislocation velocity, and plastic strain, is finished on the local coordinate system whose axes are consistent with the slip direction and normal direction of the slip plane, as shown in Figure 4a. The FEM module is constructed in the sample coordinate system, unlike that of the DDD module. To achieve the coordinate system transformation, the transformation matrix, $[R]$, is calculated by representing the vector OA in the two coordinate frames, as illustrated in Figure 4b. Then, the vectors, $\{g\}$, and tensors, $[M]$, can be transformed with rules as follows:

$$\{g'\} = [R]\{g\}, \tag{10}$$

$$[M'] = \left[R^T\right][M][R], \tag{11}$$

where $\{g'\}$ and $[M']$ are the transformed vector and tensor, respectively. Superscript T means matrix transpose.

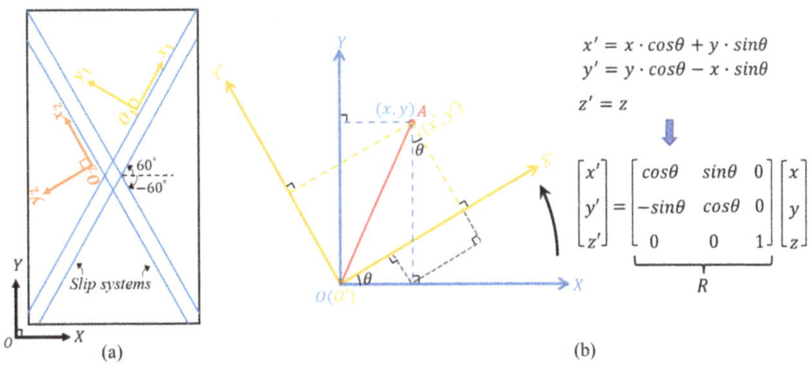

Figure 4. Schematic of two-dimensional coordinate system conversion. (**a**) (x, y) and (x', y') represent the position of point A in the coordinate system OXY and O′X′Y′, respectively. (**b**) The transformation matrix R can be obtained from the relationship of two coordinate systems.

4. Subroutines for ABAQUS

The multiscale framework is built by the secondary development of finite element software ABAQUS. In this section, the implementation details of the DCM model in ABAQUS are introduced.

4.1. User Subroutines in the Multiscale Framework

Abundant user subroutines are provided in ABAQUS, which allow advanced users to customize a variety of ABAQUS capabilities [48]. The general workflow for running the DCM using ABAQUS is illustrated in Figure 5. Three user subroutines are used for the DCM model, including URDFIL for results processing, UEXTERNALDB for external file manipulation, and UMAT for the physics-based constitutive law. The functions of these subroutines in the developed model are discussed below.

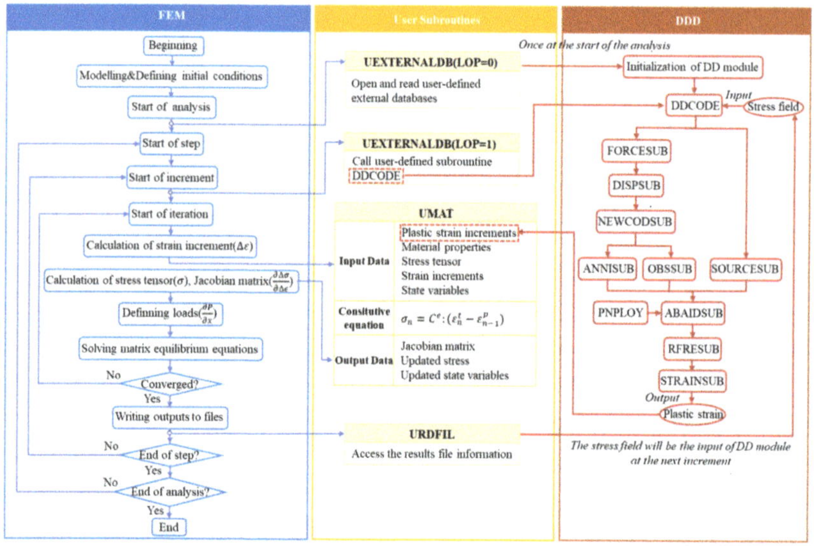

Figure 5. The workflow of the multiscale framework coupling FEM and DCM.

A user subroutine called URDFIL is written particularly as a stress relay from the FEM to DDD module. The subroutine performs two functions: it can access the results file for stress tensors of Gaussian integration points, nodes coordinates, elements, and its relative integration point numbers at the end of the increment, and it transfers this new information to the DDD module by writing them to user-defined COMMON blocks.

The DDD-relative calculations (e.g., the velocity of dislocations, dislocation multiplication, and annihilation) are implemented by calling the user subroutine UEXTERNALDB. This subroutine provides flexible points-in-time where other software or codes are able to communicate with the finite element analysis program (i.e., Abaqus/Standard), and data information stored in external files can be read in.

After user subroutine UEXTERNALDB is called, a user-supplied subroutine DDCODE is implemented to specify the DDD-relative calculations. It is constantly called at the beginning of each increment (i.e., LOP = 1) to update the state of dislocations, obstacles, and Frank–Read sources and to calculate the plastic strain induced by dislocation gliding.

There are ten self-defined subroutines, STRAINSUB, SOURCESUB, RFRESUB, ABAIDSUB, DISPSUB, NEWCODSUB, FORCESUB, OBSSUB, PNPLOY, and ANNISUB consisting of the dominant subroutine DDCODE. These subordinative subroutines implement the dislocation mechanisms by operating on the data structures of dislocations, dislocation sources, and obstacles. The data structures used in the model are introduced in the next section. The calling sequences of these subroutines are shown in Figure 5. The subroutine FORCESUB is used to calculate the force applied on dislocations using the stresses transferred from the FEM module, by Equation (1). Once the force on dislocations is obtained, their velocity and displacement can be given by Equation (2), as performed in subroutine DISPSUB. The transformation matrix and equations have been provided in Section 3.2. According to the displacement of dislocations, new coordinates of dislocations are calculated in subroutine NEWCODSUB. The subroutine OBSSUB determines if dislocations are pinned or unpinned by obstacles and it provides the dislocation numbers that are pinned. The subroutine ANNISUB decides whether the dislocation annihilates with other dislocations according to the relative positions of dislocations. The subroutine SOURCESUB, which follows the dislocation-generation rules discussed in Section 2.1, generates a dislocation dipole once all the conditions (i.e., the critical strength and nucleation time) are met. The subroutine REF-SUB eliminates the dislocations that annihilate or escape from the simulation model and re-number the remaining and newly generated dislocations. The subroutine ABAIDSUB, which calls subroutine PNPLOY, identifies which element and integration point domain the dislocations belong to. The subroutine PNPLOY is called to find whether a point is inside or outside a polygon. The last one subroutine STRAINSUB calculates plastic strain induced by each moveable dislocation and distributes them to corresponding integration points by the FE mesh-based method introduced in Section 3.2.

The plastic strain tensors at each integration point are saved in a three-dimensional array. It is important to notice that all the subroutines used in the paper is with FORTRAN language as most user subroutine interfaces in the manual of ABAQUS use it. The FORTRAN language, unlike other programming languages, stores arrays in column-major format, as shown in Figure 6a. That is to say, in a two- or multi-dimensional array, the left-most array index varies the most rapidly and the right-most array index varies the most slowly. Thus, the index of the first dimension (from left to right) represents the components of plastic strain and the component order is the same as that in UMAT, as illustrated in Figure 6b. The second and third indices are the integration point and element number where the plastic strains are distributed, respectively. This storage means of plastic strain allows the generation of simple and efficient machine code, to improve the computing and running speed. The storage format of the multi-dimensional array and the illustration of the plastic strain array are shown in Figure 7.

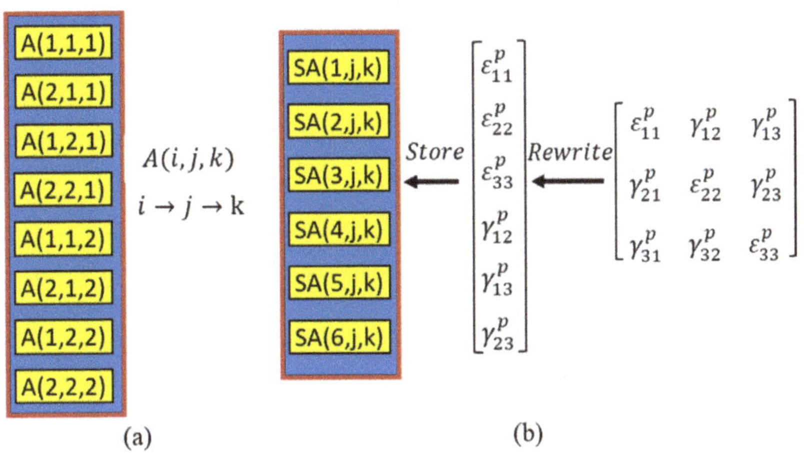

Figure 6. The storage format of arrays in Fortran (**a**), and the plastic strain tensor re-written into the vector to store in an array (**b**).

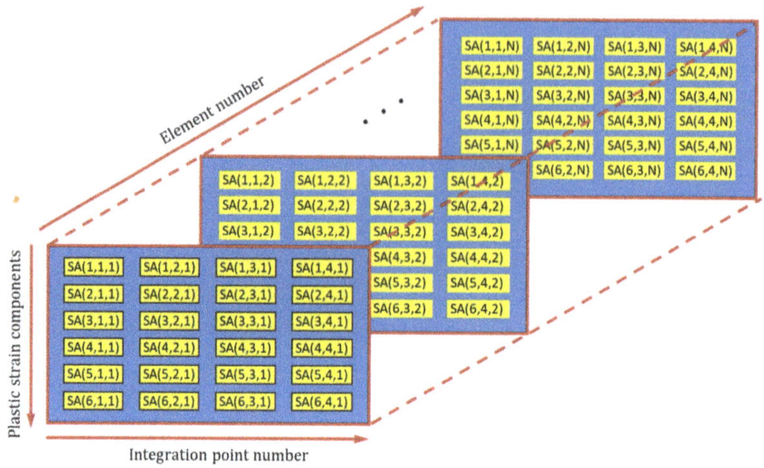

Figure 7. The data structure of plastic strain tensor at the integration points of all elements.

Moreover, the subroutine UEXTERNALDB is called once at the start of the analysis (LOP = 0). This is to obtain the reference state for the DDD module. The DDD reference state consists of the initial information of dislocation, obstacle, and Frank–Read source, such as the coordinates, critical strength, and nucleation time, before loading is applied. The data structure of the reference state is given in Section 4.2. The initial information is stored in external files. At the beginning of the analysis, subroutine UEXTERNALDB reads data from these files and uses them to initialize the DD module.

The subroutine UMAT, allowing user-defined mechanical material behavior, implements the above physics-based constitutive law (Equation (1)) by using plastic strain calculated by DDD. When the subroutine is called, the parameters such as stresses, strain increments, and state variables (SDVs) are directly passed into UMAT, while the material parameters defined in the input file are accessed by the variable PROPS. With the variables passed in, the UMAT subroutine updates the stresses and SDVs at the end of the increment and generates the Jacobian matrix of the constitutive model. An important note is that the definition of the material Jacobian affects only the convergence rate, not the simulation results obtained [48].

4.2. Data Structure for the Coupling Framework

The data structures of dislocations, obstacles, and dislocation sources are shown in Figure 8. It can be seen that the basic data structure of a dislocation consists of identity number, location information, its sign (i.e., plus or minus one, representing positive or negative dislocations, respectively), and the state information. The information of location includes the coordinate value, the located slip system, and the element and integration point number where the dislocation resides, as shown in Figure 8a. The preliminary calculation with user subroutines generates the stress, displacement, and new coordinates of the dislocation i, and then the state information (i.e., illustrating whether the dislocation is pinned, annihilated, or newly generated) is obtained. Moreover, as can be seen from Figure 8b, the data structure related to dislocation sources and obstacles is much simpler compared to the dislocations. In addition to the location information, the critical strength and time are added for sources and only strength is added for obstacles, as shown in Figure 8c. During the deformation process, these data are read and updated continuously by the self-defined subroutines mentioned in the above section. As shown in Figure 8a, for the dislocation, its relative element and integration point number are updated by the subroutine ABAIDSUB, and the driving force, displacement, and new coordinates are calculated and updated by subroutines FORCESUB, DISPSUB, and NEWCODSUB, respectively. The signs of annihilating, newly generating, and pinning are determined by subroutines ANNISUB, SOURCESUB, and OBSSUB. The data structure of sources is mainly visited and updated by subroutine SOURCESUB. Similarly, the data structure of obstacles is accessed by subroutine OBSSUB.

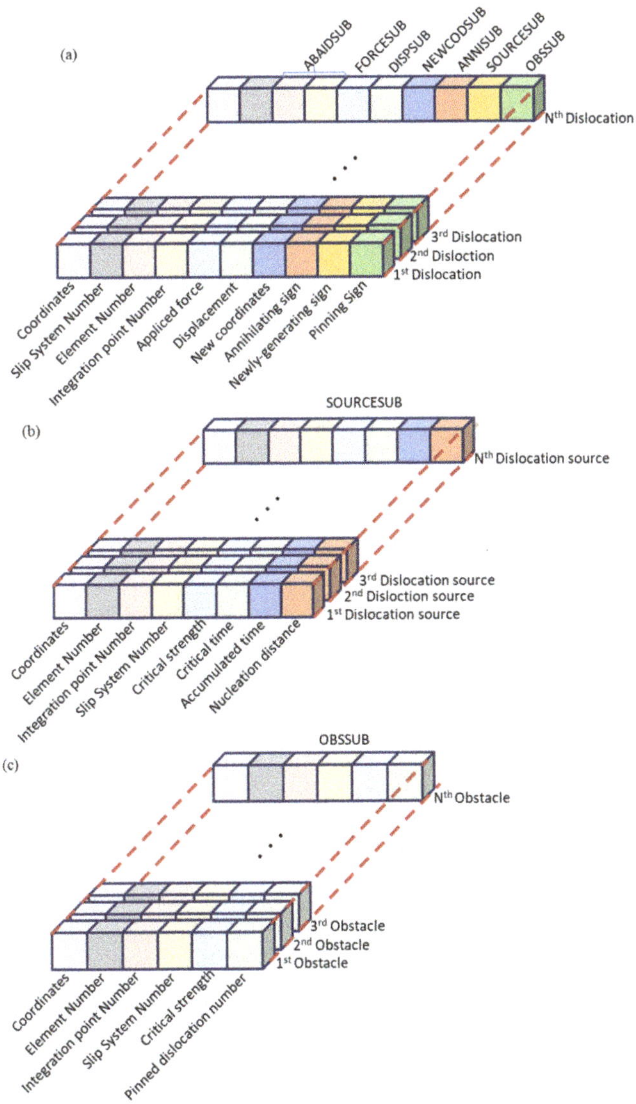

Figure 8. The data structures of dislocations (**a**), obstacles (**b**), and dislocation sources (**c**), and the relative user subroutines.

The initial data structures of dislocations, sources, and obstacles (i.e., the data structures before loading is applied) are utilized to describe the reference state of the DDD module. They are generated by MATLAB and written into text files that are visited at the start of ABAQUS/Standard analysis to initialize the DDD module.

5. Uniaxial Compression Simulation for Single Crystal Micropillar

5.1. Computational Model of Micropillars

For verification, the developed simulation model is used to study the deformation behaviors of crystal materials under uniaxial compression. As shown in Figure 9, micropillars with four different widths are built with a constant aspect ratio of height H to width L (H/L = 2). An isotropic aluminum single crystal is considered here with Poisson's ratio $v = 0.3$, shear modulus $\mu = 26$ Gpa, Burgers vector $b = 0.25$ nm, and drag coefficient $B = 10^{-4}$ Pa s [28,44,45].

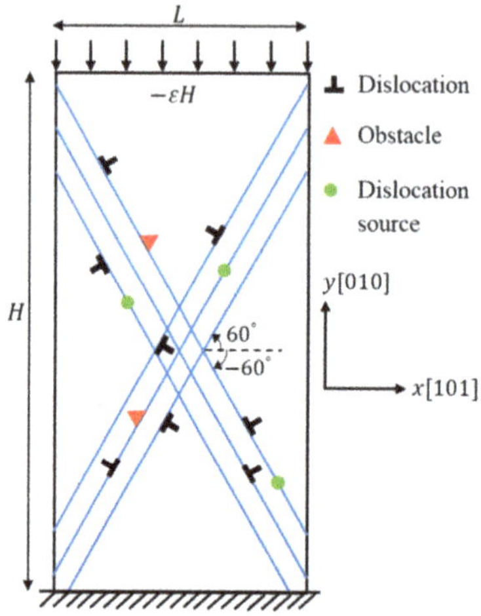

Figure 9. Sketch of micropillar compression. The black T-signs represent dislocations. The red triangle and green circle represent obstacle and source, respectively.

For face-centered crystal Al, the representative material parameters used in DDD are the same as the ones used in other research [28,30,43,45]. Two slip systems with the interaction angle of $\pm 60^{\circ}$ relative to the horizontal direction are considered with the spacing distance of two neighboring slip planes being $10b$. This approximates to the active slip systems of the face-centered crystal in plastic strain problems. The x-axis and y-axis represent the crystal directions [101] and [010], respectively. Static dislocation sources are distributed on these slip planes randomly with strength satisfying the Gaussian distribution with mean value 50 MPa and standard deviation 10 MPa. The critical time for dislocation nucleation is set as 10 ns. The obstacles in the model are with mean strength 150 MPa and 20% standard deviation.

To mimic the compression experiment, a fixed compressive strain rate $\dot{\varepsilon} = 1000/\mathrm{s}$ is imposed for all samples. Time steps of 1 ns for DD and 10 ns for the FEM are used here [49]. In all the simulations, the bottom boundaries of specimens are fixed as:

$$u_x = u_y = 0, \quad at\ Y = 0, \tag{12}$$

The upper surface is subjected to a velocity v_x such that

$$v_y = L\dot{\varepsilon}, \ at\ Y = L, \tag{13}$$

DCM simulations are performed with the boundary conditions and material parameters as described above. The simulation results are discussed below.

5.2. Effect of Sample Size on Crystal Strength

The computationally obtained $\sigma \sim \varepsilon$ curves are plotted in Figure 10 for four different simple sizes $L = \{0.3,\ 0.5, 0.8, 1.0\}$ µm. To avoid the potential effect of defect distribution (i.e., the distribution of dislocations, sources, and obstacles) on the simulation results, for each size, three simulations are carried out with different defect distributions. It is found that the smaller the sample size, the higher the strength, as observed in experimental investigation [5]. This validates that the present simulation model is capable of capturing the microscale plastic response. Due to numerous dislocations being greatly active at $L = 1.0$ μm, the stress–strain curves oscillate poorly and even fail at small strains. A similar phenomenon was also found in the compressing simulation of copper micropillars [50]. From these stress–strain curves, we compute the nominal yield stress $\sigma_{0.2}$, which is the stress value at 0.2% plastic strain offsets, and plot it in Figure 11a. The inset of Figure 11a shows the acquiring of the normal yield stress for the 0.3 μm-wide micropillar. The results indicate that the yield stress increases with decreasing pillar size. The error bars denote the range of the values of $\sigma_{0.2}$.

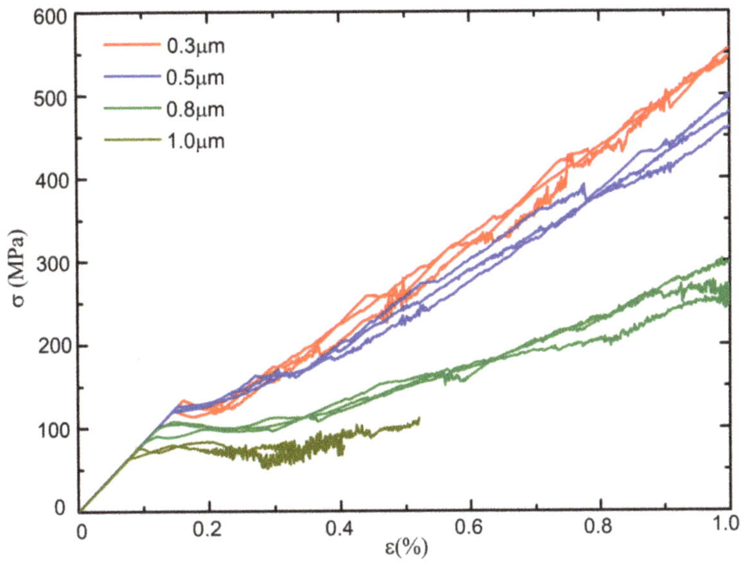

Figure 10. Stress–strain curves of micropillars with different widths $L = \{0.3, 0.5, 0.8, 1.0\}$ μm.

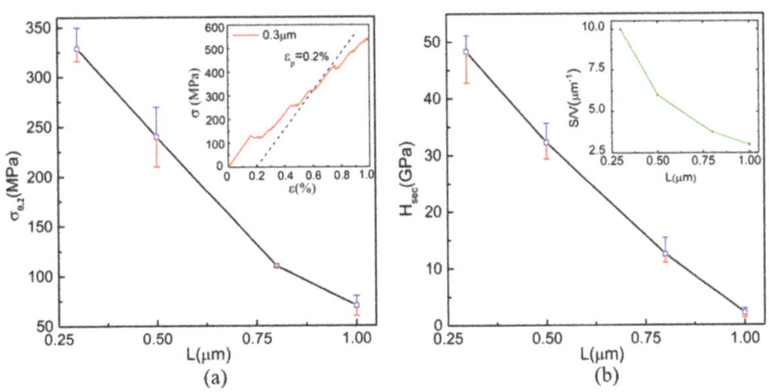

Figure 11. The yield stress at 0.2% plastic strain offsets, $\sigma_{0.2}$, (**a**) and average secant hardening modulus, H_{ave}, (**b**) versus the width of micropillars. The inset in (**a**) shows the capture of the yield stress from the one stress–strain curve, and the inset in (**b**) is the ratio of surface area to volume for each micropillar.

In addition, the secant strain hardening modulus, H_{sec}, is defined as the ratio of stress to plastic strain from $\varepsilon_p = 0.1\%$ to $\varepsilon_p = 0.3\%$, as follows:

$$H_{sec} = \frac{\sigma_{\varepsilon=0.003} - \sigma_{\varepsilon=0.001}}{0.003 - 0.001}, \quad (14)$$

Then, the average value of H_{sec} over three realizations for a particular pillar size, H_{ave}, is calculated and plotted in Figure 11b. It shows that the average strain hardening modulus decreases with sample size and that larger pillars are with the smaller scatter of H_{ave} values. In addition, the ratio of surface area to volume, which is reduced to the ratio of length to surface area for the plane strain problem, is calculated for each pillar. The inset of Figure 11b shows that the ratio decreases with sample size, although all the samples have the same length–width ratio. It indicates that dislocations in small samples have a high possibility to escape from the sample. The escape of dislocations from the free surface can result in the dislocation starvation effect [49] in small pillars and the hardening of materials.

6. Conclusions

In this paper, a concurrent multiscale simulation method is developed to shed light on the material plasticity at the microscale. In this method, the DDD module looks to the FEM module to solve the boundary conditions and the FEM module turns to DDD for plastic strain, which determines the constitutive law of the FEM.

The key issue in the DDD-FEM coupling scheme is how to reasonably regularize the discrete plastic strain to integration points of continuum mechanics. A novel FE mesh-based regularization method is proposed by using a Burgers vector density function and the characteristics of a finite element mesh in ABAQUS. It shows high accuracy and considerably improves the efficiency by avoiding the tedious calculations.

To achieve the concurrent computational model, the DDD module is programmed into Fortran codes, and then it is incorporated into the Abaqus analysis by calling user subroutines provided by Abaqus to determine the constitutive law. The stress field is stored in COMMON blocks and transferred to DDD codes. The realization of the multiscale model capitalizes on the secondary development of FEM software Abaqus, and the implementation details are presented for the first time.

The developed multiscale framework is able to capture the plastic behavior of microscale crystals. Using this multiscale method, uniaxial compression tests of micropillars are performed to study the size effect on plasticity and its underlying mechanisms. The observed secant strain hardening of the modulus and yield stress might be due to the high probability of dislocations in smaller size to escape from the free surface.

This multiscale framework can be potentially employed in studying the plastic behavior of materials with complex microstructures and deeply revealing its underlying mechanisms. In future work, the developed simulation model will consider a variety of complex dislocation motions, such as dislocation climb, interactions between dislocation and grain boundary, and dislocation junction formation.

Author Contributions: Z.Z.: Methodology, investigation, data curation, formal analysis, writing—original draft preparation. Z.T.: Conceptualization, supervision, resources, writing—reviewing and editing, funding acquisition. X.J.: Supervision, project administration, funding acquisition. All authors have read and agreed to the published version of the manuscript.

Funding: The authors would like to acknowledge the funding support from the UK's EPSRC Future Metrology Hub (Ref: EP/P006930/1), the UK's STFC Innovation Partnership Scheme (STFC-IPS) project under grant agreement No. ST/V001280/1, and the European Union's Horizon 2020 research and innovation program under grant agreement No. 767589.

Institutional Review Board Statement: Not applicable.

Informed Consent Statement: Not applicable.

Data Availability Statement: Not applicable.

Acknowledgments: The authors would like to acknowledge the scholarship received from the UK's EPSRC Future Metrology Hub (Ref: EP/P006930/1) and the China Scholarship Council (CSC) on Zhenting Zhang's PhD study.

Conflicts of Interest: The authors declare no conflict of interest.

References

1. Ferretti, E. Shape-effect in the effective laws of plain and rubberized concrete. *Comput. Mater. Contin.* **2012**, *30*, 237.
2. Xu, Z.-H.; Li, X. Sample size effect on nanoindentation of micro-/nanostructures. *Acta Mater.* **2006**, *54*, 1699–1703. [CrossRef]
3. Hudson, J.; Brown, E.; Fairhurst, C. Shape of the Complete Stress-Strain Curve for Rock. In Proceedings of the 13th Symposium on Rock Mechanics, University of Illinois, Urbana, IL, USA, 30 August–1 September 1971.
4. Brunetaud, X.; Khelifa, M.-R.; Al-Mukhtar, M. Size effect of concrete samples on the kinetics of external sulfate attack. *Cem. Concr. Compos.* **2012**, *34*, 370–376. [CrossRef]
5. Takata, N.; Takeyasu, S.; Li, H.; Suzuki, A.; Kobashi, M. Anomalous size-dependent strength in micropillar compression deformation of commercial-purity aluminum single-crystals. *Mater. Sci. Eng. A* **2020**, *772*, 138710. [CrossRef]
6. Dunstan, D.J.; Bushby, A.J. The scaling exponent in the size effect of small scale plastic deformation. *Int. J. Plast.* **2013**, *40*, 152–162. [CrossRef]
7. Abad, O.T.; Wheeler, J.M.; Michler, J.; Schneider, A.S.; Arzt, E. Temperature-dependent size effects on the strength of Ta and W micropillars. *Acta Mater.* **2016**, *103*, 483–494. [CrossRef]
8. Lavenstein, S.; El-Awady, J.A. Micro-scale fatigue mechanisms in metals: Insights gained from small-scale experiments and discrete dislocation dynamics simulations. *Curr. Opin. Solid State Mater. Sci.* **2019**, *23*, 100765. [CrossRef]
9. Yuan, S.; Huang, M.; Zhu, Y.; Li, Z. A dislocation climb/glide coupled crystal plasticity constitutive model and its finite element implementation. *Mech. Mater.* **2018**, *118*, 44–61. [CrossRef]
10. Cui, Y.; Po, G.; Ghoniem, N. Influence of loading control on strain bursts and dislocation avalanches at the nanometer and micrometer scale. *Phys. Rev. B* **2017**, *95*, 064103. [CrossRef]
11. Xie, W.; Fang, F. On the mechanism of dislocation-dominated chip formation in cutting-based single atomic layer removal of monocrystalline copper. *Int. J. Adv. Manuf. Technol.* **2020**, *108*, 1587–1599. [CrossRef]
12. Vattré, A.; Devincre, B.; Roos, A.; Feyel, F. Predicting size effects in nickel-base single crystal superalloys with the Discrete-Continuous Model. *Eur. J. Comput. Mech. Rev. Eur. Mécanique Numérique* **2010**, *19*, 65–76. [CrossRef]
13. Keralavarma, S.M.; Curtin, W.A. Strain hardening in 2D discrete dislocation dynamics simulations: A new '2.5D' algorithm. *J. Mech. Phys. Solids* **2016**, *95*, 132–146. [CrossRef]
14. Chen, G.; Ren, C.; Zhang, P.; Cui, K.; Li, Y. Measurement and finite element simulation of micro-cutting temperatures of tool tip and workpiece. *Int. J. Mach. Tools Manuf.* **2013**, *75*, 16–26. [CrossRef]
15. Schulze, V.; Autenrieth, H.; Deuchert, M.; Weule, H. Investigation of surface near residual stress states after micro-cutting by finite element simulation. *CIRP Ann.* **2010**, *59*, 117–120. [CrossRef]
16. Ji, H.; Song, Q.; Kumar Gupta, M.; Cai, W.; Zhao, Y.; Liu, Z. Grain scale modelling and parameter calibration methods in crystal plasticity finite element researches: A short review. *J. Adv. Manuf. Sci. Technol.* **2021**, *1*, 41–50. [CrossRef]
17. Herrera-Solaz, V.; Cepeda-Jiménez, C.M.; Pérez-Prado, M.T.; Segurado, J.; Niffenegger, M. The influence of underlying microstructure on surface stress and strain fields calculated by crystal plasticity finite element method. *Mater. Today Commun.* **2020**, *24*, 101176. [CrossRef]
18. Rousseau, T.; Nouguier-Lehon, C.; Gilles, P.; Hoc, T. Finite element multi-impact simulations using a crystal plasticity law based on dislocation dynamics. *Int. J. Plast.* **2018**, *101*, 42–57. [CrossRef]
19. Gurrutxaga-Lerma, B.; Balint, D.S.; Dini, D.; Sutton, A.P. A Dynamic Discrete Dislocation Plasticity study of elastodynamic shielding of stationary cracks. *J. Mech. Phys. Solids* **2017**, *98*, 1–11. [CrossRef]
20. Geada, I.L.; Ramezani-Dakhel, H.; Jamil, T.; Sulpizi, M.; Heinz, H. Insight into induced charges at metal surfaces and biointerfaces using a polarizable Lennard–Jones potential. *Nat. Commun.* **2018**, *9*, 716. [CrossRef]
21. Zhang, Y.; Cheng, Z.; Lu, L.; Zhu, T. Strain gradient plasticity in gradient structured metals. *J. Mech. Phys. Solids* **2020**, *140*, 103946. [CrossRef]
22. Xiao, X.; Yu, L. Cross-sectional nano-indentation of ion-irradiated steels: Finite element simulations based on the strain-gradient crystal plasticity theory. *Int. J. Eng. Sci.* **2019**, *143*, 56–72. [CrossRef]
23. Barboura, S.; Li, J. Establishment of strain gradient constitutive relations by using asymptotic analysis and the finite element method for complex periodic microstructures. *Int. J. Solids Struct.* **2018**, *136*, 60–76. [CrossRef]
24. Hansen, L.T.; Fullwood, D.T.; Homer, E.R.; Wagoner, R.H.; Lim, H.; Carroll, J.D.; Zhou, G.; Bong, H.J. An investigation of geometrically necessary dislocations and back stress in large grained tantalum via EBSD and CPFEM. *Mater. Sci. Eng. A* **2020**, *772*, 138704. [CrossRef]
25. Han, X. Investigate the mechanical property of nanopolycrystal silicon by means of the nanoindentation method. *AIP Adv.* **2020**, *10*, 065230. [CrossRef]
26. Li, J.; Lu, B.; Zhang, Y.; Zhou, H.; Hu, G.; Xia, R. Nanoindentation response of nanocrystalline copper via molecular dynamics: Grain-size effect. *Mater. Chem. Phys.* **2020**, *241*, 122391. [CrossRef]

27. Lemarchand, C.; Devincre, B.; Kubin, L.P. Homogenization method for a discrete-continuum simulation of dislocation dynamics. *J. Mech. Phys. Solids* **2001**, *49*, 1969–1982. [CrossRef]
28. Van der Giessen, E.; Needleman, A. Discrete dislocation plasticity: A simple planar model. *Model. Simul. Mater. Sci. Eng.* **1995**, *3*, 689–735. [CrossRef]
29. Benzerga, A.A. Micro-pillar plasticity: 2.5D mesoscopic simulations. *J. Mech. Phys. Solids* **2009**, *57*, 1459–1469. [CrossRef]
30. Huang, M.; Li, Z. Coupled DDD–FEM modeling on the mechanical behavior of microlayered metallic multilayer film at elevated temperature. *J. Mech. Phys. Solids* **2015**, *85*, 74–97. [CrossRef]
31. Cui, Y.-n.; Liu, Z.-l.; Zhuang, Z. Theoretical and numerical investigations on confined plasticity in micropillars. *J. Mech. Phys. Solids* **2015**, *76*, 127–143. [CrossRef]
32. Cui, Y.-n.; Liu, Z.-l.; Wang, Z.-j.; Zhuang, Z. Mechanical annealing under low-amplitude cyclic loading in micropillars. *J. Mech. Phys. Solids* **2016**, *89*, 1–15. [CrossRef]
33. Cui, Y.; Po, G.; Ghoniem, N. Temperature insensitivity of the flow stress in body-centered cubic micropillar crystals. *Acta Mater.* **2016**, *108*, 128–137. [CrossRef]
34. Hu, J.; Liu, Z.; Chen, K.; Zhuang, Z. Investigations of shock-induced deformation and dislocation mechanism by a multiscale discrete dislocation plasticity model. *Comput. Mater. Sci.* **2017**, *131*, 78–85. [CrossRef]
35. Hu, J.-Q.; Liu, Z.-L.; Cui, Y.-N.; Liu, F.-X.; Zhuang, Z. A New View of Incipient Plastic Instability during Nanoindentation. *Chin. Phys. Lett.* **2017**, *34*, 046101. [CrossRef]
36. Santos-Güemes, R.; Esteban-Manzanares, G.; Papadimitriou, I.; Segurado, J.; Capolungo, L.; Llorca, J. Discrete dislocation dynamics simulations of dislocation-θ′ precipitate interaction in Al-Cu alloys. *J. Mech. Phys. Solids* **2018**, *118*, 228–244. [CrossRef]
37. Huang, S.; Huang, M.; Li, Z. Effect of interfacial dislocation networks on the evolution of matrix dislocations in nickel-based superalloy. *Int. J. Plast.* **2018**, *110*, 1–18. [CrossRef]
38. Liu, F.X.; Liu, Z.L.; Pei, X.Y.; Hu, J.Q.; Zhuang, Z. Modeling high temperature anneal hardening in Au submicron pillar by developing coupled dislocation glide-climb model. *Int. J. Plast.* **2017**, *99*, 102–119. [CrossRef]
39. Zbib, H.M.; de la Rubia, T.D. A multiscale model of plasticity. *Int. J. Plast.* **2002**, *18*, 1133–1163. [CrossRef]
40. Liu, Z.L.; Liu, X.M.; Zhuang, Z.; You, X.C. A multi-scale computational model of crystal plasticity at submicron-to-nanometer scales. *Int. J. Plast.* **2009**, *25*, 1436–1455. [CrossRef]
41. Vattré, A.; Devincre, B.; Feyel, F.; Gatti, R.; Groh, S.; Jamond, O.; Roos, A. Modelling crystal plasticity by 3D dislocation dynamics and the finite element method: The Discrete-Continuous Model revisited. *J. Mech. Phys. Solids* **2014**, *63*, 491–505. [CrossRef]
42. Cui, Y.; Liu, Z.; Zhuang, Z. Quantitative investigations on dislocation based discrete-continuous model of crystal plasticity at submicron scale. *Int. J. Plast.* **2015**, *69*, 54–72. [CrossRef]
43. Song, H.; Yavas, H.; Giessen, E.V.d.; Papanikolaou, S. Discrete dislocation dynamics simulations of nanoindentation with pre-stress: Hardness and statistics of abrupt plastic events. *J. Mech. Phys. Solids* **2019**, *123*, 332–347. [CrossRef]
44. Kondori, B.; Needleman, A.; Amine Benzerga, A. Discrete dislocation simulations of compression of tapered micropillars. *J. Mech. Phys. Solids* **2017**, *101*, 223–234. [CrossRef]
45. Papanikolaou, S.; Song, H.; Van der Giessen, E. Obstacles and sources in dislocation dynamics: Strengthening and statistics of abrupt plastic events in nanopillar compression. *J. Mech. Phys. Solids* **2017**, *102*, 17–29. [CrossRef]
46. Cai, W.; Arsenlis, A.; Weinberger, C.R.; Bulatov, V.V. A non-singular continuum theory of dislocations. *J. Mech. Phys. Solids* **2006**, *54*, 561–587. [CrossRef]
47. Huang, M.; Tong, J.; Li, Z. A study of fatigue crack tip characteristics using discrete dislocation dynamics. *Int. J. Plast.* **2014**, *54*, 229–246. [CrossRef]
48. Zhou, N.; Elkhodary, K.I.; Huang, X.; Tang, S.; Li, Y. Dislocation structure and dynamics govern pop-in modes of nanoindentation on single-crystal metals. *Philos. Mag.* **2020**, *100*, 1585–1606. [CrossRef]
49. Liu, Z.; Liu, X.; Zhuang, Z.; You, X. Atypical three-stage-hardening mechanical behavior of Cu single-crystal micropillars. *Scr. Mater.* **2009**, *60*, 594–597. [CrossRef]
50. Akarapu, S.; Zbib, H.M.; Bahr, D.F. Analysis of heterogeneous deformation and dislocation dynamics in single crystal micropillars under compression. *Int. J. Plast.* **2010**, *26*, 239–257. [CrossRef]

crystals

MDPI

Article

The Reliability of SAC305 Individual Solder Joints during Creep–Fatigue Conditions at Room Temperature

Mohammed Abueed [1,*], Raed Al Athamneh [2], Moayad Tanash [2] and Sa'd Hamasha [3]

[1] Department of TD ATA Automation, Intel Corporation, Hillsboro, OR 97124, USA
[2] Department of Industrial Engineering, Faculty of Engineering, The Hashemite University, Zarqa 13133, Jordan
[3] Department of Industrial and Systems Engineering, Samuel Ginn College of Engineering, Auburn University, Auburn, AL 36849, USA
* Correspondence: mohammed.abueed@intel.com

Abstract: The failure of one solder joint out of the hundreds of joints in a system compromises the reliability of the electronics assembly. Thermal cycling is a result of both creep–fatigue mechanisms working together. To better understand the failure process in thermal cycling, it is crucial to analyze both the effects of creep and fatigue mechanisms in a methodical manner. In this work, individual solder junctions are subjected to accelerated shear fatigue testing to investigate the effects of creep and fatigue on joint dependability at room temperature. A modified fixture is used to conduct fatigue tests on an Instron 5948 micromechanical tester. SAC305 joints with an OSP surface finish were cycled under stress control at first, and then the strain was maintained for a set amount of time. In this investigation, three stress amplitudes of 16, 20, and 24 MPa are used, together with varying residence periods of 0, 10, 60, and 180 s. The fatigue life of solder junctions is described for each testing condition using the two-parameter Weibull distribution. Additionally, as a function of stress amplitude and residence time, a dependability model is created. For each testing scenario, the progression of the stress–strain loops was studied. By quantifying relevant damage metrics, such as plastic work per cycle and plastic strain at various testing circumstances, the damage due to fatigue is distinguished from creep. To investigate the relationships between plastic work and plastic strain with fatigue life, the Coffin–Manson and Morrow Energy model is used. The results indicate that using greater stress magnitudes or longer dwell periods significantly shortens fatigue life and dramatically increases plastic work and plastic strain. The housing impact is significant; in some circumstances, testing with a longer dwelling period and lower stress amplitude resulted in more damage than testing with a shorter dwelling period and higher stress levels. When illustrating the fatigue behavior of solder junctions under various stress amplitudes or dwellings, the Coffin–Manson and Morrow Energy model were both useful. In the end, general reliability models are developed as functions of plastic work and plastic strain.

Keywords: solder joints; lead-free; reliability; creep; fatigue

Citation: Abueed, M.; Al Athamneh, R.; Tanash, M.; Hamasha, S. The Reliability of SAC305 Individual Solder Joints during Creep–Fatigue Conditions at Room Temperature. *Crystals* **2022**, *12*, 1306. https://doi.org/10.3390/cryst12091306

Academic Editors: Ulrich Prahl, Sergey Guk and Faisal Qayyum

Received: 15 August 2022
Accepted: 12 September 2022
Published: 15 September 2022

1. Introduction

The robustness of electronic assemblies is determined by the mechanical integrity of solder joints against various degradation influences. Thermal cycling is one of the environmental effects responsible for degrading the fatigue life of solder joints in real-life applications. Numerous switching on–off cycles for electronics systems as in power cycling and exterior sources of elevated temperature as the heat generated from engines are examples of thermal cycling sources. In thermal cycling, the behavior of solder joints is complicated due to their suffering from several damage mechanisms, including creep, fatigue, and the interaction of both mechanisms. Creep is dominant during dwelling periods, while fatigue is influential between ramps [1,2]. Mainly, joint failure in thermal cycling failure occurs due to the mismatch between the coefficient of thermal expansions between PCB and components. This would generate cyclic shear stresses on joints as temperature alternating between extremes, named thermomechanical fatigue stress. Microstructure evaluation under thermal cycling includes stress concentration in the defected spots (such as voids) of the microstructure, then recrystallization of Sn grains, followed by intergranular crack growth among grain boundaries. In addition to thermal cycling, drop shock was found to be a vital cause in portable applications such as mobile phones and digital cameras. Researchers [3–7] found that crack initiation in drop tests started from the IMC layer, as it is brittle and considered to be the weakest. This would be enhanced if preceded with few thermal cycles and/or aging due to IMC layer growth caused by thermal aging with time.

Therefore, a joint's initial microstructure has great impact during thermal cycling, including precipitates (Ag3Sn and Cu6Sn5), Sn morphologies, and Sn grain orientations. This is due to the low number of grains in each solder joint: one to three large Sn grains, each with a number of dendrites with the same crystallographic orientation [8–13]. Generally, grain orientation correlates to a joint's reliability, as it plays a vital role in microstructure evolution [14–16] and other phenomena such as electromigration [17]. Moreover, after the reflow process, once the tin (Sn) in solder spheres comes in contact with the copper (Cu) on the solder pads, an intermetallic compound (IMC) is formed, which strengthens the mechanical bond. However, the growth of the IMC layer over time leads to weakening due to its brittle nature. A microstructure study revealed that the component side of the solder junction was where the majority of the cracks originated. There were primarily two sorts of cracks seen: those that go into the bulk solder and those that run along the IMC layer. It was discovered that the crack nearly always started next to the IMC layer, where thermal cycling causes the most stress, and eventually spread into the bulk or along the IMC layer. Other factors such as oxidation and phase change are of significant impact as well [18–22].

SnPb-based alloys show excellent performance in thermal cycling, but the industry switched to lead-free alloys after RoHS prohibited using lead in 2006. Many SAC-based alloys demonstrate competitive performance [23–25]; however, studies are still ongoing to investigate their properties comprehensively. Creep and fatigue are critical mechanical properties of actual solder joints that have been explored by many researchers. Hamasha et al. [26] studied the effect of various testing conditions of varying amplitude and strain rate on the fatigue life of actual joints of SAC105 and SAC305. Results show that SAC305 is more fatigue-resistant compared with SAC105 under varying amplitudes. Sinan et al. [27] investigated the fatigue properties of individual SAC-based doped spheres with various surface finishes. It is found that the higher the Ag content, the superior the fatigue and shear strength obtained. Xu et al. [28] explored the fatigue properties of SAC305 and 105 compared with SnPb alloys under isothermal fatigue conditions and wide strain ranges. SAC305 shows better fatigue properties than SAC105 and SnPb at specific ranges. The effect of surface finish under thermal cycling has been studied extensively. Francy et al. [29] investigated the effect of various surface finishes: OSP, ImAg, and ENIG and alloys under thermal cycling for BGAs and SMRs. ENIG outperformed other surface finishes in the case of BGAs, but with SMRs, OSP and ImAg were slightly better. Similar results were found by [30,31]. Generally, ENIC and ENIPIC were found to perform well in thermal cycling due to their Ni plating layer that blocks the growth of the IMC layer.

Others examined fatigue properties of SAC alloys under various conditions of aging durations and temperatures [28,29], stress amplitudes [30,31], strain ranges [32,33], and shear rates [34]. Many researchers utilized bulk [35,36] and dog-bone [37,38] samples in their work to examine the microstructure evolution [39,40] and fatigue properties of different alloys [40,41] under various conditions. Even though valuable results were obtained, they were not as much as utilizing actual joints due to missing the IMC layer, surface finish, and different microstructure, which impact reliability. Several researchers studied the creep effect under various testing conditions. Fahim et al. [41] examined the evolution of microstructures, including IMC of SAC-based alloys and several surface finishes at various temperatures using nanoindentation testing after aging. Significant degradation in reliability is observed after long-term aging with increasing testing temperature. The long-term aging of solder junctions at high temperatures causes considerable changes in their microstructure, including a thickening of the intermetallic compound layer (IMC) and coarsening of the precipitates. The mechanical and fatigue characteristics of solder junctions are significantly influenced by their microstructure. Different conditions of aging durations and temperatures were studied [42–46], and similar results were obtained for dominating aging temperature over duration for degradation. Others [28,47,48] explored creep properties and microstructure evolution after several thermal cycles. It is concluded that creep rate and deformation were higher when testing joints after more cycles. Doping is introduced to enhance the properties of the materials by adding small (micro) amounts of certain elements such as Ni, Bi, Cr, etc., to SAC-based alloys. The properties of doped alloys under thermal cycling and iso-thermal cycling have been studied by different researchers [34,42–52]. Doped alloys showed mechanical property improvements depending on the doping element. Various mechanical properties were investigated such as hardness [53], tensile strength [54], and modules of elasticity [55].

There are many models established to predict the fatigue life of several solder materials under thermal cycling with various testing conditions and geometries. Norris and Landzberg [56] were the ones who started early development of the reliability models of SnPb in thermal cycling, considering various temperature and frequency levels. Their model was a modified version of the Coffin–Manson equation. Engelmaier W. [57] studied the fatigue life of SnPb-based CCC joints during power cycling and generated models to predict their life based on the Coffin–Manson model. Most models in the literature were generated based on [51–53] models. Vayman S. et al. [58] created an empirical model for low-tin alloys (SnPb based) under isothermal fatigue conditions including extensive temperature range, hold times, strain ranges, frequencies, and environmental conditions. Salmela O. [59] investigated several acceleration factors for lead-free material under different thermal cycling conditions. He modified the Norris–Landzberg model with a correction factor to compromise for material and geometry. Others established empirical models for lead-free material under thermal cycling [60,61] and isothermal fatigue with aging [62–66].

According to the literature, there is extensive research that investigated the fatigue and creep properties for SnPb-based and lead-free materials. However, limited research has been performed under the combination of both effects in a systematic way. Moreover, comprehensive studies generated models for SnPb and lead-free materials under thermal cycling tests. Nevertheless, there is no study that created a model for actual SAC-based solder joint material under a combination of creep and fatigue. In this study, expanded research of our previous work related to the combined effects of fatigue and creep explored at room temperature [65], and various iso-thermal temperatures [8], is implemented. Moreover, reliability models as a function of dwell time for fatigue life based on Morrow Energy and Coffin–Manson models are generated for SAC305 actual solder joints. These mathematical/empirical models seem to be sophisticated enough to predict life as it was generated according to defined procedures, including identifying the failure mechanism and modes and developing/modifying models based on physics or pervious models considering many factors such as temperature, dwell time, etc.

2. Materials and Methods

2.1. Test Vehicle

The test vehicle is a coupon of PCB made of FR-4 glass epoxy with 10 mm × 10 mm × 1 mm dimensions, as shown in Figure 1. Each testing coupon consists of a full array of nine SAC305 solder joints. Each 30 mil-diameter joint is reflowed after being attached to a 22 mil SMD Cu pad. The pitch between joints is 3 mm to assure enough space among joints during testing. OSP is the only surface finish utilized in this work.

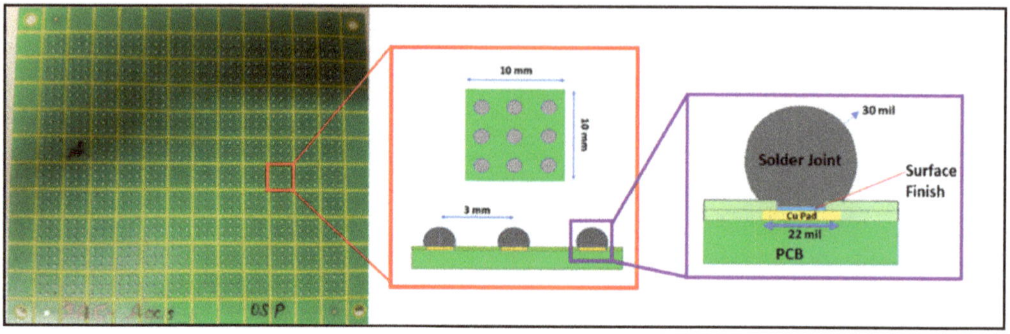

Figure 1. Test Vehicle.

Typical SAC305 solder paste reflow profile is utilized for reflowing the solder joints. The profile starts with a preheating phase to 150–225 °C/200 s, followed by an increase to a peak temperature of 245 °C with less above liquid period than typical SAC305 paste. This is due to the direct exposure of joints to convection heat, unlike assemblies with components. Finally, the temperature is cooled down with a 4 °C/s rate. The whole reflow process is performed within a nitrogen environment. A detailed reflow profile is shown in Figure 2.

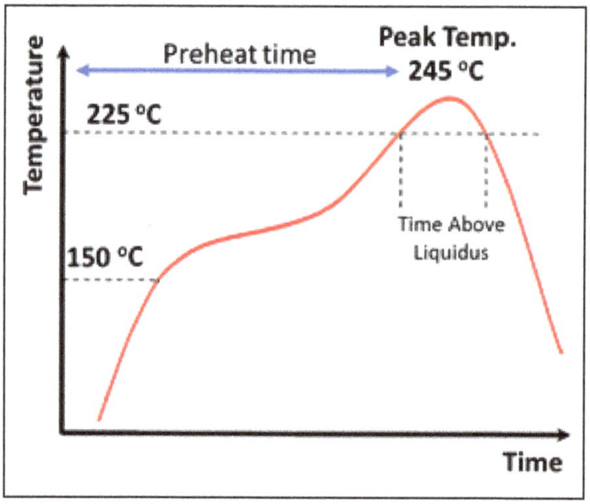

Figure 2. The Reflow profile.

2.2. Experimental Set-Up

Instron 5948 micromechanical tester engaged with a customized fixture is used to implement an accelerated shear fatigue testing on individual solder joints, as shown in Figure 3. The fixture is a circular tip 1 mm in diameter utilized to cycle the joints between stress extremes. The X–Y stage, with the help of a load cell, is used to fit the joints in the middle of the tip and assure the proper tip stand-off range according to the JEDEC standard. A schematic diagram is shown in Figure 4. The displacement and the load are measured by the machine and load cell, respectively. These measurements are recorded continuously by the data acquisition system.

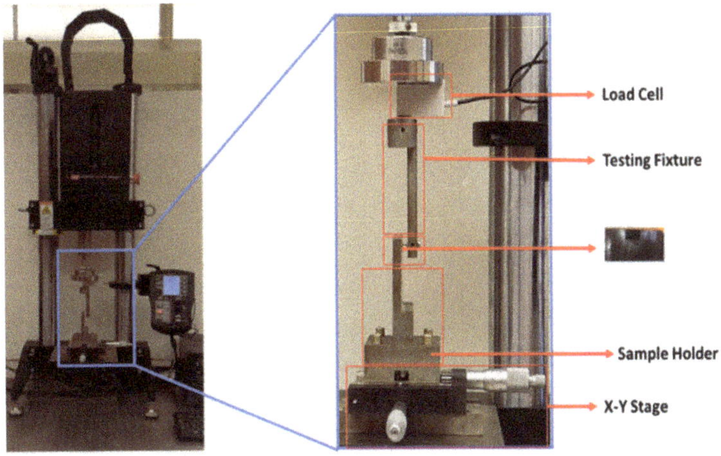

Figure 3. The Instron 5948 Micro-Tester.

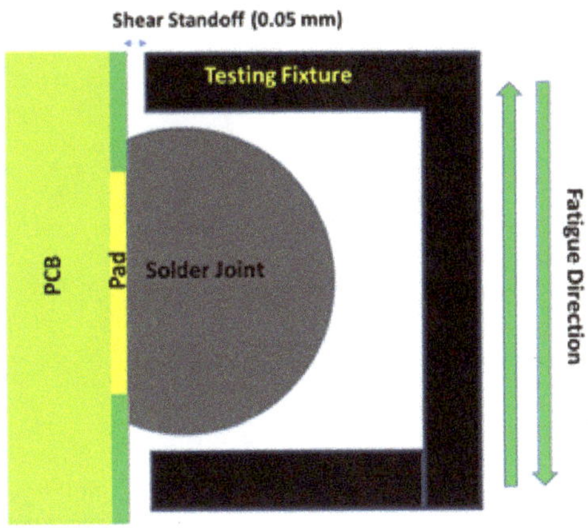

Figure 4. Schematic of a customized testing fixture.

2.3. Test Profile

Two profiles are employed to explore the effects of creep and fatigue in addition to quantifying them. First, a stress-controlled cyclic fatigue test is implemented to examine fatigue effect only, as shown in Figure 5. Cyclic-defined stress is applied at a constant ramp rate of 0.1 mm/s on the solder joint and switched between stress extremes (positive and negative) until complete joint failure (no dwelling conditions). The second profile is of two parts: a stress-controlled followed by a strain-controlled condition, as shown in Figure 6. As soon as the predefined value of stress is reached (stress-controlled) at a positive extreme, the strain is held constant (strain-controlled) for a certain dwelling period. During this part, the material suffers from stress relaxation and/or creep; therefore, it is described as creep–relaxation. Once the dwelling is over, the joint is cycled (switched) to the negative stress extreme with the same conditions of stress and dwelling. The second test is utilized to explore the combined effects of creep and fatigue. The maximum dwelling was 180 s because negligible stress drop (damage) was observed after this duration.

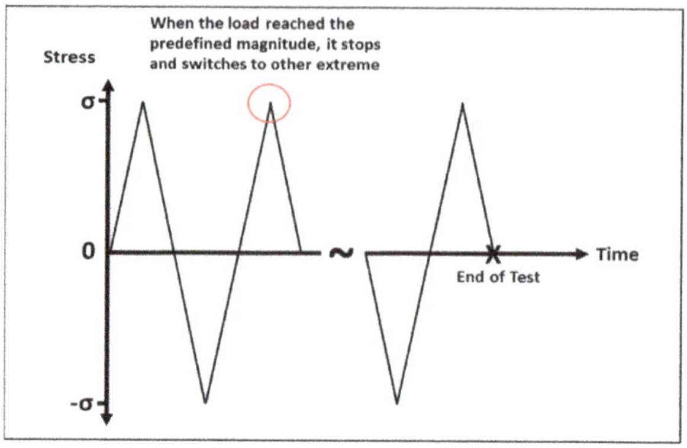

Figure 5. Mechanical fatigue testing profile.

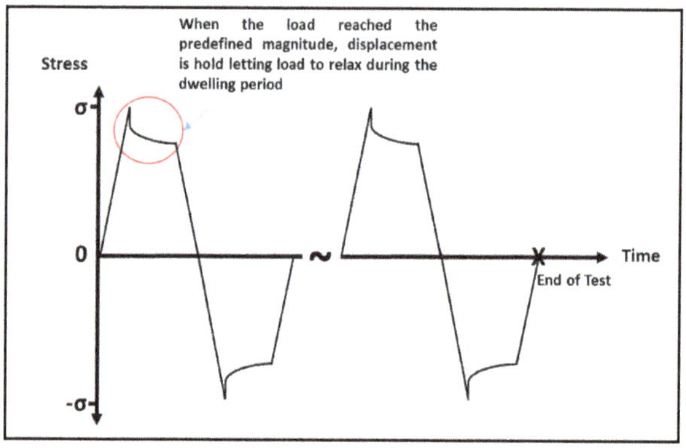

Figure 6. Combined creep–fatigue testing profile.

All experiments were performed at room temperature under several stress magnitudes (16, 20, and 24 MPa) and various dwelling periods (0, 10, 60, and 180 s). A constant ramp rate was employed of 0.1 mm/s for all replicates where seven joints were the sample size for each combination. Picking shear rate is critical in these tests due to viscoplastic behavior on SAC alloys. Faster ramp rate will cause the joints to resist more and does not give it enough time to creep. However, lower rates would cause the creep to be the dominant without allowing any effect for the fatigue. So, the defined shear rate was picked carefully based on previous studies [9–11,20,21] at such types of joints. The test matrix is shown in Table 1.

Table 1. Test Matrix.

Load Amplitude (Mpa)	Fatigue Only Test	Creep–Fatigue Test		
	0 s Dwell	10 s Dwell	60 s Dwell	180 s Dwell
16 MPa	7 samples	7 samples	7 samples	7 samples
20 MPa	7 samples	7 samples	7 samples	7 samples
24 MPa	7 samples	7 samples	7 samples	7 samples

3. Results and Discussions

3.1. Weibull Plots Analysis and Prediction Modeling

3.1.1. Mechanical Fatigue Condition

Weibull Plots

Two parameter Weibull plots are generated for each testing combination to describe the fatigue behavior of solder joints. To demonstrate the degradation in fatigue life or joint reliability, a two-parameter Weibull equation [48] is applied as shown in Equation (1).

$$R(t) = \mathrm{e}^{-\left(\frac{t}{\theta}\right)^{\beta}} \tag{1}$$

where $R(t)$ is the reliability at time t (the probability of not fail), t is the time or number of cycles, θ is the scale parameter or characteristic life, and β is the shape parameter. Figure 7 shows the Weibull plot for joints cycled according to the mechanical fatigue profile (no dwelling) with various stress amplitudes. A significant reduction in joints reliability is observed at higher stress levels. Weibull distribution plots were constructed using Minitab Software for each stress level, considering the maximum likelihood estimation method as the parametric estimation method. The variability of data is low according to the shape (β) parameter values.

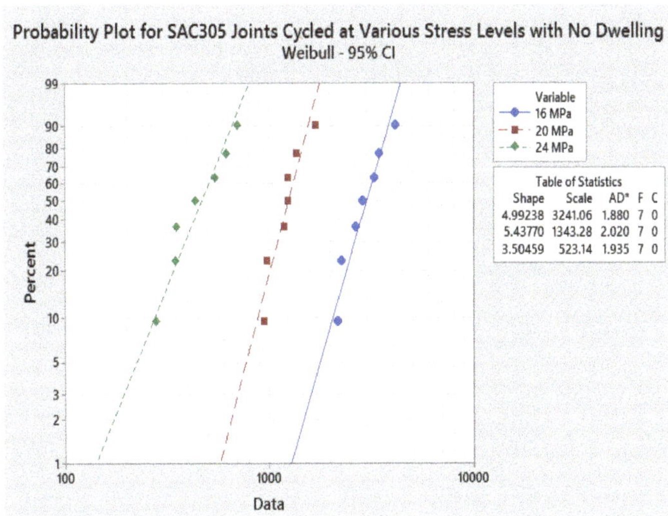

Figure 7. Weibull distributions for SAC305 joints cycled at different stress amplitudes at no dwelling.

Prediction Modeling

The characteristic life and stress amplitude are related according to power equation [48] shown in Equation (2).

$$N_{63} = C \times P^{-n} \tag{2}$$

where N_{63} is the characteristic life, P is the stress magnitude, C, n are material constants. n is called the ductility factor where lower value implies higher ductility. Figure 8 illustrates the fatigue life as a function of stress amplitude. Seven solder joints are cycled until complete failure with each stress level. The fatigue life is reduced drastically at higher stress levels.

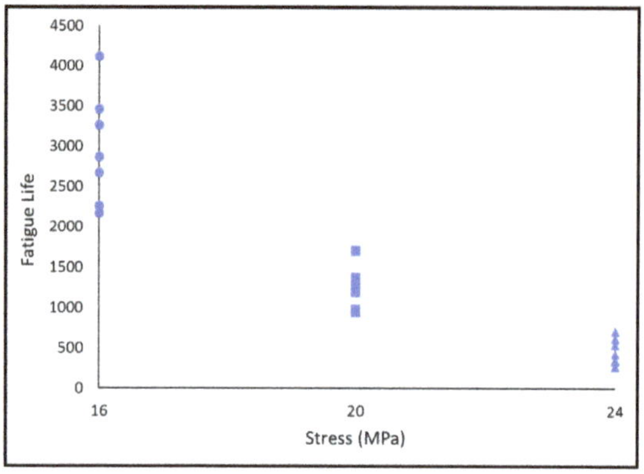

Figure 8. The fatigue life of SAC305 solder joints as a function of stress amplitude.

Characteristic life is a vital parameter to define reliability as a function of stress level according to Equation (1). Therefore, the characteristic life of SAC305 joints as a function of stress amplitude at the "no dwelling" condition is plotted, as shown in Figure 9. Data were fitted to a power equation which was found satisfactory to describe the characteristic life of joints as a function of stress amplitude according to Equation (2). It is concluded that the power value for the generated power equation is −4.34, which reflects the fatigue ductility exponent coefficient of the material. Moreover, increasing the stress value by a factor of 2 will lead to life reduction by a factor of 19. The ductility index (n) represents the material ductility, where the higher value of n means lower ductility. Similar results were found by others [21,66]. C constant in the stress life equation works as a scale parameter of the relationship between the fatigue life and the stress level, and n coefficient provides the shape of this relationship.

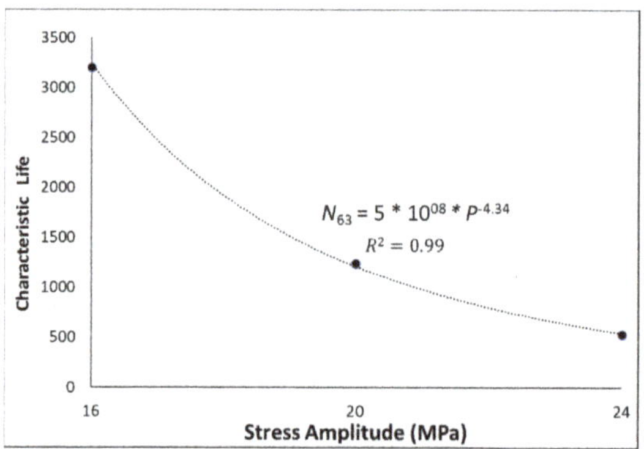

Figure 9. Characteristic life as a function of stress amplitude for SAC305 solder joints at no dwelling.

3.1.2. Dwelling (Creep–Fatigue) Condition

Weibull Plots

In order to examine the effects of dwelling on the fatigue life of the solder joints at different stress conditions, individual solder joints were tested at amplitudes of 16, 20, and 24 MPa with different dwell times of 0, 10, 60, and 180 s at 25 °C. The shape and scale parameters of the Weibull distribution were obtained for all combinations. In Figure 10, the Weibull plots for 10 s of dwelling level are generated. At certain dwelling times, increasing the stress amplitude leads to substantial fatigue life reduction. The same trend is found for other dwelling periods of 60 and 180 s. Moreover, at a certain stress level, the fatigue life is decreased extremely with dwellings of 10 s. Life continues to decrease with longer dwellings but in a smaller amount, as shown in Figure 11. Based on the preliminary life data analysis, only one data point was found as an outlier for the solder joints that were cycled at the 16 MPa stress level with a 180 s dwell time. This data point was eliminated from the reliability analysis. In Figure 12, characteristic life as a function of dwell time for various stress amplitudes is plotted. Creep effect (due to dwelling) was found to be substantial. There are some results for less life of lower stress and higher dwelling conditions than ones cycled with higher stress and shorter dwellings.

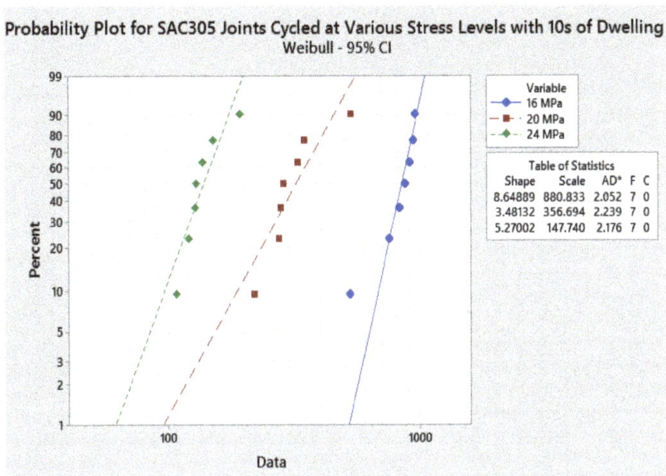

Figure 10. Weibull distributions for SAC305 joints cycled at different stress amplitudes at 10 s dwelling.

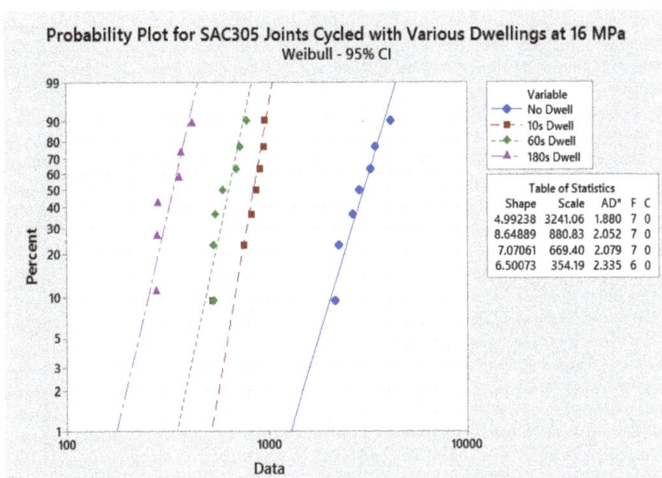

Figure 11. Weibull distributions for SAC305 joints cycled at different dwellings with 16 MPa stress level.

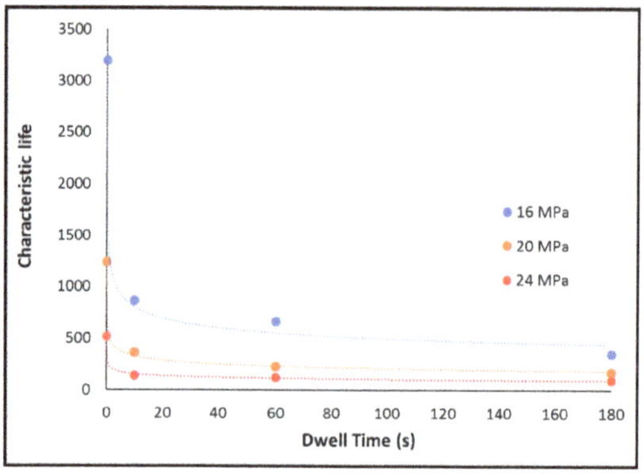

Figure 12. Characteristic life as a function of dwell time for SAC305 solder joints cycled with various stress levels.

Prediction Modeling

To characterize the reliability model as a function of dwelling, the relationship between characteristic life and stress amplitudes for joints cycled with various dwellings are plotted as shown in Figure 13, where data are fitted to power equations for each dwelling period. A decreasing trend for the power value (the material ductility exponent) is identified when the dwell time is increased. However, the constant C was observed to decrease with dwelling. The R-square values are above 99% for all fitting lines.

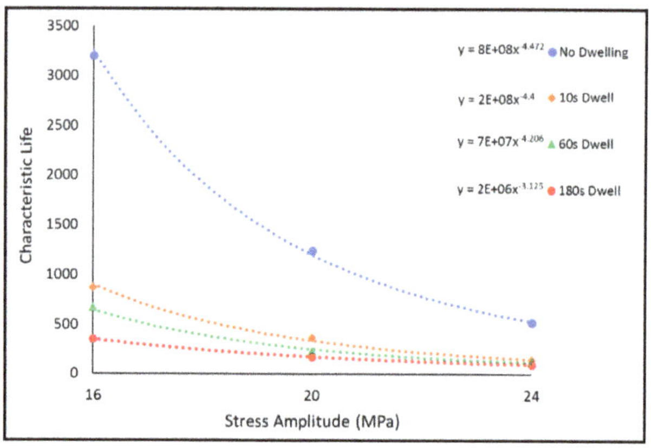

Figure 13. Characteristic life as a function of stress amplitude for SAC305 solder joints at various dwelling times.

To predict the reliability as a function of dwell time and stress amplitude, the correlation between n and C values as a function of dwell time (t_d) must be identified. The correlations to predict the power values of n and C value as a function of dwell time are shown in Figures 14 and 15. From Equation (2), the characteristic life is predicted as a function of stress amplitude and dwell time as shown in Equation (3).

$$N_{63} = 5 \times 10^8 \times e^{(-0.03t_d)} \times P^{-(-0.0075t_d + 4.5188)} \tag{3}$$

where N_{63} is the characteristic life, P is the stress amplitude, and t_d is the dwell time. To find out the general reliability model of solder joints as a function of dwell time, parameters in Equation (1) must be determined. In our case, there is no observed trend for the shape parameter of the Weibull plot at different dwell times and stress amplitudes. The shape parameter values were found between 3.5 and 14 with an average of 6.66. For the scale parameter (θ); the finding of characteristic life from Equation (3) is substituted in Equation (1). As a result, the general reliability model as a function of dwell time is established as shown in Equation (4).

$$R(t) = e^{-\left(\frac{t}{5\times10^{8}\times e^{(-0.03\times t_d)}\times P^{(0.0075\times t_d+4.5188)}}\right)^{6.66}} \quad (4)$$

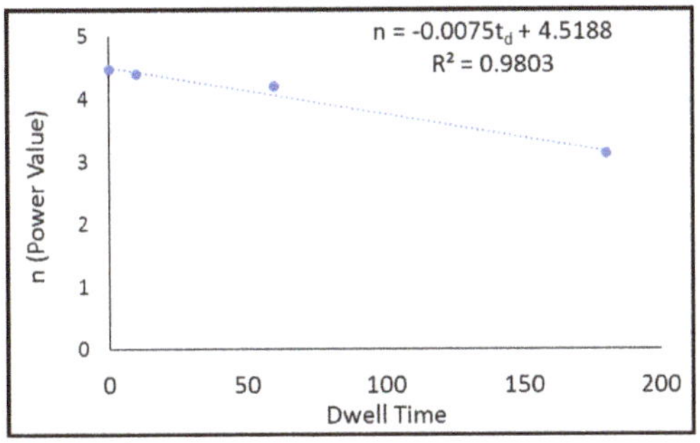

Figure 14. Power value (n) as a function of dwell time.

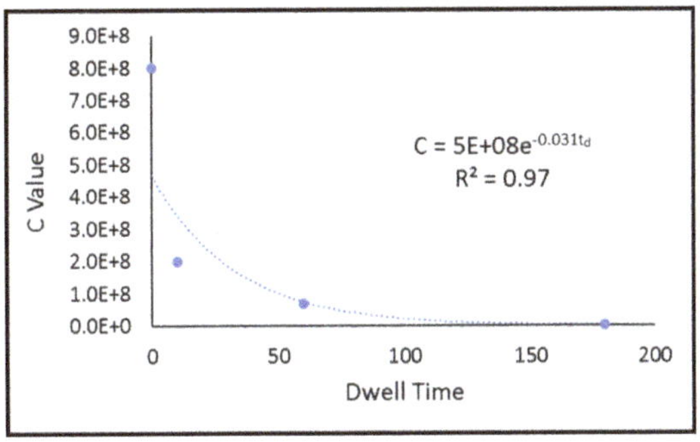

Figure 15. Constant (C) as a function of dwell time.

3.2. Stress–Strain Analysis

Hysteresis loops or stress–strain loops are essential to determine the damage parameters for each cycle represented by the inelastic work and plastic strain per cycle. These parameters are directly related to the accumulated damage. The loop area represents the inelastic (damage) work per cycle, where its width along the x-axis represents plastic strain. Figure 16 shows the average hysteresis loops for joints cycled with various stress levels at no dwelling. Both inelastic work (area inside the loop) and the plastic strain are obviously increased with higher stress amplitude. In Figure 17, average hysteresis loops are generated for 10 s of dwelling periods. At a specific dwelling, the same trend is observed; the hysteresis loop is enlarged drastically at higher stress levels. Moreover, more stress–relaxation is noticed with higher stress magnitudes. The enlargement of hysteresis loops is due to a massive increase in acculturated damaged work due to the creep effect in addition to fatigue damage. Consequently, evolution in hysteresis loops is generated for various dwelling times at certain stress levels of 24 MPa, as shown in Figure 18. Similar behavior is noticed for the plastic strain parameter. The higher stress or more prolonged dwelling would cause more plastic strain to be accumulated at a particular dwelling or stress levels, respectively.

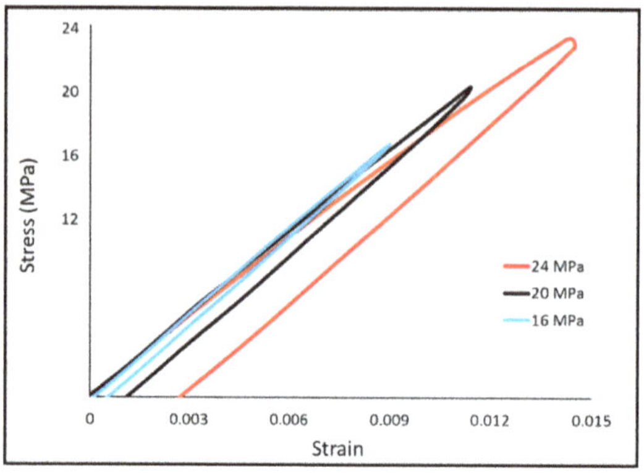

Figure 16. The hysteresis loops for SAC305 joints cycled with different stress amplitudes at no dwelling.

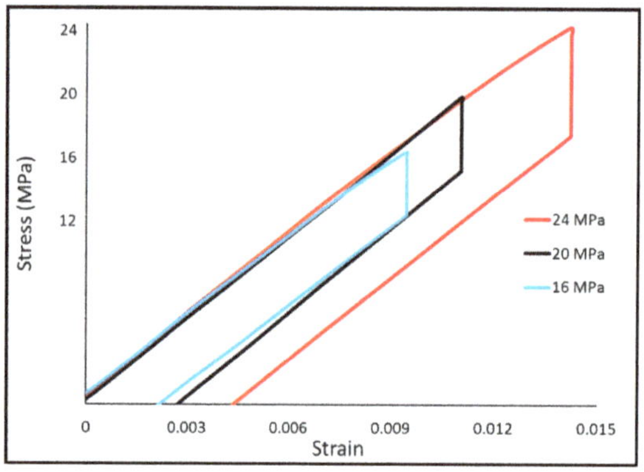

Figure 17. The hysteresis loops for SAC305 joints cycled with different stress amplitudes at 10 s dwelling.

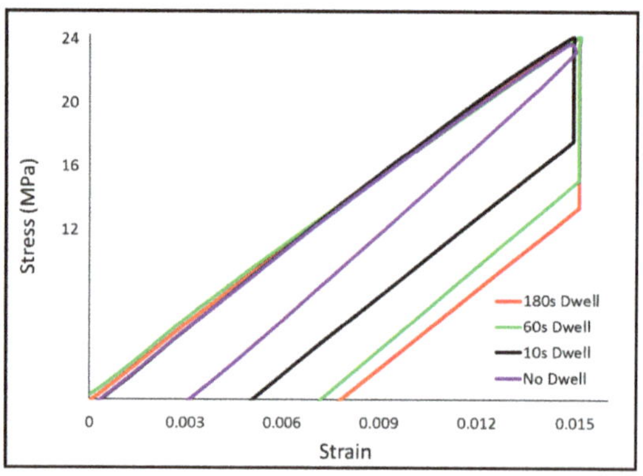

Figure 18. The hysteresis loops for SAC305 joints cycled at various dwellings with 24 MPa stress level.

The evolution in the hysteresis loop is directly related to the damage parameters, inelastic work, and plastic strain per cycle. Therefore, it is essential to examine the progression of such parameters during the joint lifetime to explain the evolvement of stress–strain loops. Figure 19 represents the typical development of inelastic work during the joint's life. This would provide an understanding of such behavior. The evolution is divided into three regions; the first region is the strain hardening, and it lasts for a few cycles. The second region includes the constant or steady-state region, which represents most of the joint's life. The last phase is the crack initiation and propagation. The same behavior is noticed for all combinations of dwellings and stress levels, as shown in Figures 20 and 21.

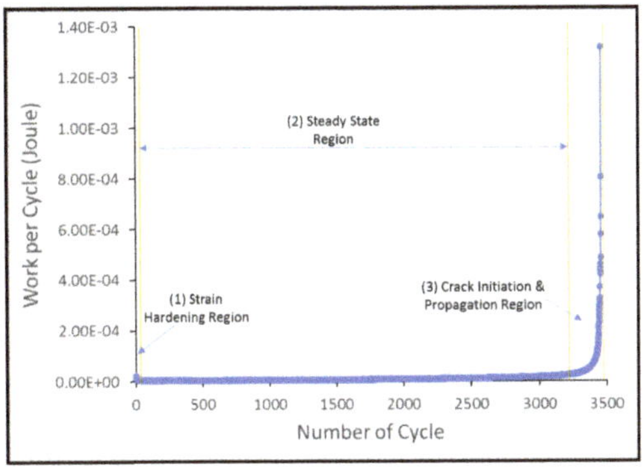

Figure 19. Inelastic work vs. the number of cycles for SAC305 solder joints cycled at 16 MPa stress amplitude until failure.

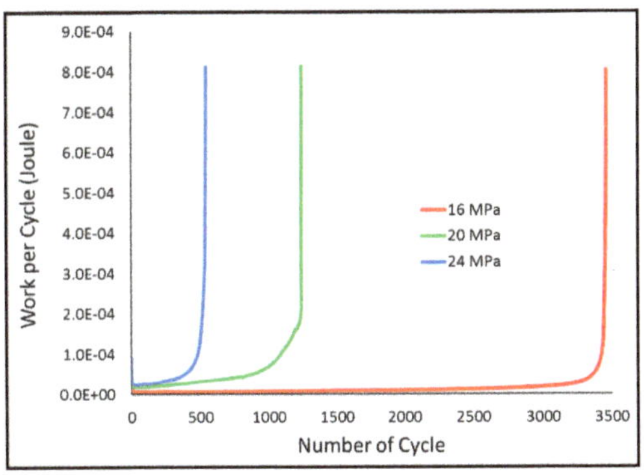

Figure 20. Inelastic work vs. the number of cycles for SAC305 solder joints cycled at various stress amplitudes at no dwelling until failure.

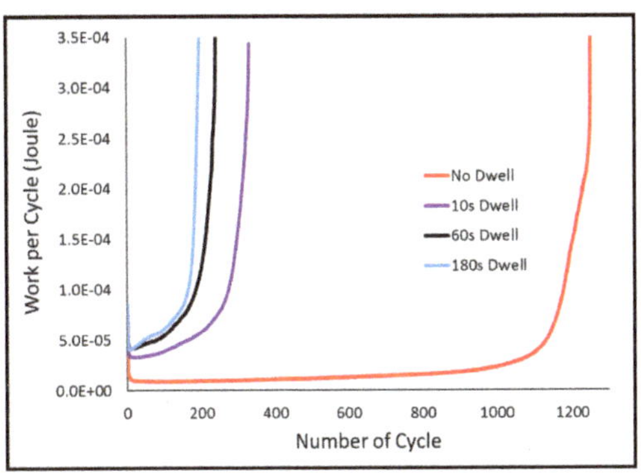

Figure 21. Inelastic work vs. the number of cycles for SAC305 solder joints cycled at 20 MPa stress amplitudes at various dwellings until failure.

3.3. Creep Effect

In this study, the creep effect is demonstrated by evaluating the damaged work per cycle and comparing it for both conditions of no and 10 s of dwelling. At the no dwelling conditions, the damaged work is related to cyclic fatigue only, since the only effect is mechanical fatigue. The creep effect is considered negligible due to the fast ramp rate employed. This would not allow a considerable creep effect to take place between ramps. However, the damage due to creep is dominant during dwelling, and the damage due to fatigue is acquired between ramps in the case of a combined creep–fatigue test. According to the above, it is assumed that inelastic (damaged) work is caused by fatigue only in the cyclic fatigue test with no dwelling, and inelastic (damaged) work due to creep–fatigue is generated due to the combined effects of creep–fatigue at dwelling conditions. Consequently, the damaged work due to creep is determined approximately by subtracting damage due to fatigue only (at no dwelling) from the one generated in the dwelling experiment. Figure 22, for example, shows the hysteresis loops for both conditions of no dwelling (green-colored) and with 10 s of dwelling (yellow-colored) in addition to a bar chart summarizing the damaged work generated for both cases. For more clarification, the damaged work due to fatigue is colored as green in the bar graph, and the damage due to creep is added accordingly for each case with more extended dwelling periods. Results show that work due to creep increases with longer dwelling times, where the related damage on a life reduction basis is reflected obviously. Despite the amount of inelastic work due to creep being almost similar to fatigue in the case of 10 s dwelling, the reflection on life reduction is massive by reducing life by a factor of 3. The same trends for other stress levels with higher damaged work are noticed due to more damaged being created due to higher stress.

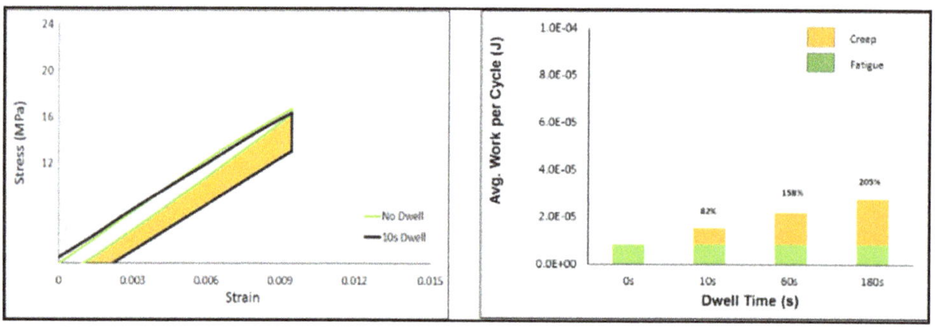

Figure 22. Hysteresis loops (**right**) with no dwelling and with 10 s dwelling cycled at 16 MPa, and a bar chart (**left**) for average inelastic work for both cases.

Figure 23 illustrates the average accumulated work until complete failure for SAC305 joints as a function of the dwell times time for different stress levels. For a certain stress level, it is noticed that increasing dwelling periods generates lower accumulated work until complete failure. This can be explained by the smaller amount of cycles observed until complete failure for extended dwellings which are accompanied with less accumulated work. Accumulated work includes other types than damaged or plastic work, which might be work due to friction or generated heat accompanied with each cycle. The trend is different during the dwellings. At no dwelling, 10 s, and 60 s of dwellings, the lower stress level generates more accumulated work than higher stress ones due to the significant additional cycles (accompanied with more accumulated work) observed until complete failure. On the other hand, the non-significant difference in the number of cycles until complete failure at 16 MPa stress level compared with the other stress levels of 20 and 24 MPa causes less accumulated work at 180 s of dwelling. Moreover, the number of cycles until complete failure for 20 and 24 MPa are close together, but more accumulated work is generated in the case of 24 MPa due to the higher average damage work during the 180 s dwelling. That is why it shows more accumulated damage work in the case of 24 MPa.

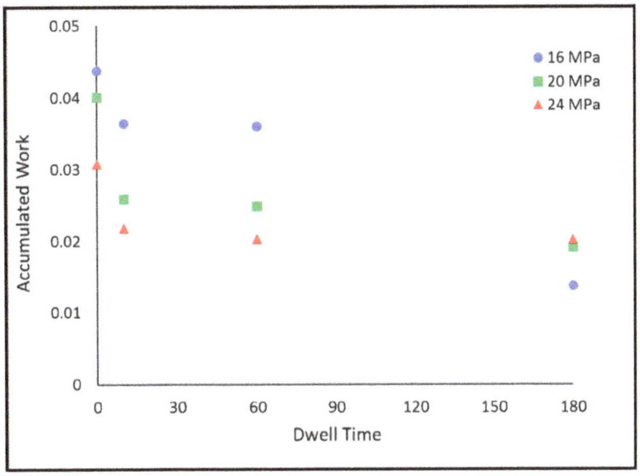

Figure 23. Accumulated work until complete failure vs. dwell time at different stress amplitudes.

Figures 24 and 25 illustrate the average inelastic work per cycle and plastic strain as a function of dwell time, respectively. Both the average work per cycle and the plastic strain were observed to increase with higher stress levels at specific dwellings drastically. This can be explained by more damage accumulating with higher stress levels at fixed dwelling and generating more inelastic work and plastic strain. Moreover, the longer dwelling would produce more inelastic work and plastic strain at a specific stress level. This is due to creep damage accumulated with extended dwellings. However, the creep effect might be substantial on both quantities of inelastic work and plastic strain compared with stress level. In many cases, inelastic work and plastic strain are higher at lower stress levels with longer dwellings than higher stress amplitude with shorter dwell times.

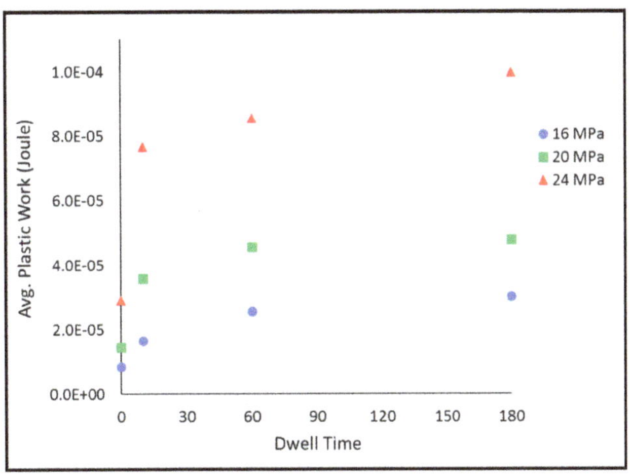

Figure 24. Average work per cycle as a function of dwelling times at various stress levels.

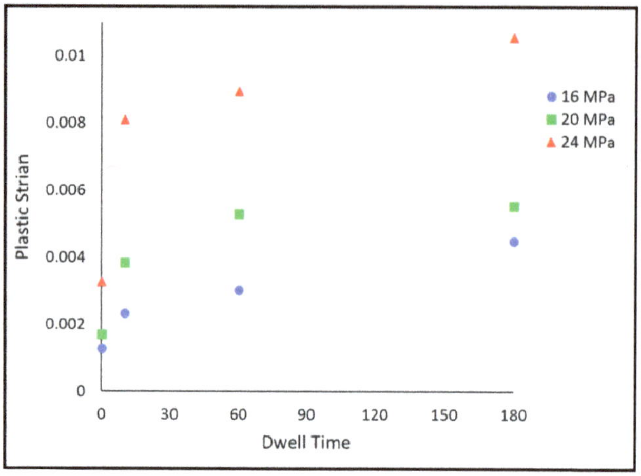

Figure 25. Plastic strain per cycle as a function of dwelling times at various stress levels.

3.4. The Coffin–Manson and Morrow Energy Models

Coffin–Manson Model

Figure 26 describes the plastic strain as a function of dwell time. It is clearly shown that there is a trend of plastic strain as a function of dwell time at various stress amplitudes. With a high *R-square* for all curves, data are fitted to a power equation, so the plastic strains can be predicted as a function of dwell time according to Equation (5).

$$PS = D \times {t_d}^{0.119} \tag{5}$$

where PS is the plastic strain, D is constant, and t_d is the dwell time. The D-value could also be formulated as a function of stress amplitude according to Figure 27. Thus, Equation (6) can be expressed as shown in Equation (6). The dwell time's impact on the value of the plastic strain is depicted by the D coefficient in terms of its magnitude.

$$PS = 0.0003\, e^{0.1248\, P} \times {t_d}^{0.119} \tag{6}$$

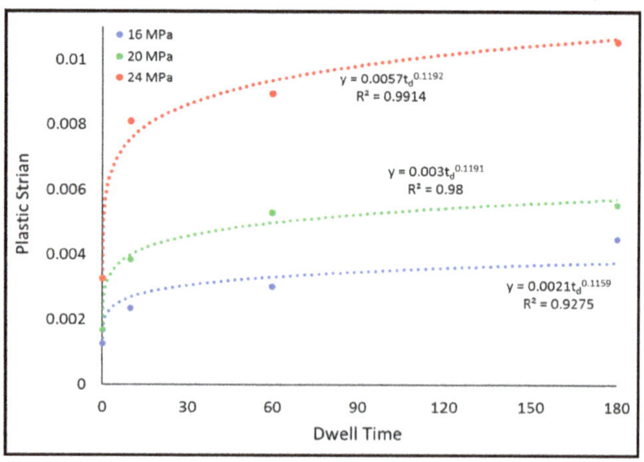

Figure 26. Plastic strain per cycle as a function of dwelling time curves at various stress levels.

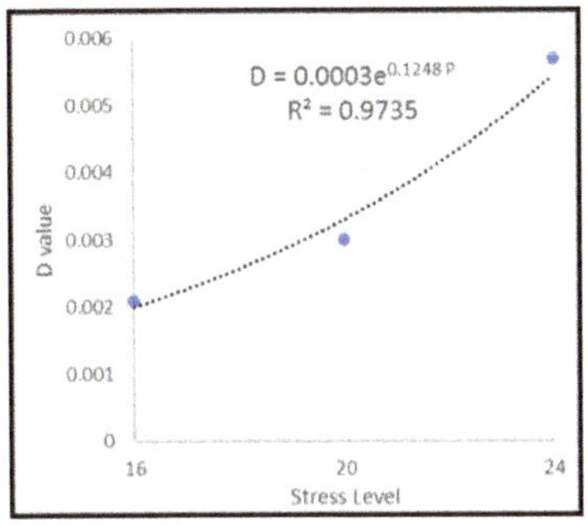

Figure 27. D-value as a function of stress amplitude.

Coffin–Manson is one of the most common models employed for fatigue life prediction as a function of plastic strain. The correlation between fatigue life and plastic strain is illustrated in Equation (7).

$$N_{63} = PS^{\frac{1}{R}}\ Z^{\frac{-1}{R}} \tag{7}$$

where N_{63} is the characteristic fatigue life, Z is the fatigue ductility coefficient, PS is the plastic strain, and R is the fatigue exponent. Based on the Coffin–Manson equation, the general reliability model as a function of plastic strain could be developed under certain conditions. If there is no clear trend for the coefficient of fatigue ductility (Z) and the fatigue exponent (R) at various dwellings, this implies that dwelling does not affect the Coffin–Manson equation. Moreover, data points for all conditions should have a similar trend (slope) to be fitted to a global Coffin–Manson equation no matter what the dwelling time is. To establish such a model, the characteristic life as a function of plastic strain (Equation (7)) must be obtained. Then, the new equation is substituted in Equation (1) to obtain the reliability model. To examine the model applicability in our case, the above-stated conditions must be checked. Figure 28 demonstrates the characteristic life as a function of plastic strain for various dwelling periods. Data points have the same trend (slope) and could be fitted to a global Coffin–Manson equation. The values for Coffin–Manson equation constants at various dwellings are generated accordingly, as shown in Table 2. It is obviously shown that there is no clear trend in these constants regardless of the dwelling time. This means dwelling has no effect on the Coffin–Manson model, and a general model could be developed. The values of the coefficient of fatigue ductility (Z) and the fatigue exponent (R) for the global equation are 0.19 and 0.646, respectively. Moreover, the global Coffin–Manson model is illustrated in Figure 29.

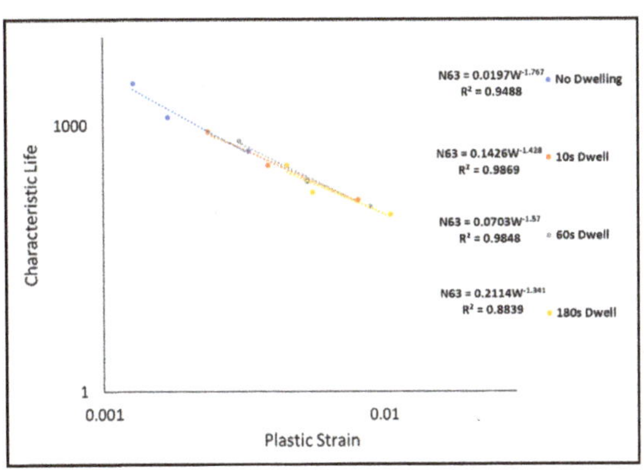

Figure 28. Characteristic life vs. plastic strain for SAC305 joints at various dwelling times on a log–log scale.

Table 2. Fatigue ductility coefficient and fatigue exponent of the Coffin–Manson model.

Dwelling Time	Fatigue Ductility (*Z*)	The Fatigue Exponent (*R*)
0	0.108	0.565
10	0.255	0.7
60	0.185	0.636
180	0.313	0.745
Global	0.19	0.646

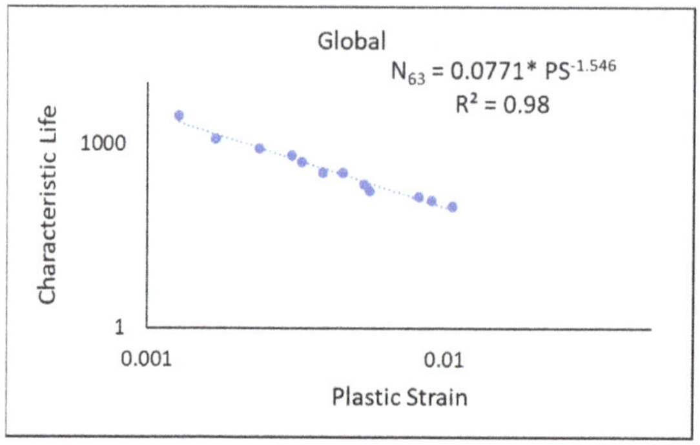

Figure 29. Characteristic life (global Coffin–Manson equation) vs. plastic strain for SAC305 joints at various dwelling times on a log–log scale.

Finally, the characteristic life equation as a function of plastic strain (Equation (8)) must be obtained and substituted in the reliability equation (Equation (1)) to develop the reliability model as a function of plastic strain. The characteristic life as a function of plastic strain is obtained from Figure 29 as shown in Equation (8).

$$N_{63} = 0.0771 \times PS^{-1.546} \tag{8}$$

Substituting Equation (8) in the reliability equation (Equation (1)) and considering the average shape parameter for all combinations as the defined shape parameter, the general reliability model is developed as shown in Equation (9).

$$R(t) = e^{-\left(\frac{t}{0.0771 \times PS^{-1.456}}\right)^{6.66}} \tag{9}$$

where PS is the plastic strain, and t is the number of cycles.

3.5. Morrow Energy Model

Inelastic work is another parameter considered to measure damage, where the Morrow Energy equation is the common related model. Figure 30 illustrates the inelastic work as a function of dwell time for various stress levels. It shows a trend in inelastic work as a function of dwell time at various stress amplitudes. With high R-square for all curves, data are fitted to a power equation, so the inelastic work is predicted as a function of dwell time according to Equation (10)

$$W = H \times {t_d}^{0.12} \tag{10}$$

where W is the inelastic work, H is constant, and t_d is the dwell time. On the other hand, the H-value could also be formulated as a function of stress amplitude according to Figure 31.

Thus, Equation (10) can be expressed as shown in Equation (11). The H coefficient illustrates the magnitude of the dwell time impact on the value of the inelastic work per cycle.

$$W = (-5.0 \times 10^{-6} \times P - 7 \times 10^{-5})\ t_d^{0.12} \tag{11}$$

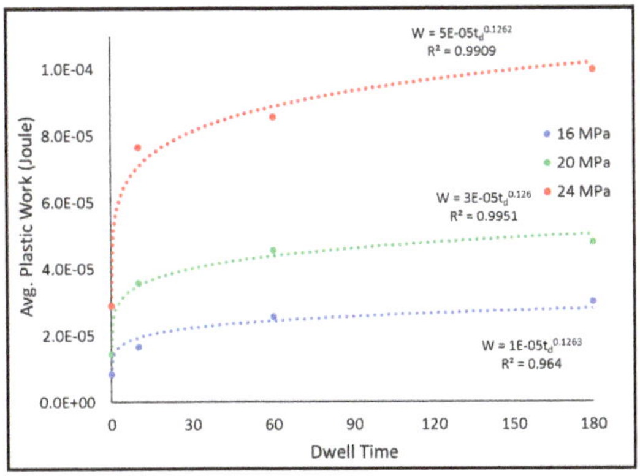

Figure 30. Inelastic work per cycle as a function of dwelling time curves at various stress levels.

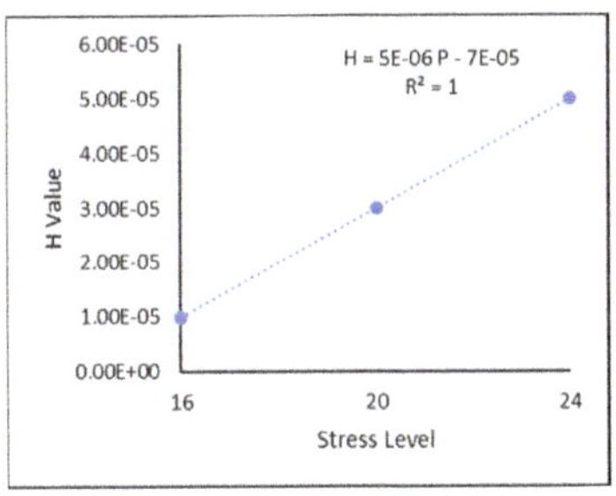

Figure 31. *H*-value as a function of stress amplitude.

Morrow Energy is a common model used to predict life as a function of inelastic work as shown in Equation (12).

$$N_{63} = G^{\frac{1}{m}}\ W^{\frac{-1}{m}} \tag{12}$$

where N_{63} is the characteristic fatigue life, G is the fatigue ductility coefficient, W is the inelastic work, and m is the fatigue exponent. In the same way as the Coffin–Manson model, the reliability model as a function of inelastic work based on the Morrow Energy model could be established under similar circumstances defined above. Our data show that fatigue ductility and fatigue exponent have no clear trend at various dwellings, as shown in Figure 32. Furthermore, the data points on a log–log scale demonstrate having a similar trend (slope) and could be fitted to the global Morrow Energy equation. The fatigue ductility and the fatigue exponent constants for all dwellings are specified accordingly, as shown in Table 3. It is obviously shown that there is no clear trend in these constants regardless of the dwelling time. This means dwelling has no effect on the Morrow Energy model, and the global model could be developed. Figure 33 shows the global model for Morrow Energy equation considering that the global constants for fatigue ductility coefficient and fatigue exponent are 0.0025 and 0.737, respectively.

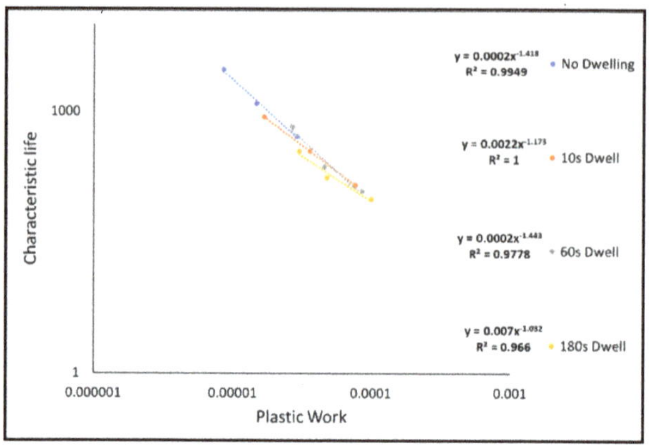

Figure 32. Characteristic life vs. inelastic work for SAC305 joints at various dwelling times on a log–log scale.

Table 3. Fatigue ductility coefficient and fatigue exponent of the Morrow Energy model.

Dwelling Time	Fatigue Ductility (G)	The Fatigue Exponent (m)
0	0.0023	0.713
10	0.0055	0.85
60	0.0028	0.69
180	0.0085	0.96
Global	0.0025	0.737

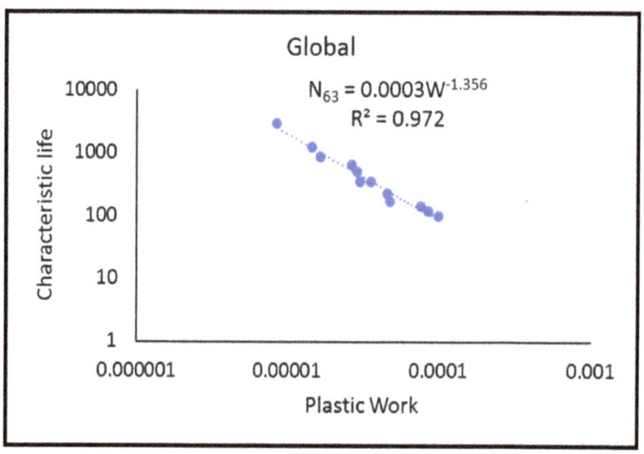

Figure 33. Characteristic life (global Morrow Energy equation) vs. plastic work for SAC305 joints at various dwelling times on a log–log scale.

To generate the reliability model (Equation (1)) as a function of inelastic work, the characteristic life equation as a function of plastic work (Equation (12)) must be attained. Therefore, the characteristic life as a function of plastic work is developed from Figure 33 as shown in Equation (13).

$$N_{63} = 0.0003\ W^{-1.356} \tag{13}$$

Finally, substituting Equation (13) in the reliability equation (Equation (1)) and considering the average shape parameter for all combinations as the defined shape parameter, the general reliability model is developed as shown in Equation (14).

$$R(t) = e^{-\left(\frac{t}{0.0003 \times W^{-1.356}}\right)^{6.66}} \tag{14}$$

where W is the plastic work, and t is the number of cycles.

3.6. Microstructure Analysis

The failure mode for samples at room temperature was determined by studying SEM images for tested joints after various dwelling times. This would provide an idea about the effect of dwelling on material evolution under various dwelling periods. Results show that failure is located within the bulk region among all cases of fatigue and creep–fatigue tests as shown in Figure 34.

Figure 34. SEM images for tested joints under various dwelling periods compared with no dwelling condition (most left).

4. Conclusions

The damage mechanism under thermal cycling is complicated and not very well understood. Creep and fatigue failure mechanisms related to several factors of dwell time and stress level are detected to have a major degradation effect on fatigue life. It was found that increasing the dwell time and/or the stress magnitude will reduce life. Moreover, at certain stress magnitudes, life is substantially decreased when increasing dwell time. The relation between life and stress amplitude is explored according to the life–stress equation. Results found to fit a power equation for all cases, and empirical models to predict the reliability as a function of dwell time and stress level, were generated accordingly. Furthermore, increasing the dwell time and/or stress level leads to more damage per cycle, which means more dissipated work per cycle and plastic strain. Degradation models were generated as a function of plastic strain and inelastic work per cycle based on Coffin–Manson and Morrow Energy models, respectively.

Author Contributions: Conceptualization, S.H. and M.A.; methodology, R.A.A., S.H. and M.A.; validation, R.A.A. and M.A.; formal analysis, M.A.; investigation, M.A. and M.T.; resources, M.A.; data curation, S.H. and M.A.; writing—original draft preparation, S.H. and M.A.; writing—review and editing, M.T. and S.H.; visualization, S.H.; supervision, S.H.; project administration, R.A.A., M.T. and S.H. All authors have read and agreed to the published version of the manuscript.

Funding: This research received no external funding.

Institutional Review Board Statement: Not applicable.

Informed Consent Statement: Not applicable.

Data Availability Statement: The datasets generated during and/or analyzed during the current study are available from the corresponding author on reasonable request.

Conflicts of Interest: The authors declare no conflict of interest.

References

1. Lau, J.H. Electronics packaging technology update: BGA, CSP, DCA and flip chip. *Circuit World* **1997**, *23*, 22–25. [CrossRef]
2. Murty, G.S. Stress relaxation in superplastic materials. *J. Mater. Sci.* **1973**, *8*, 611–614. [CrossRef]
3. Kim, Y.M.; Kim, Y.H. Drop impact reliability of Cu-Zn/Sn-Ag-Cu/Cu-Zn solder joints. In Proceedings of the 2008 International Conference on Electronic Materials and Packaging, Taipei, Taiwan, 22–24 October 2008; IEEE: Piscataway, NJ, USA, 2008; pp. 236–238. [CrossRef]
4. Waidhas, B.; Proschwitz, J.; Pietryga, C.; Wagner, T.; Keser, B. Study of the board level reliability performance of a large 0.3 mm pitch wafer level package. In Proceedings of the 2019 IEEE 69th Electronic Components and Technology Conference (ECTC), Las Vegas, NV, USA, 28–31 May 2019; IEEE: Piscataway, NJ, USA, 2019; pp. 1159–1164. [CrossRef]
5. Xu, L.; Pang, J.H. Effect of intermetallic and Kirkendall voids growth on board level drop reliability for SnAgCu lead-free BGA solder joint. In Proceedings of the 56th Electronic Components and Technology Conference, San Diego, CA, USA, 30 May–2 June 2006; IEEE: Piscataway, NJ, USA, 2006; p. 8. [CrossRef]
6. Lee, J.H.; Kumar, S.; Kim, H.J.; Lee, Y.W.; Moon, J.T. High thermo-mechanical fatigue and drop impact resistant Ni-Bi doped lead free solder. In Proceedings of the 2014 IEEE 64th Electronic Components and Technology Conference (ECTC), Orlando, FL, USA, 27–30 May 2014; IEEE: Piscataway, NJ, USA, 2014; pp. 712–716. [CrossRef]
7. Xu, L.; Pang, J.H. Effect of thermal and electromigration exposure on solder joint board level drop reliability. In Proceedings of the 2006 8th Electronics Packaging Technology Conference, Singapore, 6–8 December 2006; IEEE: Piscataway, NJ, USA, 2006; pp. 570–575. [CrossRef]
8. Abueed, M.; Athamenh, R.; Hamasha, S.; Suhling, J.; Lall, P. Effect of Fatigue on Individual SAC305 Solder Joints Reliability at Elevated Temperature. In Proceedings of the 2020 19th IEEE Intersociety Conference on Thermal and Thermomechanical Phenomena in Electronic Systems (ITherm), Orlando, FL, USA, 21–23 July 2020; IEEE: Piscataway, NJ, USA, 2020; pp. 1043–1050. [CrossRef]
9. Arfaei, B.; Cotts, E. Correlations between the microstructure and fatigue life of near-eutectic Sn-Ag-Cu Pb-free solders. *J. Electron. Mater.* **2009**, *38*, 2617–2627. [CrossRef]
10. Bieler, T.R.; Jiang, H.; Lehman, L.P.; Kirkpatrick, T.; Cotts, E.J.; Nandagopal, B. Influence of Sn grain size and orientation on the thermomechanical response and reliability of Pb-free solder joints. *IEEE Trans. Compon. Packag. Technol.* **2008**, *31*, 370–381. [CrossRef]
11. Arfaei, B.; Xing, Y.; Woods, J.; Wolcott, J.; Tumne, P.; Borgesen, P.; Cotts, E. The effect of Sn grain number and orientation on the shear fatigue life of SnAgCu solder joints. In Proceedings of the 2008 58th Electronic Components and Technology Conference, Lake Buena Vista, FL, USA, 27–30 May 2008; IEEE: Piscataway, NJ, USA, 2008; pp. 459–465. [CrossRef]
12. Bieler, T.R.; Zhou, B.; Blair, L.; Zamiri, A.; Darbandi, P.; Pourboghrat, F.; Lee, T.-K.; Liu, K.-C. The role of elastic and plastic anisotropy of Sn in recrystallization and damage evolution during thermal cycling in SAC305 solder joints. *J. Electron. Mater.* **2012**, *41*, 283–301. [CrossRef]
13. Lehman, L.P.; Xing, Y.; Bieler, T.R.; Cotts, E.J. Cyclic Twin Nucleation in Tin-Based Solder Alloys. *Acta Mater.* **2010**, *58*, 3546–3556. [CrossRef]
14. Lee, T.K.; Zhou, B.; Bieler, T.R. Impact of isothermal aging and Sn grain orientation on the long-term reliability of wafer-level chip-scale package Sn–Ag–Cu solder interconnects. *IEEE Trans. Compon. Packag. Manuf. Technol.* **2012**, *2*, 496–501. [CrossRef]
15. Telang, A.U.; Bieler, T.; Lucas, J.P.; Subramanian, K.N.; Lehman, L.; Xing, Y.; Cotts, E.J. Grain-boundary character and grain growth in bulk tin and bulk lead-free solder alloys. *J. Electro. Mater.* **2004**, *33*, 1412–1423. [CrossRef]
16. Mondal, D.; Haq, M.A.; Suhling, J.C.; Lall, P. Effects of β-Sn Crystal Orientation on the Deformation Behavior of SAC305 Solder Joints. In Proceedings of the 2022 IEEE 72nd Electronic Components and Technology Conference (ECTC), San Diego, CA, USA, 31 May–3 June 2022; IEEE: Piscataway, NJ, USA, 2022; pp. 1658–1667. [CrossRef]
17. Fu, X.; Zhou, B.; Yao, R.; En, Y.; Chen, S. Effect of grain orientation and microstructure evolution on electromigration in flip-chip solder joint. In Proceedings of the 2019 IEEE 69th Electronic Components and Technology Conference (ECTC), Las Vegas, NV, USA, 28–31 May 2019; IEEE: Piscataway, NJ, USA, 2019; pp. 1324–1327. [CrossRef]
18. Etris, S.; Lieb, K.; Sisca, V.; Moore, I.; Batik, A.; Rathore, H.; Yih, R.; Edenfeld, A. Fatigue behavior of solders used in flip-chip technology. *J. Test. Eval.* **1973**, *1*, 170–178. [CrossRef]
19. Hamasha, S.; Jaradat, Y.; Qasaimeh, A.; Obaidat, M.; Borgesen, P. Assessment of solder joint fatigue life under realistic service conditions. *J. Electron. Mater.* **2014**, *43*, 4472–4484. [CrossRef]
20. Hamasha, S.; Akkara, F.; Su, S.; Ali, H.; Borgesen, P. Effect of cycling amplitude variations on SnAgCu solder joint fatigue life. *IEEE Trans. Compon. Packag. Manuf. Technol.* **2018**, *8*, 1896–1904. [CrossRef]

21. Chowdhury, M.; Hoque, M.A.; Fu, N.; Suhling, J.C.; Hamasha, S.; Lall, P. Characterization of material damage and microstructural evolution occurring in lead free solders subjected to cyclic loading. In Proceedings of the 2018 IEEE 68th Electronic Components and Technology Conference (ECTC), San Diego, CA, USA, 29 May–1 June 2018; IEEE: Piscataway, NJ, USA, 2018; pp. 865–874. [CrossRef]
22. Su, S.; Jian, M.; Wei, X.; Akkara, F.J.; Hamasha, S.; Suhling, J.; Lall, P. Effect of Surface Finish on the Fatigue Behavior of Bi-based Solder Joints. In Proceedings of the 2019 18th IEEE Intersociety Conference on Thermal and Thermomechanical Phenomena in Electronic Systems (ITherm), Las Vegas, NV, USA, 28–31 May 2019; IEEE: Piscataway, NJ, USA, 2019; pp. 1155–1159. [CrossRef]
23. Kim, K.S.; Huh, S.H.; Suganuma, K. Effects of intermetallic compounds on properties of Sn–Ag–Cu lead-free soldered joints. *J. Alloy. Compd.* **2003**, *352*, 226–236. [CrossRef]
24. Yoon, J.W.; Kim, S.W.; Jung, S.B. IMC morphology, interfacial reaction and joint reliability of Pb-free Sn–Ag–Cu solder on electrolytic Ni BGA substrate. *J. Alloy. Compd.* **2005**, *392*, 247–252. [CrossRef]
25. Yu, D.Q.; Wang, L. The growth and roughness evolution of intermetallic compounds of Sn–Ag–Cu/Cu interface during soldering reaction. *J. Alloy. Compd.* **2008**, *458*, 542–547. [CrossRef]
26. Hamasha, S.D.; Borgesen, P. Effects of strain rate and amplitude variations on solder joint fatigue life in isothermal cycling. *J. Electron. Packag.* **2016**, *138*, 021002. [CrossRef]
27. Su, S.; Akkara, F.J.; Abueed, M.; Jian, M.; Hamasha, S.; Suhling, J.; Lall, P. Fatigue properties of lead-free doped solder joints. In Proceedings of the 2018 17th IEEE Intersociety Conference on Thermal and Thermomechanical Phenomena in Electronic Systems (ITherm), San Diego, CA, USA, 29 May–1 June 2018; IEEE: Piscataway, NJ, USA, 2018; pp. 1243–1248. [CrossRef]
28. Xu, H.; Lee, T.K.; Kim, C.U. Fatigue properties of lead-free solder joints in electronic packaging assembly investigated by isothermal cyclic shear fatigue. In Proceedings of the 2014 IEEE 64th Electronic Components and Technology Conference (ECTC), Orlando, FL, USA, 27–30 May 2014; IEEE: Piscataway, NJ, USA, 2014; pp. 133–138. [CrossRef]
29. Akkara, F.J.; Zhao, C.; Ahmed, S.; Abueed, M.; Su, S.; Hamasha, S.D.; Suhling, J.; Lall, P. Thermal cycling reliability of newly developed lead-free solders for harsh environments. In Proceedings of the SMTA International, Rosemont, IL, USA, 22–26 September 2019.
30. Hamasha, S.; Akkara, F.; Abueed, M.; Rababah, M.; Zhao, C.; Su, S.; Suhling, J.; Evans, J. Effect of Surface Finish and High Bi Solder Alloy on Component Reliability in Thermal Cycling. In Proceedings of the 2018 IEEE 68th Electronic Components and Technology Conference (ECTC), San Diego, CA, USA, 29 May–1 June 2018; pp. 2032–2040. [CrossRef]
31. Coyle, R.; Parker, R.; Longgood, S.; Sweatman, K.; Howell, K.; Arfaei, B. iNEMI Pb-Free Alloy Characterization Project Report: Part VI-The Effect of Component Surface Finishes and Solder Paste Composition on Thermal Fatigue of SN100C Solder Balls. In Proceedings of the SMTA International Conference, Fort Worth, TX, USA, 13–17 October 2013.
32. Al Athamneh, R.; Abueed, M.; Hani, D.B.; Hamasha, S.D. Effect of aging on SAC 305 solder joints reliability in accelerated fatigue shear test. In Proceedings of the SMTA International Conference, Rosemont, IL, USA, 14–18 October 2018.
33. Al Athamneh, R.; Abueed, M.; Hani, D.B.; Su, S.; Hamasha, S.; Suhling, J.; Lall, P. Effect of aging on the fatigue life and shear strength of SAC305 solder joints in actual setting conditions. In Proceedings of the 2019 18th IEEE Intersociety Conference on Thermal and Thermomechanical Phenomena in Electronic Systems (ITherm), Las Vegas, NV, USA, 28–31 May 2019; IEEE: Piscataway, NJ, USA, 2019; pp. 1146–1154. [CrossRef]
34. Yang, C.; Chan, Y.S.; Lee, S.R.; Ye, Y.; Liu, S. Comparison of thermal fatigue reliability of SnPb and SAC solders under various stress range conditions. In Proceedings of the 2009 International Conference on Electronic Packaging Technology & High Density Packaging, Beijing, China, 10–13 August 2009; IEEE: Piscataway, NJ, USA, 2009; pp. 1119–1123. [CrossRef]
35. Osterman, M.; Dasgupta, A.; Han, B. A strain range based model for life assessment of Pb-free SAC solder interconnects. In Proceedings of the 56th Electronic Components and Technology Conference, San Diego, CA, USA, 30 May–2 June 2006; IEEE: Piscataway, NJ, USA, 2006; p. 7. [CrossRef]
36. Dudek, R.; Kaulfersch, E.; Rzepka, S.; Rollig, M.; Michel, B. FEA based reliability prediction for different Sn-based solders subjected to fast shear and fatigue loadings. In Proceedings of the 2008 International Conference on Electronic Packaging Technology & High Density Packaging, Shanghai, China, 28–31 July 2008; IEEE: Piscataway, NJ, USA, 2008; pp. 1–7. [CrossRef]
37. Hasan, S.M.K.; Fahim, A.; Suhling, J.C.; Hamasha, S.; Lall, P. Evolution of the Mechanical Behavior of Lead Free Solders Exposed to Thermal Cycling. In Proceedings of the 2019 18th IEEE Intersociety Conference on Thermal and Thermomechanical Phenomena in Electronic Systems (ITherm), Las Vegas, NV, USA, 28–31 May 2019; IEEE: Piscataway, NJ, USA, 2019; pp. 1332–1341. [CrossRef]
38. Fahim, A.; Hasan, K.; Ahmed, S.; Suhling, J.C.; Lall, P. Mechanical Behavior Evolution of SAC305 Lead Free Solder Joints under Thermal Cycling. In Proceedings of the 2019 18th IEEE Intersociety Conference on Thermal and Thermomechanical Phenomena in Electronic Systems (ITherm), Las Vegas, NV, USA, 28–31 May 2019; IEEE: Piscataway, NJ, USA, 2019; pp. 734–744. [CrossRef]
39. Zhao, X.; Watte, P.; de Vries, H.; van Hees, G. Challenges in Predicting the Solder Interconnect Lifetime of High Power Electronics. In Proceedings of the 2018 19th International Conference on Electronic Packaging Technology (ICEPT), Shanghai, China, 8–11 August 2018; IEEE: Piscataway, NJ, USA, 2018; pp. 1511–1517. [CrossRef]
40. Kumar, P.M.; Gergely, G.; Horváth, D.K.; Gácsi, Z. Investigating the microstructural and mechanical properties of pure lead-free soldering materials (SAC305 & SAC405). *Powder Metall. Prog.* **2018**, *18*, 49–57. [CrossRef]
41. Fahim, A.; Hasan, S.K.; Suhling, J.C.; Lall, P. Nanoindentation Testing of SAC305 Solder Joints Subjected to Thermal Cycling Loading. In Proceedings of the International Electronic Packaging Technical Conference and Exhibition, Anaheim, CA, USA, 7–9 October 2019; American Society of Mechanical Engineers: New York, NY, USA, 2019; Volume 59322, p. V001T08A005. [CrossRef]

42. Pang, J.H.; Low, T.H.; Xiong, B.S.; Luhua, X.; Neo, C.C. Thermal cycling aging effects on Sn–Ag–Cu solder joint microstructure, IMC and strength. *Thin Solid Films* **2004**, *462*, 370–375. [CrossRef]
43. Fahim, A.; Ahmed, S.; Suhling, J.C.; Lall, P. Mechanical characterization of intermetallic compounds in SAC solder joints at elevated temperatures. In Proceedings of the 2018 17th IEEE Intersociety Conference on Thermal and Thermomechanical Phenomena in Electronic Systems (ITherm), San Diego, CA, USA, 29 May–1 June 2018; IEEE: Piscataway, NJ, USA, 2018; pp. 1081–1090. [CrossRef]
44. Yin, L.; Meilunas, M.; Arfaei, B.; Wentlent, L.; Borgesen, P. Effect of microstructure evolution on Pb-free solder joint reliability in thermomechanical fatigue. In Proceedings of the 2012 IEEE 62nd Electronic Components and Technology Conference, San Diego, CA, USA, 29 May–1 June 2012; IEEE: Piscataway, NJ, USA, 2012; pp. 493–499. [CrossRef]
45. Fahim, A.; Ahmed, S.; Chowdhury, M.R.; Suhling, J.C.; Lall, P. High temperature creep response of lead free solders. In Proceedings of the 2016 15th IEEE Intersociety Conference on Thermal and Thermomechanical Phenomena in Electronic Systems (ITherm), Las Vegas, NV, USA, 31 May–3 June 2016; IEEE: Piscataway, NJ, USA, 2016; pp. 1218–1224. [CrossRef]
46. Teo, J.R. Thermal cycling aging effect on the reliability and morphological evolution of SnAgCu solder joints. *IEEE Trans. Electron. Packag. Manuf.* **2007**, *30*, 279–284. [CrossRef]
47. Zhang, J.; Hai, Z.; Thirugnanasambandam, S.; Evans, J.L.; Bozack, M.J.; Zhang, Y.; Suhling, J.C. Thermal aging effects on the thermal cycling reliability of lead-free fine pitch packages. *IEEE Trans. Compon. Packag. Manuf. Technol.* **2013**, *3*, 1348–1357. [CrossRef]
48. Basit, M.M.; Motalab, M.; Suhling, J.C.; Hai, Z.; Evans, J.; Bozack, M.J.; Lall, P. Thermal cycling reliability of aged PBGA assemblies-comparison of Weibull failure data and finite element model predictions. In Proceedings of the 2015 IEEE 65th electronic components and technology conference (ECTC), San Diego, CA, USA, 26–29 May 2015; IEEE: Piscataway, NJ, USA, 2015; pp. 106–117. [CrossRef]
49. Akkara, F.J.; Zhao, C.; Athamenh, R.; Su, S.; Abueed, M.; Hamasha, S.; Suhling, J.; Lall, P. Effect of solder sphere alloys and surface finishes on the reliability of lead-free solder joints in accelerated thermal cycling. In Proceedings of the 2018 17th IEEE Intersociety Conference on Thermal and Thermomechanical Phenomena in Electronic Systems (ITherm), San Diego, CA, USA, 29 May–1 June 2018; IEEE: Piscataway, NJ, USA, 2018; pp. 1374–1380. [CrossRef]
50. Yang, L.; Fenglian, S.; Hongwu, Z.; Yang, W. Influence of Ag, Cu and additive Bi elements on the thermal property of low-Ag SAC solder alloys. In Proceedings of the 2011 6th International Forum on Strategic Technology, Harbin, China, 22–24 August 2011; Volume 1, pp. 72–75. [CrossRef]
51. Akkara, F.J.; Zhao, C.; Gordon, S.; Su, S.; Abueed, M.; Hamasha, S.; Suhling, J.; Lall, P. Effect of Aging on Component Reliability in Harsh Thermal Cycling. In Proceedings of the 2019 18th IEEE Intersociety Conference on Thermal and Thermomechanical Phenomena in Electronic Systems (ITherm), Las Vegas, NV, USA, 28–31 May 2019; IEEE: Piscataway, NJ, USA, 2019; pp. 717–723. [CrossRef]
52. Akkara, F.J.; Zhao, C.; Su, S.; Hamasha, S.; Suhling, J. Effects of mixing solder sphere alloys with bismuth-based pastes on the component reliability in harsh thermal cycling. In Proceedings of the SMTA International, Rosemont, IL, USA, 14–18 October 2018.
53. Chen, H.; Wang, L.; Han, J.; Li, M.; Liu, H. Microstructure, orientation and damage evolution in SnPb, SnAgCu, and mixed solder interconnects under thermomechanical stress. *Microelectron. Eng.* **2012**, *96*, 82–91. [CrossRef]
54. Fu, N.; Suhling, J.C.; Hamasha, S.; Lall, P. Evolution of the cyclic stress-strain and constitutive behaviors of SAC305 lead free solder during fatigue testing. In Proceedings of the 2017 16th IEEE Intersociety Conference on Thermal and Thermomechanical Phenomena in Electronic Systems (ITherm), Orlando, FL, USA, 30 May–2 June 2017; IEEE: Piscataway, NJ, USA, 2017; pp. 1353–1360. [CrossRef]
55. Erinc, M.; Schreurs, P.J.G.; Geers, M.G.D. Intergranular thermal fatigue damage evolution in SnAgCu lead-free solder. *Mech. Mater.* **2008**, *40*, 780–791. [CrossRef]
56. Norris, K.C.; Landzberg, A.H. Reliability of controlled collapse interconnections. *IBM J. Res. Dev.* **1969**, *13*, 266–271. [CrossRef]
57. Engelmaier, W. Fatigue life of leadless chip carrier solder joints during power cycling. *IEEE Trans. Compon. Hybrids Manuf. Technol.* **1983**, *6*, 232–237. [CrossRef]
58. Vayman, S.; Fine, M.E.; Jeannotte, D.A. Isothermal fatigue of low tin lead based solder. *Metall. Trans. A* **1988**, *19*, 1051–1059. [CrossRef]
59. Salmela, O. Acceleration factors for lead-free solder materials. *IEEE Trans. Compon. Packag. Technol.* **2007**, *30*, 700–707. [CrossRef]
60. Solomon, H. Fatigue of 60/40 solder. *IEEE Trans. Compon. Hybrids Manuf. Technol.* **1986**, *9*, 423–432. [CrossRef]
61. Pan, N.; Henshall, G.A.; Billaut, F.; Dai, S.; Strum, M.J.; Benedetto, E.; Rayner, J. An acceleration model for Sn-Ag-Cu solder joint reliability under various thermal cycle conditions. In Proceedings of the SMTAI, Chicago, IL, USA, 25 September 2005; pp. 876–883.
62. Al Athamneh, R.; Hamasha, S. Fatigue behavior of SAC-Bi and SAC305 solder joints with aging. *IEEE Trans. Compon. Packag. Manuf. Technol.* **2019**, *10*, 611–620. [CrossRef]
63. Motalab, M.; Cai, Z.; Suhling, J.C.; Zhang, J.; Evans, J.L.; Bozack, M.J.; Lall, P. Improved predictions of lead free solder joint reliability that include aging effects. In Proceedings of the 2012 IEEE 62nd Electronic Components and Technology Conference, San Diego, CA, USA, 29 May–1 June 2012; IEEE: Piscataway, NJ, USA, 2012; pp. 513–531. [CrossRef]
64. Al Athamneh, R.; Hani, D.B.; Ali, H.; Hamasha, S. Fatigue life degradation modeling of SnAgCu solder joints after aging. *IEEE Trans. Compon. Packag. Manuf. Technol.* **2020**, *10*, 1175–1184. [CrossRef]

65. Abueed, M.; Alathamneh, R.; Suhling, J.; Lall, P. Effect of Creep And Fatigue on Individual SAC305 Solder Joint Reliability in Iso-Thermal Cycling. In Proceedings of the SMTA International, Rosemont, IL, USA, 22–26 September 2019; pp. 308–314.
66. Won, Y.; Cho, J.; Agonafer, D.; Asheghi, M.; Goodson, K.E. Fundamental cooling limits for high power density gallium nitride electronics. *IEEE Trans. Compon. Packag. Manuf. Technol.* **2015**, *5*, 737–744. [CrossRef]

crystals

MDPI

Article

Comparison of Structural, Microstructural, Elastic, and Microplastic Properties of the AAAC (A50) and ACSR (AC50/8) Cables after Various Operation Periods in Power Transmission Lines

Aleksandr A. Levin [1,*], Maria V. Narykova [1], Alexey I. Lihachev [1], Boris K. Kardashev [1], Andrej G. Kadomtsev [1], Nikita D. Prasolov [1], Andrei G. Panfilov [1], Roman V. Sokolov [1], Pavel N. Brunkov [1], Makhsud M. Sultanov [2], Alexander V. Strizhichenko [2] and Ilia A. Boldyrev [2]

[1] Ioffe Institute, Politekhnicheskaya ul. 26, St. Petersburg 194021, Russia
[2] "Moscow Power Engineering Institute", National Research University, Volzhsky Branch, Lenina pr. 69, Volzhsky 404110, Russia
* Correspondence: aleksandr.a.levin@mail.ioffe.ru

Abstract: In modern economic infrastructure, Al cables of overhead power transmission lines are used both without and with a steel core (respectively, all aluminum alloy conductor (AAAC) and aluminum conductor steel reinforced (ACSR) cables). In this article, the changes in structural, microstructural, and elastic-microplastic properties have been analyzed for the outer wires of the AAAC (A50) and ACSR cables (AC50/8 cables with a steel core of ~8 mm^2 cross-section, hereinafter referred to as AC50) with the cross-section of the stranded conductor of ~50 mm^2, which were in operation for 0–20 years in the Volgograd region of Russia. Using the techniques of X-ray diffraction, electron backscattered diffraction, densitometry, and the acoustic method, the structural and microstructural features of the wires have been compared and found to be correlated with their elastic-microplastic properties. It has been ascertained that the presence of a steel core in AC50 leads to a decrease in the defectiveness of the near-surface layer of their aluminum wires. Compared with A50 cables, the development of void defects in the near-surface layer of Al-wires of AC50 cables slows down (by ~1 year with a service life of ~10 years and by ~3 years with a service life of ~20 years).

Keywords: aluminum wires; overhead power transmission lines; XRD; EBSD; densitometry; elastoplastic properties; density; near-surface layer

Citation: Levin, A.A.; Narykova, M.V.; Lihachev, A.I.; Kardashev, B.K.; Kadomtsev, A.G.; Prasolov, N.D.; Panfilov, A.G.; Sokolov, R.V.; Brunkov, P.N.; Sultanov, M.M.; et al. Comparison of Structural, Microstructural, Elastic, and Microplastic Properties of the AAAC (A50) and ACSR (AC50/8) Cables after Various Operation Periods in Power Transmission Lines. *Crystals* **2022**, *12*, 1267. https://doi.org/10.3390/cryst12091267

Academic Editors: Ulrich Prahl, Sergey Guk and Faisal Qayyum

Received: 12 August 2022
Accepted: 1 September 2022
Published: 6 September 2022

Publisher's Note: MDPI stays neutral with regard to jurisdictional claims in published maps and institutional affiliations.

1. Introduction

Uninsulated (bare) cables A50 and AC50/8 (hereinafter also referred to as AC50) are mainly used for transmitting electrical energy in overhead electrical networks on land in all macroclimatic regions with either mild or cold climates. The main difference between AC50 and A50 cables is the presence of a steel core in the AC50 ones. Comparative characteristics of the investigated types of cables according to the manufacturer's certificates [1] are given in Table 1. It is obvious that the presence of a steel core in the AC50 cables of the ACSR (aluminum conductor steel reinforced) type significantly increases the bearing capacity to withstand higher loads than the capabilities of the A50 cable belonging to the AAAC (all aluminum alloy conductor) type. The advantages of the A50 cable include its low weight (since it is completely made of aluminum) and a higher current conductivity (slightly higher than AC50 with a close full diameter of the cable conductor).

Table 1. Characteristics of A50 and AC50 overhead power-line cables according to Ref. [1].

Cable Type	Number of Al Wires in Cable	Number of Steel Wires in Cable	Individual Wire Diameter, mm	Cable Section, mm^2	Estimated Cable Weight, kg/km	Cable Breaking Strength (≤), N
A50	7	0	3.0	49.5	135	8198
AC50	6	1	3.2	56.24 [a]	195	17,112

[a] Al/Fe section is of 48.2 mm^2/8.04 mm^2.

Overhead power lines are mainly constructed of supports, cables, and insulators, with cables (wires in them) being the most weak ("vulnerable") element of this system. The expected service life of the cables listed in Table 1 is at least 45 years [1], but it is not limited to this period and, rather, is determined by the technical condition of the wires. The study of the factors that cause the destruction of wires and the prediction of their serviceability is a vital area of research to develop ways to improve their normal operational properties.

There are several reasons for the premature failure of cables of overhead power lines; therefore, let us briefly analyze some of them. The first reason is wire damage and defects, which can occur at any stage of manufacturing of the cables and processing of the cable wires, including continuous casting, extrusion, wire drawing, or stranding. In the papers [2,3], various defects were investigated and the reasons for their appearance were explained. In particular, one of the most common causes is the presence of oxide particles in aluminum conductors (the starting material for wires) and shrinkage porosity. Such defects (nano and micropores, inclusions such as oxides, carbides, etc.) can indeed be stress concentrators and lead to the formation of microcracks with further destruction [4,5] that reduces their service life. Compliance with the required standards in the manufacture of cables is an important and necessary step to ensure a long service life.

Another reason for the premature failure of cables may be a significant increase in the mechanical load due, for example, to the action of a strong wind or ice formations [6,7]. Furthermore, one of the most dangerous types of loads is high-frequency vibrations due to the laminar wind regime, which induces fatigue fretting (fretting deterioration in between conductor strands) [8]. Fretting deterioration residues oxidize and become harder. These hard particles increase the run-off of conductors [2,3,9,10]. For instance, a study to assess the causes of premature fatigue failure of AAAC cables (Brazil) showed that vibration reduction led to a more than threefold increase in residual service life [11]. Fretting fatigue studies have also been performed in laboratory conditions for AAAC cables (e.g., [12–14]) and for ACSR ones as well (e.g., [15,16]).

Environmental conditions are of great importance during the operation of cables. As is known, overhead power lines are exposed to atmospheric corrosion [17–19]. The aluminum wires that make up the outer coil of A50 and AC50 cables have high corrosion resistance due to the formation of a resistant amorphous oxide film up to ~10 nm thick on the aluminum surface in ambient air [18–21]. Galvanized steel is used as the central element of the AC50 cable inasmuch as zinc coating is an effective way to protect the steel core from direct contact with the aluminum part of the cable [20]. However, existing operating conditions can lead to the destruction of the zinc layer and the development of galvanic corrosion between steel and aluminum. It has been found that the pH of the aqueous medium (water film, air atmosphere) is of decisive importance for the zinc coating corrosion process since it controls the dissolution of the passive oxyhydroxide surface [22]. Therefore, external conditions are of particular importance here, such as the relative humidity of the air, the amount of precipitation, the presence of industrial enterprises, etc. [19,23]. The results of studies of ACSR cables after operation show different results in assessing the integrity of the zinc coating of the steel core and its effect on the degradation of the structure of the aluminum part of the cables. In fact, some studies have shown that no obvious corrosion is found on a steel core under mild climate conditions [24,25]. On the contrary, a significant violation of the zinc coating and corrosion damage of aluminum wires were observed in some studies, although the mechanical strength of these cables (with steel cores) still remained at a fairly high level [26–28]. In [29], data on cables with steel cores are adduced, according to which the electrical resistance is significantly increased as a result of galvanic corrosion, so that the local corrosion leads to the failure of power lines. A comparative study showed that the rate of galvanic corrosion in ASCR cables was significantly higher than in AAAC ones [30]. Given the various results of studying the effect of galvanic corrosion on the performance properties of cables, various methods are proposed to improve the solution to this problem, from taking the geographical zoning into account [31] to various methods for assessing its impact [32] (because of the peculiarity of the location of the development of galvanic corrosion and the impossibility of its determination during visual inspection).

Ref. [33] discussed the advantages of using AAAC cables (vs. ASCR) due to the lack of galvanic corrosion, lower resistance, and better nominal current in it. As with ASCR cable corrosion, the main challenge is knowing where to start looking in order to identify incipient damage in time for repair. In [34], AAAC and ACSR conductors were studied in real-life conditions, and it has been concluded that the use of AAAC conductors is more efficient than ACSR, since the latter have a noticeably greater weight, which limits their use (when used on relatively weak structures of wooden supports). However, a comparative study of the fatigue characteristics of AAAC and ACSR cables in laboratory conditions showed that the ASCR type had a fatigue life fivefold greater than the AAAC type, which is an important parameter in evaluating the performance of cables [13,35]. A separate study evaluating the fatigue strength of ASCR cables in a dry environment and in a NaCl corrosive environment showed that wear damage was more severe in a corrosive environment [36]. A similar conclusion about the criticality of corrosion abrasion of cables with a steel core was made in [2].

Analysis of the available scientific data has shown that AAAC and ASCR cables have their advantages and disadvantages depending on the operating conditions. In laboratory studies, it is unlikely to be possible to achieve real operating conditions that take the whole range of factors which provoke early destruction of the conductor into account. The complexity of predicting the behavior of wires during their use is due to the fact that it depends on many factors: the structural state of wire elements, the state of the surface layer, environmental conditions, mechanical stress, friction, the presence/absence of galvanic corrosion, and a number of other conditions. Previously [10,37,38], we studied changes in the structure and microstructure of the individual outer wires of A50 cables (AAAC type) as a function of their service life, correlating with changes in their elastic-plastic properties and surface electrical resistance. In this paper, the methods used to study those cables (namely, the acoustic method for measuring elastoplastic properties and complementary methods of energy-dispersion microanalysis (EDX), scanning electron microscopy (SEM), electron backscattered diffraction (EBSD), X-ray diffraction (XRD), and densitometry) are applied to similar individual outer wires from AC50 cables (ACSR type). The comparison of two types of cables, A50 and AC50, carried out in this work is of interest because the results obtained can be analyzed, taking into account the identity of their operating conditions, particularly the temperature and environment of use (in the Volgograd region of Russia), scale factor (wire/cable diameter), type of stress state, and service lifetime. The main difference is the presence of a steel core in an AC50 cable, which, as the analysis of literature data shows, has its advantages and disadvantages. In this regard, the identification of the characteristics that change most strongly with time may make it possible to identify the parameters that are precursors of destruction. In this article, complementing [10,37], we contribute to increasing the available data on the analysis and predicting the performance of overhead power line cables by considering wires from AC50 cables compared to wires from A50 cables with comparable service lives of 0 to ~20 years.

2. Materials and Methods

2.1. Samples

To compare the structural, microstructural, elastic, and microplastic properties of two types of uninsulated cables, namely, A50 (AAAC) and AC50 (ACSR), individual aluminum wires of overhead power-line cables were studied after different periods of operation, see Table 2. Hereinafter, wires of the A50-type cables will be referred to as "A50-type wires" or "A50 wires" for short, and wires of the AC50-type cables will be referred to as "AC50-type wires" or, for short, "AC50 wires".

Table 2. Samples of cables of overhead power lines with different service life, which were selected for research (samples A50 are from [10,37]) [1].

Sample, *N*	**5-2**	**5**	**8**	**2-2**	**7**	**6**
Type	A50	AC50	A50	AC50	A50	AC50
Service life t, years	0	0	10	8	18	20
Wire diameter d, mm	3.02	3.20	2.85	3.24	3.03	3.07

[1] Additionally, samples of the A50 type that do not have an AC50 counterpart in service life (N3, 35 years of service life (same as in Refs. [10,37]) and N10, 54 years) were investigated in some experiments.

The nominal characteristics of the A50 and AC50 types of cables according to the manufacturer's passport (Interstate Technical Standard GOST 839-2019) [1]) are summarized in Table 1. The A50 cable consists of seven separate Al wires and, according to the cable passport, has a cable conductor cross-section of 49.5 mm^2, the diameter of individual aluminum wires is 3 mm, the estimated cable weight is 135 kg/km. Steel-aluminum cables designated as AC50 were also investigated. They contain six Al wires with a diameter of 3.2 mm, whirled around a steel core with the same diameter of 3.2 mm. The cross-sections of the AC50 cable according to the passport are 48.2 mm^2 (Al) and 8.04 mm^2 (steel), the estimated weight of the cable is 195 kg/km. It should be noted that the values of the diameters of the wires studied, measured experimentally, differ from the nominal values and range from 2.85 mm to 3.03 mm and from 3.20 mm to 3.24 mm for A50 and AC50 wires, respectively (Table 2).

In manufacturing the aluminum wires of the above cables, cold-drawn aluminum of grade A7E (GOST 11069) [39] was used. Table 3 shows the chemical composition of the aluminum alloy according to the manufacturer's quality certificate.

Table 3. Chemical composition of A7E-grade aluminum used by the manufacturer (wt.%) according to Ref. [39]. The lower limit of the Al weight content and the upper limits of possible impurities in the wire material are indicated.

Al	**Fe**	**Si**	**Zn**	**Ga**	**Mg**	**Cu**	**Ti + V + Cr + Mn**	**Other**
99.59	0.20	0.08	0.04	0.03	0.02	0.01	0.01	0.02

All the Al wires are manufactured using the same technology from aluminum of the same grade, and the results of XRD studies [37] revealed that the unexploited A50 and AC50 wires (respectively, N5-2 and N5) showed almost the same structural and microstructural parameters. That is why, in the study by different methods of a series of A50-type samples of different length service lives, the results obtained on N5 AC50-type wire were considered as related to the initial state (service life of 0 years), unless otherwise indicated.

Samples of the optimal length for the studies will be specified in the description of each method. They were cut from the outer wire of the tested cables and washed in an ultrasonic acetone bath to remove external impurities and protective grease. Under the term "outer wires", we mean Al wires under the influence of the atmosphere and also other wires (either exclusively Al wires in the case of A50 cables or, in addition, steel core in the case of AC50 ones).

It should be noted that from the available samples, wires of A50 and AC50 types with the maximum possible service life were selected for comparison. However, there is a difference of 2 years between the service life of the samples after operation (10 and 8 years or 18 and 20 years for samples A50 and AC50, respectively, see Table 2). Taking into account the long expected service life of 45 years in accordance with the manufacturer's standard [1], it can be expected that this difference will be insignificant when comparing their properties.

2.2. Experimental Details of Measurements and Analysis

All details of the experiment and analysis of the results of EDX, SEM, EBSD, XRD measurements, and densitometric and acoustic investigations are detailed in [10,37]. Therefore, this article provides only a brief description of the experimental equipment, measurements, and their analysis.

2.2.1. Experimental Details of EDX, SEM, and EBSD

To study samples by SEM, EDX, and EBSD by use of a JSM 7001F scanning electron microscope (JEOL, Tokyo, Japan) equipped with an INCA PentaFETx3 system for EDX microanalysis and with an HKL Nordlys detector (Oxford Instruments, Abingdon-on-Thames, UK) to obtain EBSD maps, polished cross sections of wire samples with a diameter of 2.85–3.24 mm were prepared (see Table 2). For the preparation of polished cross sections, a MultiPrep8 machine (Allied, San Francisco, CA, USA) was used with a gradual reduction in the abrasive grain for mechanical grinding of the section. The final polishing of the section surface was carried out with an argon-ion beam using a 1061 SEM Mill system of ion milling and polishing (Fischione, Export, PA, USA). The EBSD maps were taken with a step of 0.5 μm in areas of 100 × 100 μm^2 in size, approximately in the center of the cross sections and near their edges at a distance of ~150 μm from the outer surface of the wires, i.e., in positions corresponding to the internal bulk and near-surface layers of wires, respectively. The EDX analysis was carried out on an area of 500 × 500 μm^2 in cross-section to obtain averaged values of the sample composition. During the analysis, the spectrum was continuously accumulated while the area was scanned by an electron beam.

2.2.2. Experimental Details of Densitometric Measurements

The densities of wire samples were determined by hydrostatic weighing on a Shimadzu AUW 120D analytical balance using an SMK-301 attachment (Shimadzu Corporation, Kyoto, Japan). This method, called the densitometric method, employs weighing a sample in air and in a liquid as well as determining the density of the liquid used at a given temperature. This method makes it possible to obtain the integral densities ρ_d of the samples (hereinafter also referred to as "densitometric density" or "integral density"). To accurately determine the density, samples of aluminum wires 80 mm long and weighing about 1.5 g were used, and distilled water was used as the liquid. The dependence of the density on the temperature of such a liquid is known with the required accuracy. So, the relative error $\delta\rho_d/\rho_d$ in determining the density did not exceed $1 \cdot 10^{-4}$. To study the surface-layer influence, the wires were investigated after different periods of service life, and the distribution of the change in density over the cross-section of the wires (density defect $\Delta\rho_{dL}/\rho_{dL}$) was obtained. Chemical etching was carried out in a 20% NaOH aqueous solution. During etching, the thickness of the removed layer was determined as [10]:

$$T_{\text{etch}} \approx R_0 \cdot \left(1 - \sqrt{\frac{m_i}{m_0}}\right), \quad (1)$$

where R_0 and m_0 are the radius and mass of the sample in the form of a cylindrical rod before polishing, respectively, and m_i is the mass of the sample after polishing the ith layer.

The layer density (ρ_{dL}) and the value of the density defect $\Delta\rho_{dL}/\rho_{dL}$ in polished layers were calculated with the formulas [10]:

$$\rho_{dL} = \frac{(m_0 - m_i) \cdot \rho_{di} \cdot \rho_{d0}}{m_0 \cdot \rho_{di} - m_i \cdot \rho_{d0}}, \quad (2)$$

$$\frac{\Delta\rho_{dL}}{\rho_{dL}} = \frac{\rho_{dL} - \rho_{d0}}{\rho_{dL}}, \quad (3)$$

where ρ_{d0} is the density of the sample before polishing, while ρ_{di} is that after polishing of the ith layer.

The thickness of each removed layer ranged from a few microns near the wire surface to ~5 μm after the thickness of the removed layer reached $T_{etch} \approx 20$ μm. This provided information on the distribution of the density defect over the cross-section of the sample with a decrease in its radius.

2.2.3. Experimental Details of XRD Measurements and Analysis

Samples for XRD measurements were cut in the form of a cylinder ~25 mm long and ~3–3.2 mm (Table 2) in diameter and glued onto a low-background single-crystal Si(119) holder. XRD measurements were carried out on a D2 Phaser powder diffractometer (Bruker AXS, Karlsruhe, Germany) in the Bragg–Brentano θ-θ geometry with Cu-$K_{\alpha1,2}$ radiation from a copper anode after a K_{β}-filter in the form of a Ni-foil and a linear position-sensitive semiconductor X-ray detector LYNXEYE (Bruker AXS, Karlsruhe, Germany) with an opening angle of 5°. The temperature in the chamber where the holder with the sample was installed was kept equal to 314 ± 1 K during the measurements. The measurements were carried out using the method of θ-2θ scanning in the range of diffraction angles 2θ = 6–141° with a step of $\Delta 2\theta_{step}$ = 0.02°. The regimes were used either with sample rotation around an axis coinciding with the axis of the diffractometer goniometer or without rotation, when the X-ray beam was incident on the long side of a cylinder-like sample (see illustration in [10]). Owing to XRD patterns obtained with rotation, the influence of the averaged effects of preferential orientation was estimated. XRD patterns measured without rotation were used to obtain the parameters of XRD reflections (observed Bragg angle ($2\theta_{obs}$), observed full width at half-maximum intensity ($FWHM_{obs}$), maximum (I_{max}), and integral (I_{int}) intensities of a reflection) for quantitative analysis of the structural and microstructural parameters of the Al material of the studied samples. Using the *EVA* program [40], the XRD patterns were corrected for the contribution of the background and Cu-$K_{\alpha2}$ radiation, and the values of the above reflection parameters were determined. The obtained observed Bragg angles $2\theta_{obs}$ of the reflections were adjusted with the angular corrections $\Delta 2\theta_{zero}$ (detector zero shift) and $\Delta 2\theta_{displ}$ (displacement due to the mismatch of the sample surface with the focal plane of the diffractometer; see [10]), which were determined from additional XRD measurements of the samples, immersed to surface level in NaCl powder, certified with XRD powder standard Si640f (NIST, Gaithersburg, MD, USA). X-ray phase analysis was performed by means of the *EVA* program using the Powder Diffraction File-2 (PDF-2) powder database [41].

Using the Bragg angles $2\theta_B$ obtained after angular corrections and the Miller indices *hkl* of the reflections, the parameters a of the cubic unit cell of the wire Al material were calculated, corresponding to each individual XRD reflection. In this case, the estimated standard deviation (e.s.d.) δa of the parameter a was calculated analytically from $\delta 2\theta_B$ (e.s.d. of the Bragg angle), which was taken as half of the step $\Delta 2\theta_{step}$ with which the XRD pattern was measured (i.e., $\delta 2\theta_B$ = 0.01°). The average value of the parameter a was determined using the *Celsiz* program [42], which performs refinement using the least squares method.

The density of the Al material ρ_x (g/cm^3) determined from XRD data (hereinafter "X-ray density" or "XRD density") was calculated from the ratio of the mass of the unit cell of the crystalline Al phase to the volume of the unit cell as:

$$\rho_x = \frac{C_{a.m.u.} Z \cdot A_r}{V_{cell} \cdot 10^{-24}}, \quad (4)$$

where V_{cell} (Å^3) = a^3 is the volume of the Al cubic unit cell, Z = 4 is the number of formula units in the Al unit cell, A_r = 26.9815384 a.m.u. (atomic mass units) is the tabular value of the atomic mass of Al, and $C_{a.m.u.}$ = 1.660539066 10^{-24} g/a.m.u. is the conversion factor from a.m.u. to grams.

Using the expression:

$$\frac{\Delta p}{p} = \frac{p - p_0}{p}, \tag{5}$$

the defect of the unit cell parameter, $\Delta a/a$, and the X-ray density defect, $\Delta \rho_x/\rho_x$ were also calculated. In these calculations, in expression (5), $p = a$ or ρ_x for any reflection from the sample (either the average value of the unit cell parameter or density of a sample), and p_0 is either the value of the unit cell parameter a_0 or the density ρ_{x0} of the Al material in the bulk of a new sample (0 years of service).

The determination of microstructural parameters (average size D of coherent scattering regions, which are also called crystallites, and absolute value of average microstrain ε_s in them) was carried out using the *SizeCr* program [43] in accordance with the method described in [10,37] and analyzed in detail in [43]. Using the *SizeCr* program, based on the observed ratio $FWHM_{obs}/B_{int}$ (where $B_{int} = I_{int}/I_{max}$ is the integral width of a reflection), the reflection's type (Gaussian, or Lorentzian, or pseudo-Voigt (pV)) was determined for each reflection, and, depending on the type of reflection, a correction of $FWHM_{obs}$ for instrumental broadening was made, see [10,37,43]. As a rule, the observed XRD reflections were pV-type ones. That is, they were characterized by the ratio $0.637 < FWHM_{obs}/B_{int} < 0.939$. The values $FWHM_{corr}$ ($FWHM_{obs}$ corrected for instrumental broadening according to reflection type) were used to calculate the microstructural parameters. The microstructural characteristics of Al wire materials (average crystallite size D and absolute value of average microstrain ε_s) were determined by the graphical methods WHP (Williamson–Hall plot) [44] and SSP (Strain–Size plot) [45] adapted to the observed pV type of XRD reflections. The points of the WHP and SSP graphs, corresponding to every XRD reflection *hkl*, were calculated using the *SizeCr* program, utilizing the coefficients $K_{Scherrer} = 0.94$ and $K_{strain} = 4$ of the Scherrer and Wilson–Stokes equations, connecting the corresponding $FWHM_{corr}$ components with crystallite size D^{hkl} and microstrain $\varepsilon_s{}^{hkl}$ values, respectively (see [43–45]). In the absence of microstrains (model $\varepsilon_s = 0$), the *SizeCr* program calculated the sizes $D^{hkl}{}_0$ of crystallites for each observed reflection as well as the mean-square-root value D_0 for all reflections. When setting a fixed value of D^{hkl}, the values of the microstrain $\varepsilon_s{}^{hkl}$ corresponding to each observed XRD reflection *hkl* were calculated.

The penetration depth T^{hkl}_{pen} of X-rays for each reflection with Miller indices *hkl* (see [46] and illustration in [10]) was estimated as:

$$T^{hkl}_{pen} = \frac{\sin(\theta B)}{2 \cdot \mu_l \cdot \rho_x}, \tag{6}$$

where μ_l = 48.657 cm^2/g is the linear absorption coefficient of Al in the case of Cu-$K_{\alpha 1}$ radiation (after correcting for the Cu-$K_{\alpha 2}$ contribution), and θ_B is half of the $2\theta_B$ Bragg angle (after accounting for angular corrections). The structural and microstructural characteristics of the sample material, which were calculated for each reflection with Miller indices *hkl*, correspond to the material's state averaged along the crystallographic direction [*hkl*] over the volume of the near-surface layer with a thickness equal to the penetration depth T^{hkl}_{pen}. As a result, analysis of different reflections detected at different Bragg angles $2\theta_B$ makes it possible to obtain profiles of changes in structural and microstructural parameters with depth from the surface.

2.2.4. Experimental Details of Acoustic Measurements

Acoustic resonant methods are based on the analysis of steady oscillations of samples in the form of rods, the length of which is much greater than their transverse dimensions. The determination of elastic and microplastic properties (elastic modulus E and decrement δ of elastic vibrations and microplastic deformation diagrams $\sigma(\varepsilon_d)$) makes it possible to study microprocesses that can take place in samples when external conditions change.

Acoustic properties were studied with the composite oscillator technique applied to aluminum samples in the form of rods ~25 mm long. This length corresponds to the resonant frequency of a polycrystalline aluminum sample of about 100 kHz. Young's modulus E (also called the "modulus of elasticity" or "elastic modulus") and elastic vibration decrement δ were measured in the range of amplitudes of vibrational deformation ε from $\sim 1 \cdot 10^{-6}$ to $\sim 3 \cdot 10^{-4}$. The modulus of elasticity was determined using the formula [47]:

$$E = 4\rho \cdot l^2 \cdot f^2, \quad (7)$$

where l is the sample length, ρ is its density (for calculations of E, the density ρ_d determined by the densitometric method was used), and f is the oscillation frequency of the rod-shaped wire samples close to 100 kHz. Upon studying the $E(\varepsilon)$ dependence, microplastic deformation diagrams $\sigma(\varepsilon_d)$ were constructed, which made it possible to evaluate the properties of the material in the "stress—inelastic strain" coordinates customary in mechanical tests, when the value of the amplitudes of vibrational stresses $\sigma = E \cdot \varepsilon$ (Hooke's law) is plotted along the ordinate axis, and the nonlinear inelastic deformation $\varepsilon_d = \varepsilon \cdot (\Delta E/E)_h$ is plotted along the abscissa axis, where $(\Delta E/E)_h = (E - E_i)/E_i$ is the amplitude-dependent defect of Young's modulus. The quantities E_i and δ_i measured at small amplitudes ε when both E and δ are yet independent of ε are called amplitude-independent Young's modulus and amplitude-independent decrement of elastic vibrations, respectively.

Note that the measurements were performed at moderate amplitudes, i.e., the dislocation structure of the material did not change.

3. Results

3.1. SEM and EDX Results

SEM and EDX results for the N8 and N7 outer wires with service lifetimes of 10 and 18 years, respectively, taken from A50-type cables (i.e., without steel cores) as well as for new (0 years of service) wires of an AC50-type cable were discussed earlier in [10,37]. As in the case of A50-type wires, SEM images have shown that the prepared outer-wire cross-sections of AC50-type steel core cables are quite suitable for obtaining the EDX spectra and EBSD mapping. The EDX spectra of cross-sectional surfaces from N2-2 and N6 AC50-type samples with service lives of 8 and 20 years, close to the service lives of wires of A50-type cables mentioned above, are shown in Figure 1. According to the results of the analysis of EDX spectra, the chemical composition of the elements of wire samples of both types of different service life from 0 to 20 years has demonstrated no significant difference and is consistent with the composition of the certificate data (Table 3). As expected, the Al content of ~99 wt% prevails in all samples. The EDX spectra show peaks corresponding to the Fe and Si atoms, which, according to the certificate data (Table 3), are the most abundant among impurity atoms, as well as peaks attributed to O atoms due to Al oxidization [10,37]. The EDX spectra also show the presence of peaks of extraneous elements, namely, Ar as a result of argon ion polishing of the surface of the samples and Ni from the walls of the microscope chamber.

3.2. EBSD Results

Distribution EBSD maps of Euler orientation angles φ_1, Φ, and φ_2 (correspondingly, angles of intrinsic rotation, nutation, and precession, see Ref. [48] for definition) with superimposed grain boundaries are shown in Figure 2. As grains, regions of the crystal structure were considered the misorientation within which did not exceed 2°. The colors of the grains on the map correspond to a combination of Euler angles that describe the orientation of the crystal lattice in a given grain.

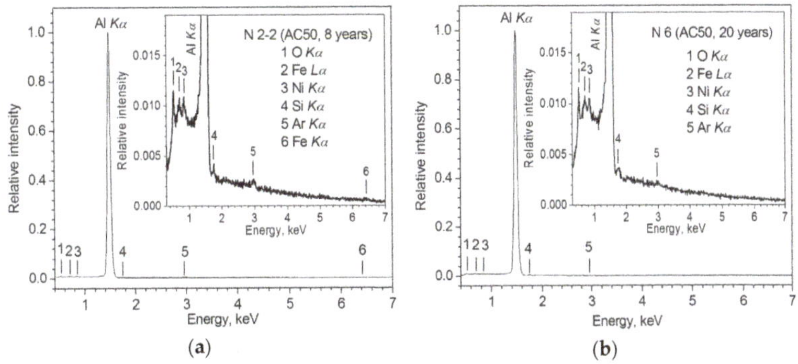

Figure 1. EDX spectra of the AC50 samples (**a**) N2-2 (8 years of service life), (**b**) N6 (20 years of service life).

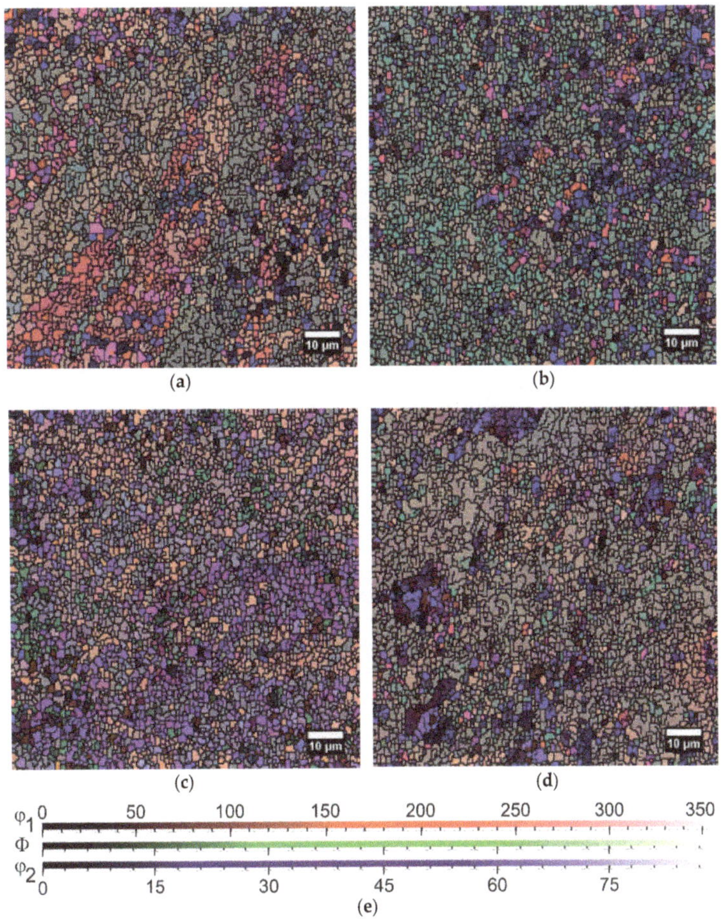

Figure 2. Distribution maps of Euler angles at the center (**a**,**b**) and at the edge (**b**,**c**) for cross-sections of AC50 samples after (**a**,**c**) 8 (sample N2-2) and (**b**,**d**) 20 (N6) years of service life. The legend for maps (**a**–**d**) is shown in (**e**). Scale for the Euler angle Φ is the same as for φ_2.

As a result of the analysis of maps of the distribution of Euler orientation angles in the center and at the edge of the studied thin sections of the cross-sections of the AC50-type samples after operation for 8–20 years (N2-2 and 6), we have constructed histograms of the distribution of grains by size (Figure 3), dependences of the area occupied by grains on their size (Figure 4a), grain-aspect-ratio distribution histogram that is ratio of the maximum grain size to the minimum (Figure 5), and distribution of grain boundary misorientation angle (Figure 6). For the non-exploited AC50 sample (N5, 0 years of service life) and the A50 sample N8 after 10 years of operation, the distribution maps of the Euler orientation angles in the center and at the edge and the distribution histograms constructed for them are presented in [10], while the dependence of the area occupied by grains vs. grain size for these A50 type specimens is plotted in Figure 4b. For the central regions of wire cross-sections of A50-type long-service life (up to 62 years) cables, similar EBSD studies were carried out in [37].

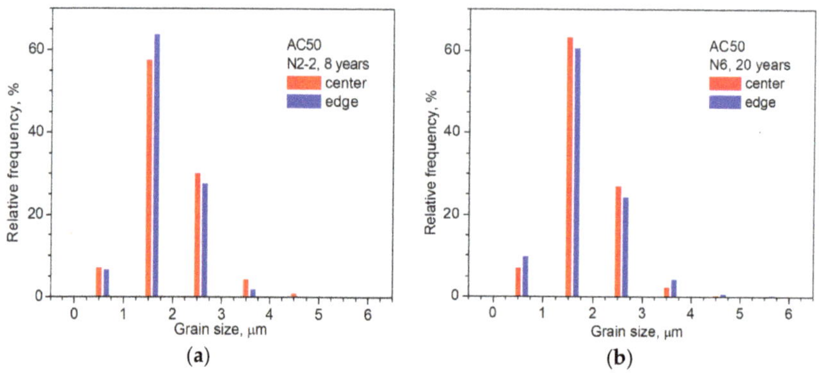

Figure 3. Histograms of grain size distribution for the centers and edges of cross-sections of outer wires from the AC50 cables after (**a**) 8 years of service (N2-2) and (**b**) 20 years of service (N6). For better visualization, the histograms for sample edges are shifted along the abscissa axes in (**a**,**b**).

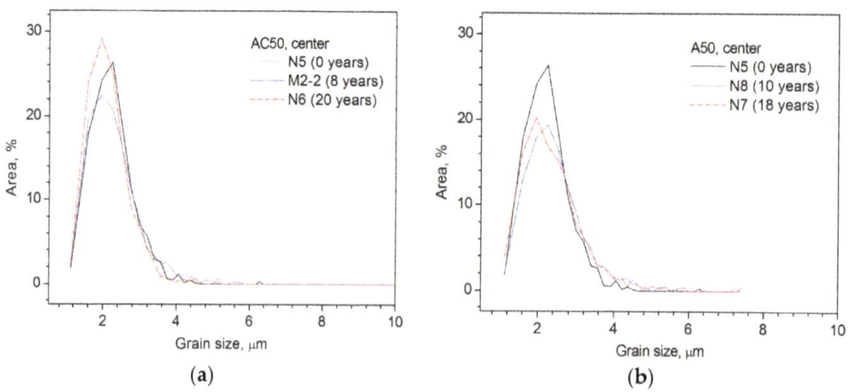

Figure 4. Dependencies of the relative area occupied by grains vs. their size for the cross-sections of outer wires from the (**a**) AC50 and (**b**) A50 cables at the centers of the samples. All dependencies are given on the same scale for comparison. Numbers of samples and their service lives are indicated in the legends of the Figures.

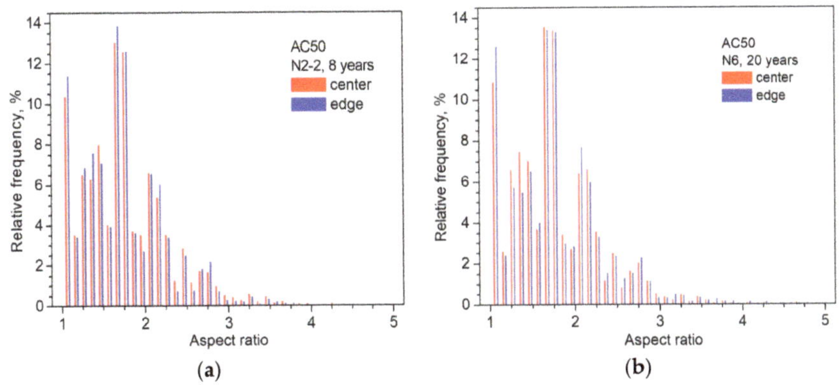

Figure 5. Histograms of grain aspect ratio distribution for the centers and edges of cross-sections of outer wires from the AC50 cables after (**a**) 8 years of service (N2-2) and (**b**) 20 years of service (N6). For better visualization, the histograms for sample edges are shifted along the abscissa axes in (**a**,**b**).

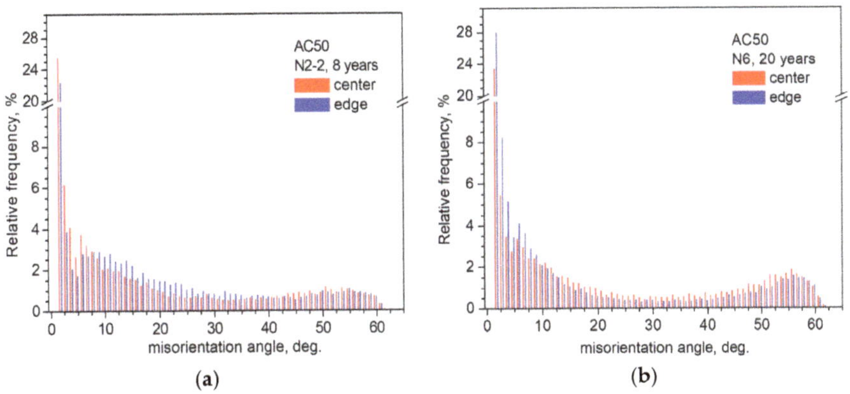

Figure 6. Histograms of grain boundary misorientation angle distribution for the centers and edges of cross-sections of outer wires from the AC50 cables after (**a**) 8 years of service (N2-2) and (**b**) 20 years of service (N6). For better visualization, the histograms constructed for the edges of the samples are shifted along the abscissa axes in (**a**,**b**).

One can see from Figure 3a,b that, as in the case of A50-type cables [10], the general nature of the grain size distribution in the center and at the edge of the cross-section of samples of different service life periods does not change, and in all cases, the most frequently occurring grain sizes are of ~2 μm. At the same time, in the entire size range from 0.5 to 6 μm, insignificant differences are observed in the relative frequency of occurrence of grains in the center and near the edge of the samples.

With an increase in the service life of 0 to 18–20 years, the relative area occupied by grains with sizes of ~2 μm changes slightly (within ~±5% of ~26% in N5 wire without operation) in the bulk of AC50 and A50 wires (Figure 4a,b). At the same time, the area occupied by grains with sizes 4–6 μm increases slightly from ~0.5–1% to ~1–2%.

As for the A50-type samples [10,37], the grain aspect ratio distribution histograms in the AC50 type samples (Figure 5a,b) weakly vary depending on the period of operation and position, remaining practically unchanged in the center and at the edges of the cross-sections of the wires and after various periods of operation. This invariance of the aspect ratio shows that the grain shape in the A50- and AC50-type samples is practically the same inside the cables and in their near-surface layers and remains the same during operation.

Recently [37], angle-distribution histograms of grain-boundary misorientation, which were constructed for the bulk regions of A50 wires (centers of cross-sections) with different service lives, have shown that the number of high-angle grain boundaries with misorientation in the range of 50–60° decreases when the service life is long (down to 0.5–0.9% in the sample with a service life of 62 years from 1.5–2.5% in the unexploited sample N5 (0 years)). This decrease is accompanied by an increase in the number of low-angle boundaries with a misorientation angle of less than 15° (up to 1.9–28% in the sample with a service life of 62 years from 1.7–22% for N5 (0 years) wire). It has been suggested in [37] that such a distribution of the misorientation angle of grain boundaries is caused by the alignment of the crystal lattice of grains along one common direction. Meanwhile, at shorter service lives of A50 wires, a similar process of reducing the number of high-angle boundaries with a misorientation in the range of 50–60° is initially observed after 10 years of operation (sample N8, 0.4–0.8%) [37] in comparison with the non-operated cable (sample N5, 1.5–2.5%). With a further increase in the service life to 18 years (A50 sample N7), the number of high-angle boundaries with a misorientation of 50–60° increases again, up to 0.8–1.2%, although not recovering to the value in the N5 wire (0 years) [37]. This growth, respectively, can be associated with the destruction of grain alignment along the general direction.

The grain boundary misorientation angle distribution histograms for the bulk regions (centers of cross-sections) of the outer wires of AC50 cables (Figure 6 for samples with a service life of 8 and 20 years, see also [10] for a non-exploited sample) show an evolution similar to A50-type wires with comparable service lives up to the 18 years described above. Initially, with an increase in the service life from 0 years (sample N5 [10,37]) to 8 years (N2-2), there is a decrease in the number of high-angle boundaries with a misorientation in the range of 50–60°, although somewhat less than in the A50 wire after 10 years of operation (from 1.5–2.5% in sample N5 (service life of 0 years) down to 0.75–1.2% in sample N2-2 (8 years, AC50) and 0.4–0.8% in sample N8 (10 years, A50)), i.e., the grains probably tend to line up along one direction. With a further increase in the service life of AC50-type wires up to 20 years (sample N6), there is an increase in the number of high-angle boundaries with a misorientation of more than 15°, in particular, with a misorientation in the range of 50–60° to 1.1–1.8% (in comparison to 0.8–1.2% in the A50 wire N7 (service life of 18 years)), which presumably indicates an increase in the number of grains with an arbitrary orientation of the crystal lattice.

The grain boundary misorientation angle distribution histograms for near-surface layers of wires (edges of cross sections) of the outer wires of AC50 cables (for samples with a service life of 8 and 20 years, see Figure 6; for an unused sample, see [10]) are similar to the histograms obtained for the centers of cross-sections. At the edges, there is only a slight decrease in the number of high-angle boundaries with a misorientation in the range of 50–60° compared to the centers of the sections (respectively, for the edge and center, 1.0–1.9% and 1.5–2.5% in sample N5 (0 years), 0.75–1.0% and 0.75–1.2% in sample N2-2 (8 years), and 0.95–1.6% and 1.1–1.8% in sample N6 (20 years)), i.e., the misorientation of the grains practically does not change either in the bulk or in the near-surface layers of the AC50 wires. A completely opposite situation is observed in the wires of the A50-type wires. In addition to the drop in high-angle grain boundaries after 10 years of service, as discussed above, the absence of a rigid steel core in A50 cables results in a noticeable difference in histograms obtained from the center and from the edge of the cross-section. At the edge of the section, i.e., in the near-surface layer of the A50 wires, the number of high-angle boundaries with misorientations in the range of 50–60° is noticeably reduced (respectively, ~0.1% and 0.4–0.8% at the edge and in the center of the cross-section of sample N8 (10 years) [10]), which probably indicates that in the near-surface layer of the A50 wires, the crystal lattice of grains tends to align along one common direction more strongly than in the bulk regions of the wires.

Thus, the analysis of EBSD maps obtained from the centers and edges of the cross-sections of the samples (Figure 2 and Refs. [10,37]) shows that the distribution histograms of the misorientation angle of the grain boundaries (Figure 6) are the most sensitive to changes in the microstructure, whereas the grain-size distribution histograms (Figure 3) are less sensitive. Compared with AC50 wires, in wires taken from A50 type cables without a steel core in a layer at a distance of ~150 μm from the surface, there is a tendency towards greater grain alignment. The presence of a steel core in AC50-type cables leads to a significant change in grain misorientation in AC50-type wires compared to A50 wires after a comparable service life of up to 20 years in power lines. Grain misorientation in near-surface layers (near the edge of the wire cross-section) and in the bulk of AC50 wires varies slightly for samples of the same service life, although depending on the duration of operation in a manner similar to the A50-type wires (initially, there is a tendency to align the crystal lattices of grains with a service life of 8 to 10 years, with an increase in the duration of operation up to 18–20 years, the arbitrariness of grain orientation increases). In the wires of both types of cables, the aspect ratios of grains in near-surface layers and in the bulk of the wires practically do not change (Figure 5), indicating the invariance of the grain shape in both regions of the samples.

3.3. Results of Densitometric Measurements

Systematic determination of the density ρ_d of wires from used A50- and AC50-type cables of approximately the same service life, carried out by densitometry, has shown that the density of the aluminum part of the AC50 cable is somewhat higher than that of the A50 one, despite the fact that the service life of AC50 wires is somewhat longer. For instance, the integral density ρ_d of aluminum wire of sample N6 (AC50) is 0.05% higher than that of N7 (A50), with a service life of 20 and 18 years, respectively.

To elucidate the possible reasons for this difference, we determined the density ρ_{dL} of the near-surface layer of the samples under study (see Formula (2)). Figure 7 shows examples of such distributions of the true density defect (cf. Formula (3), in %) over the cross-section for the studied samples of A50 wires (from [10]) and for AC50 ones after 18 and 20 years of operation, respectively, where T_{etch} is the thickness of the removed layer, determined by Formula (1), and $\Delta\rho_{dL}/\rho_{dL}$ is the value of the defect in the density of the near-surface layer relative to the density of the entire cross-section of the wire. As follows from the analysis of dependences, the absolute value of the density defect decreases as much as the near-surface layer is removed, i.e., the deviation of the integral density ρ_{dL} of the surface layer decreases with respect to the density in the bulk of the wire. The main change in the density defect in wires of both types is observed in a relatively thin layer about 10 μm thick, that is, the lowest value of the layer density is observed in a narrow near-surface layer. Such a change in density indicates that defects of a void nature (nano and micropores, microcracks) are concentrated just in this narrow near-surface layer. After the removal of this layer (~10 μm) in both types of wires, an insignificant decrease in the absolute value of $\Delta\rho_{dL}/\rho_{dL}$ continues until a layer of ~30 μm is removed from the surface.

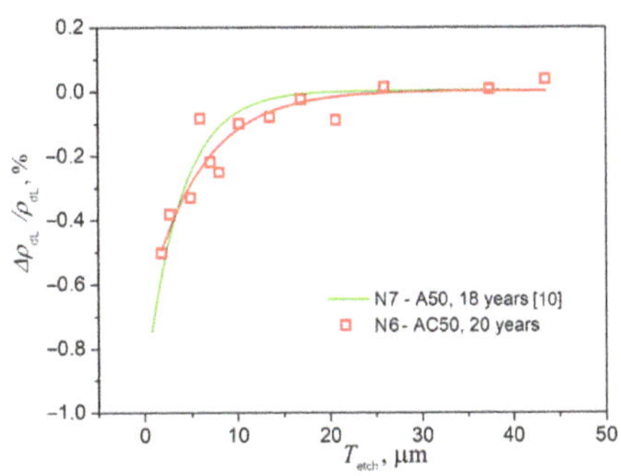

Figure 7. Dependence of the value of the density defect in the near-surface layer of samples N7 (A50, service life of 18 years) [10] and N6 (AC50, service life of 20 years).

Thus, within the statistical spread of the experimental values $\Delta\rho_{dL}/\rho_{dL}$, both types of wires show the presence of two characteristic values of the thickness of near-surface defect layers (NSDLs), in which there appears to be a noticeable decrease in density compared to the bulk of the wires. In narrow near-surface layers less than ~10 μm thick, the density is the lowest due to the high concentration of void defects and demonstrates the greatest changes with distance from the surface in depth, and at depths from the surface of ~10 to ~30 μm, it weakly increases until stabilization, apparently due to the reduction in the number of defects. It is also worth noting that the absolute value of the density defect of the near-surface layer in wires from cables with a steel core (AC50) is somewhat less (by ~0.1–0.2%) than for A50 wires, and, as follows from Figure 7, this difference is most noticeable in a narrow near-surface layer ~5–10 μm thick. The smaller absolute value of the NSDL-density defect in AC50 wires probably reflects a smaller number of void defects in these wires due to the softening effect of the steel core.

Additionally, the densitometry method was used to study the relative change in the integral density (defect $\Delta\rho_d/\rho_d$ of the integral density) of A50 samples in the range from 0 to 54 years (Figure 8). The values of the integral density $\Delta\rho_d$ were determined for the wires after 0, 8, 18, 35, and 54 years of operation (see Table 2 for sample numbers). It has been established that the most dramatic change in the integral density ρ_{dL} is observed in the range from 0 to 20 years of operation. An increase in service life of more than 20 years also leads to a further slight decrease in the density of the wire material. However, when analyzing the data, it should also be taken into account that, as was shown in [10,37], a noticeable amount of Al_2O_3 oxides are formed on the surface of the wires, the density of which (~3.7 g/cm^3) is significantly higher than the density of aluminum (~2.7 g/cm^3). That is why it can be assumed that the true absolute value of the integral defect of aluminum wire without the aluminum oxide contribution is greater due to the formation of void microdefects, i.e., with a service life of more than 20 years, the density fitting curve without taking into account the contribution of aluminum oxides will go lower in the graph of Figure 8.

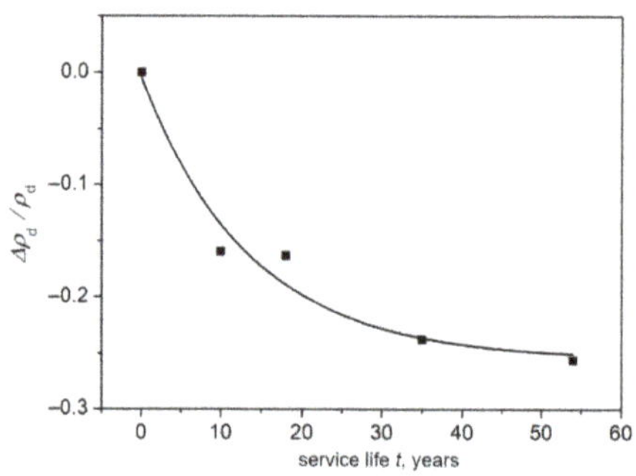

Figure 8. Dependence of the integral value of the density defect in A50 wires on the service life during operation up to 54 years.

3.4. Results of XRD Investigations

Figure 9a–c presents the measured XRD patterns of AC50 samples with service lives from 0 to 20 years. For A50-cable wires with comparable service lifetimes of up to 18 years, the XRD patterns can be found in [10,37]. The XRD reflections of the cubic Al crystalline phase at angular 2θ positions corresponding to Bragg angles are schematically shown in Figure 9d by vertical bars whose heights are proportional to the intensities of the reflections according to the PDF-2 card 01-073-9843 [49].

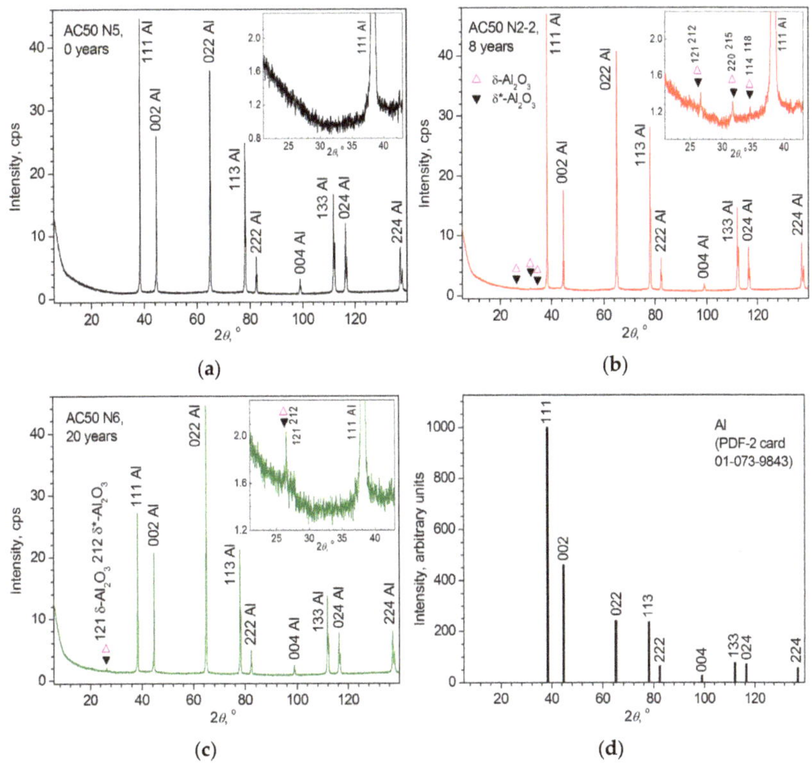

Figure 9. XRD patterns detected from wires of AC50 cables after (**a**) 0 (N5), (**b**) 8 (N2-2), and (**c**) 20 (N6) years of exploitation. (**d**) schematically presents the XRD pattern of Al according to PDF-2 card 01-073-9843 (the height of each bar represents the intensity of the corresponding XRD reflection according to the PDF-2 card). Insets in (**a**–**c**) present the 2θ diffraction angle range 21–43° on a larger scale, exhibiting the weak reflections attributed to δ*- and δ-Al_2O_3. The Bragg angle positions of the δ*- and δ-Al_2O_3 reflections are shown according to PDF-2 cards 00-056-1186 and 00-046-1215, respectively, using different symbols. The Miller indices *hkl* of the observed reflections are indicated.

As in the case of wires from the A50-type cables [37], the XRD patterns of the AC50 wires contain all the Al reflections with an enlarged 022 reflection intensity that indicates the polycrystalline nature of the wires with a noticeable effect of preferential orientation along the [011] crystallographic direction (see Figure 9a–d for comparison). As for the A50 wires, the intensity of the 111 reflection for the AC50 wires with a service life of up to 8 years remains the highest despite the presence of a preferential orientation along the [011] direction (Figure 9a,b), whereas the preferential orientation along [011] develops more strongly for wires after 20 years of operation, the 022-reflection intensity becomes the highest.

Quantitative ratios of the intensities of reflections 022 to 111 and 002 to 022, which characterize the preferential orientation of Al wires from AC50 and A50 cables, are given in Table 4 and are graphically shown in Figure 10. As one can see, both types of wires show the identical tendency to grow for the preferential orientation along the [011] crystallographic direction. Nevertheless, in AC50 type wires (with a steel core), the preferential orientation is more pronounced at all service lives, although the difference in preferential orientation between the AC50 and A50 wires decreases as the service life increases to 20 years. At the same time, the ratio of maximum intensities of XRD Al 022 and 111 reflections ($I_{max}^{022}/I_{max}^{111}$) increases by ≈2.2 times in A50 wires with an increase in service life from 8 to 20 years compared with ≈2.0 times in AC50 wires with an increase in service life from 10 to 18 years. Such a decrease in the rate of amplification of the effects of preferential orientation can be attributed to the stabilizing effect of the steel core in AC50 wires. In addition, it is likely that the stronger influence of the effects of preferential orientation in AC50 wires is due to the fact that in the starting state of AC50 wire (0 years of operation), the preferential orientation along [011] is already more pronounced than in A50 wire without operation, which is probably due to the peculiarity of the technological process in the production of aluminum wire of various batches for A50 and AC50 wires.

Table 4. XRD analysis results of A50 and AC50 wires (temperature of measurements is T = 314 ± 1 K). Sample numbers and sample service life (years) are given according to Table 2. Table data from PDF-2 for crystalline Al phase are shown for comparison.

		Observed Preferential Orientation		WHP	SSP
Sample *N*/Years	[*hkl*]	$I_{max}^{022}/I_{max}^{111}$, % $I_{max}^{002}/I_{max}^{022}$, %	a, Å/ρ_x, g/cm^3 D_0, nm ($\varepsilon_s = 0$)	D, nm ε_s, %	D, nm ε_s, %
			A50 type		
5-2/0	[011]	42.0(3) 76.4(7)	4.05026(12)/2.6973(2) 109(16)	111(14) 0.010(14)	109(16)0
8/10	[011]	69.0(4) 66.1(4)	4.0515(5)/2.6949(11) 139(16)	302(54) 0.031(2)	298(26) 0.031(2)
7/18	[011]	153.8(1.6) 55.6(6)	4.0525(9)/2.6927(17) 126(33)	246(55) 0.033(3)	252(32) 0.034(3)
			AC50 type		
5/0	[011]	75.8(9) 68.4(9)	4.05032(10)/2.6972(2) 138(16)	141(17) 0.007(11)	138(16) 0
2-2/8	[011]	90.9(6) 38.8(4)	4.0511(4)/2.6956(7) 136(29)	212(28) 0.026(3)	219(19) 0.026(3)
6/20	[011]	180.7(2.0) 42.7(5)	4.0525(8)/2.6929(15) 120(23)	187(27) 0.029(3)	167(13) 0.025(3)
			Table data for crystalline Al powder		
PDF-2 card 01-071-4008, [50] T = 312.3 K	no	- -	4.050694/2.69642 -	-	-
PDF-2 card 01-073-9843, [49] T = 298 K	no	24.0 191.7	4.04932(2)/2.6992(4) -	- -	- -

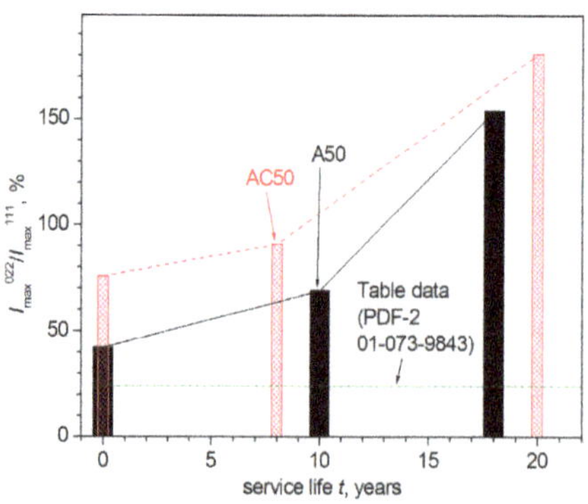

Figure 10. Dependence of the $I_{max}^{022}/I_{max}^{111}$ ratio on the service life t of the Al wires from the AC50 and A50 cables. Lines connecting the experimental points are guides to the eye only.

As in the case of the A50 wires [37], the XRD patterns of the AC50 wires after operation (Figure 9b,c) show the formation of very weak reflections attributed to δ*- and/or δ-modifications of Al_2O_3 [51,52] (δ-phase, space group $P4_12_12$ (92), PDF-2 card 00-056-1186; δ*-phase, space group *P*222 (16), PDF-2 card 00-046-1215). With a service life of up to 20 years, δ*-/δ-Al_2O_3 reflections in AC50 wires after operation develop more strongly compared to those in A50 ones (without a steel core), apparently due to the oxidizing effect of the steel core, which is enhanced by possible damage to aluminum wires when rubbing against the steel core. Whereas the ratios *q* of the integrated intensity I_{int} of the 121 δ*- (and/or 212 δ-) Al_2O_3 reflection to I_{int} of the Al reflection with the highest intensities (with Miller indices *hkl* = 111 for service life up to 10 years and 022 after 18 years) in A50 wires are, respectively, 0.21(5)% and 0.5(1)% for samples N8 (service life of 10 years) and N7 (18 years), this value *q* amounts to 0.53(3)% and 0.9(1)% for the AC50 wires N2-2 (10 years) and N6 (20 years), respectively, see Figure 11.

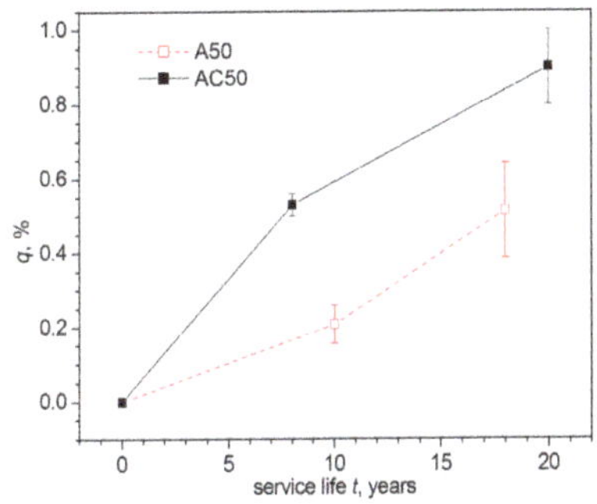

Figure 11. The fraction *q* of the Al_2O_3 crystalline phases in wires from A50 and AC50 cables with an exploitation duration of up to 20 years, which is estimated as $q = I_{int}^{Al_2O_3}/I_{int}^{Al} \cdot 100\%$, where $I_{int}^{Al_2O_3}$ is the integral intensity of the 121 δ*-(212 δ-) Al_2O_3 reflection and I_{int}^{Al} is the integral intensity of the strongest Al reflection.

A thorough investigation of the XRD patterns measured from samples of various life time periods showed that reflections corresponding to the same Miller indices *hkl* are systematically shifted, which corresponds to a change in the cubic unit cell parameter of the wire Al material, even after applying the angular corrections for zero shift and displacement, as described in Section 2.2.3. As an example, Figure 12 shows a reflection with a Miller index of *hkl* = 133 for the studied A50 and AC50 wires with a service life of 0 to 18–20 years.

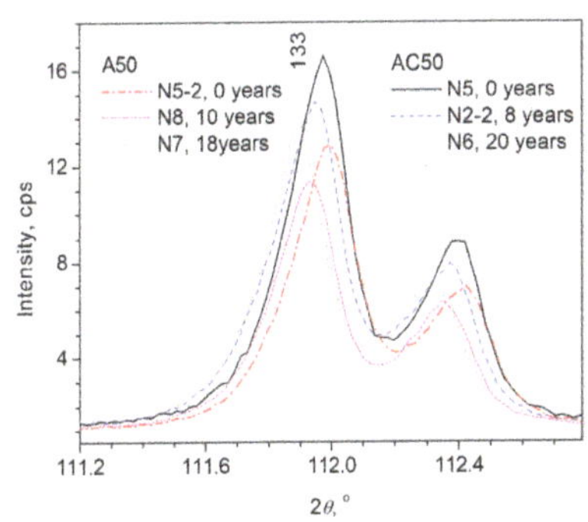

Figure 12. Part of the XRD patterns in the vicinity of the *hkl* = 133 reflection for several A50 and AC50 samples.

Cubic unit cell parameters a of the Al material of the A50 and AC50 wires, averaged over all observed reflections and calculated by the *Celsiz* [42] program using the least-squares method, are presented in Table 4 and graphically shown in Figure 13a against the service life duration. The corresponding X-ray densities ρ_X and density defects $\Delta\rho_X/\rho_X$ of the Al material of A50 and AC50 wires, which are calculated with Formulas (4) and (5), respectively, are shown in Table 4 and Figure 13b.

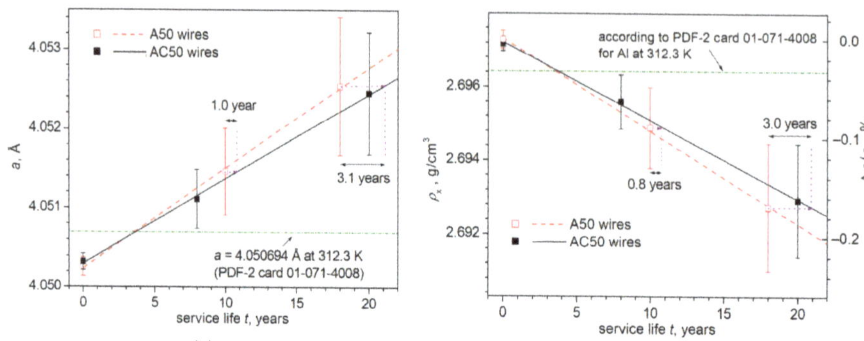

(**a**) (**b**)

Figure 13. The dependences of (**a**) the Al cubic crystal unit cell parameter a and (**b**) density ρ_X calculated from the XRD data on the service life t of the Al wires of overhead power transmission lines. At the right axis of (**b**), the scale of the corresponding density defect (reduction of density) $\Delta\rho_X/\rho_X$ (where $\Delta\rho_X = \rho_X - \rho_{X\,0\,years}$, $\rho_{X\,0\,years}$ is the mean density of the A50 wire N5-2 (service life of 0 years)) of the Al wires is shown. The horizontal line in (**a**) indicates the tabulated value of the Al cubic unit cell parameter a according to the PDF-2 card 01-071-4008 [50] and that in (**b**) exhibits the corresponding calculated ρ_X value. The lines running through the experimental points are guides to the eye only. In the graphs (**a**,**b**), the duration of the delay of changes of the a and ρ_X parameter values for A50 and AC50 wires are shown for two service life points.

As one can see from Table 4 and Figure 13a,b, the unit cell parameter and density Al of the material of the wires from the A50 and AC50 cables in the initial state (service life of 0 years) are almost the same (within e.s.d. to within the fourth decimal place, namely, a = 4.05026(12) Å and 4.05032(10) Å and ρ_X = 2.6973(2) g/cm^3 and 2.6972(2) g/cm^3, respectively, at an XRD measurement temperature of 314 ± 1 K). The values of these parameters are noticeably differ from the tabulated Al values (a = 4.05069 Å and ρ_X = 2.6964 g/cm^3) at a temperature of 312.3 K, which is close to the temperature of the XRD measurements in this work. As previously noted [37], a decrease in the average unit cell parameter and, accordingly, an increase in the Al density of the wire material can be associated with the incorporation of a small number of Fe and Si atoms into the Al structure, which, according to the manufacturer's passport [39] and EDX results (Figure 1a,b and Ref. [37]), are present in the wire composition in amounts up to 0.20 wt.% and 0.08 wt.% (see Table 3), respectively.

When the service life t of cables of both types in overhead power lines increases, both the cubic unit cell parameter a and the X-ray density ρ_X of the wire Al material alter almost linearly over the studied time intervals up to 20 years (Figure 13a,b), with lattice parameter a increasing and ρ_X decreasing for wires from cables of both types. Although the nature of the tendencies of changes in the structure parameters of wires from AAAC (A50) and ACSR (AC50) cables over time is the same, yet the rate of temporal change of the characteristics of A50 and AC50 wires is different. The presence of a steel core in AC50 cables leads to a decrease in the stretching rate of the unit cell parameter a of the Al wire material from 1.26(4) $\cdot$ 10^{-4} Å/year in A50 wires down to 1.07(3) $\cdot$ 10^{-4} Å/year in AC50 wires with a service life of up to 20 years. At the same time, the rate of decrease in the density ρ_X of the A50 and AC50 wires (i.e., the rate of degradation of Al wire due to the formation of void defects) decreases in absolute value from $-2.52(8) \cdot 10^{-4}$ g/cm^3/year to $-2.13(7) \cdot 10^{-4}$ g/cm^3/year, respectively. As a result, after about 11 years of service, the unit cell parameter a and density ρ_x of the Al material of AC50 type wires are the same as in Al wires from A50 type cables after 10 years. With a long service life of ~20 years of operation, the gain for AC50 wires is already ~3 years (see Figure 13a,b), i.e., the structure parameters of AC50-wire Al material after about 23 years of service are the same as those of A50-wire Al material after 20 years. Thus, the presence of a steel core in an ACSR (AC50) cable delays the structural degradation of the Al material of the cable wires.

The average sizes D_0 of crystallites, averaged over all D^{hkl} values obtained from the $FWHM_{corr}$ of each observed Al reflection with the Miller indices hkl according to the Scherrer equation in the framework of the model of zero contribution of microstrains (model $\varepsilon_s = 0$), reveal close values in wires of both types, namely, within D_0 = 109(16)–138(16) nm for unused wires (0 years of service life), 139(41)–136(29) nm and 126(33)–120(23) nm for A50/AC50 wires after 10/8 years and 18/20 years of operation, respectively (Table 4, see also Supplementary Materials (SM) Figure S1).

However, the XRD-data analysis by means of the WHP and SSP techniques (see WHP and SSP plots in Figure S2 in SM for AC50 samples and Ref. [37] for A50 wires) shows that, whereas average microstrain ε_s is zero or very close to zero for unused wires of both types (service life of 0 years) and the average crystallite sizes D are slightly larger in unused AC50 wires (in A50 (N5-2) wires, ε_s = 0.010(14)%, D = 111(14) nm and ε_s = 0%, D = 109(6) nm according to WHP and SSP, respectively, compared with ε_s = 0.007(11)%, D = 141(17) nm (WHP) and ε_s = 0%, D = 138(16) nm (SSP) in AC50 (N5) wires), microstrains in the wires after operation evolve accompanied with a notable increase in the size of crystallites (see Table 4 and Figure 14). Moreover, in wires from cables without a steel core (A50 type cables), absolute values of average microstrain ε_s and average crystallite sizes D grow after 10–18 years of operation, noticeably larger than in wires from AC50 cables (with a steel core) after 8–20 years of service in overhead power lines (cf. ε_s = 0.031(2)% and 0.034(3)%, D = 246(55) nm and 302(54) nm according to WHP and SSP techniques, respectively, for A50 wires and ε_s = 0.025(3)% and 0.029(3)%, D = 167(13) nm and 219(19) nm according to WHP and SSP techniques, correspondingly, for AC50 wires).

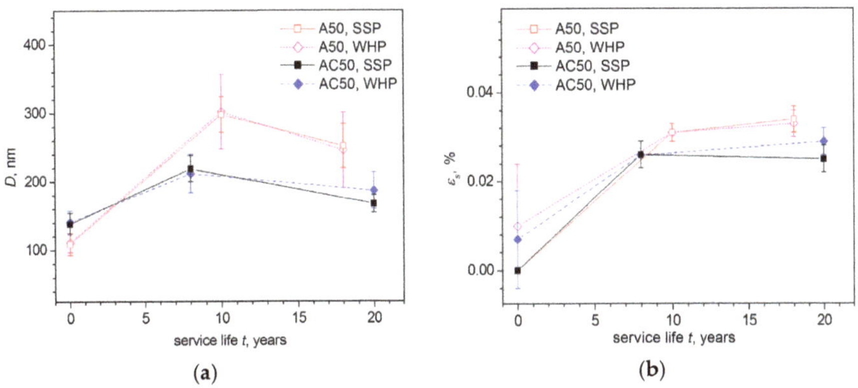

Figure 14. Comparison of dependences of (**a**) average crystallite size D and (**b**) absolute value of average microstrain ε_s, calculated by the WHP and SSP techniques, on the service-life duration t for Al wires from the cables of the AC50 and A50 types.

Thus, the use of a steel core in the AC50-type cable (the cross-sectional area of the aluminum components of the cable remains practically the same, ~50 mm^2, as in A50-type cable without the steel core) slows down the changes in the average parameters of not only the structure (the cubic unit cell parameter a of the Al material and its density ρ_x), but also of the microstructure (the average size D of crystallites and absolute value of average microstrain ε_s in them) of the Al material of cable wires.

It is to be expected that the steel core will affect not only the average characteristics of the wires but also their NSDLs. In order to investigate how the presence of a steel core in a cable affects the state of NSDLs of wires from that cable, XRD profiling of the structural and microstructural characteristics of AC50 wires in comparison with A50 wires was carried out by means of the method described in [10]. As noted in experimental Section 2.2.3, the essence of XRD profiling is that each Bragg angle $2\theta_B$ of the reflection with Miller indices hkl corresponds to the X-ray penetration depth T_{pen}^{hkl} in accordance with Formula (6), determined by the linear absorption coefficient μ_l and the X-ray density ρ_x of the material (Al in the case of the wires under study). If the samples were powder or there was no profile dependence on the depth of penetration of X-rays, then the structural and microstructural characteristics obtained from the analysis of reflections, i.e., at all depths from the sample surface, would have close values statistically disordered for different reflections within e.s.d.

Figure 15a,b presents a comparison of the distributions of the values of the cubic unit cell parameter a and the XRD mass density ρ_x of the wire Al material, which are calculated from the structural data with Formula (4), along with the penetration depth $T = T_{pen}^{hkl}$ of X-rays, which is estimated with the Formula (6) from the Bragg angles $2\theta_B$ of the observed reflections, for wires from AAAC (A50) and ACSR (AC50) cables.

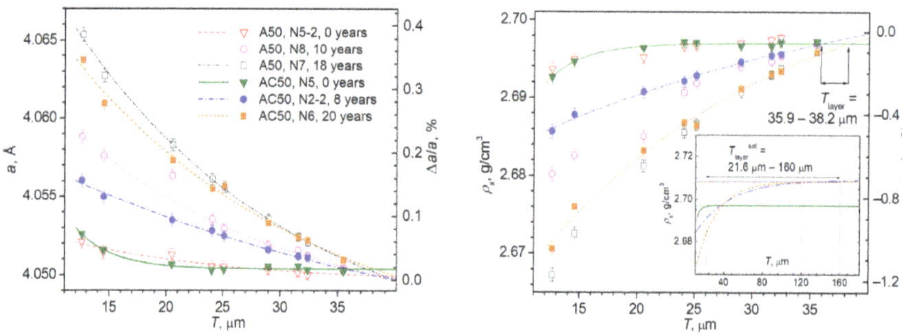

(a) (b)

Figure 15. Distribution of (**a**) the cubic unit cell parameter $a(T)$ and (**b**) the XRD mass density $\rho_x(T)$ of the wire Al material along the depth T from the surface of the A50 and AC50 wires. Samples are numbered according to Table 2 and their service lifetimes are shown in (**a**). The symbols for (**b**) are the same as shown in (**a**). At the right sides of (**a**,**b**), the axes are shown corresponding, respectively, to the lattice defect $\Delta a/a$ and the density defect $\Delta\rho_X/\rho_X$, which are estimated with respect to the bulk of the non-exploited sample of A50 type (N5-2, 0 years of operation). The approximation lines in (**a**,**b**) are drawn according to the exponential decay law. In (**b**), the thicknesses T_{layer} of the NSDL are estimated for AC50 wires from the intersection of the distribution curves for AC50 samples of different non-zero service lives with the curve corresponding to a non-used sample (N5-2, service life of 0 years). The inset in (**b**) presents the extrapolations of the distribution curves of AC50 wires to depths of 200 μm from the surface for samples. The shown estimates of the total thickness $T_{layer}{}^{sat}$ of the NSDL for wires are obtained from the intersection with tangents drawn at the points when the distribution curves reach a plateau (when ρ_x reaches the value $\rho_x{}^{sat}$ = 99.99% of the density estimated from the distribution curve at a depth of 200 μm).

As one can see from Figure 15a,b, the cubic unit cell parameter a and the density ρ_x of the Al material of wires with different service life lengths from cables of both types are not the same at different depths T from the surface of the wires, but show smooth dependences $a(T)$ and $\rho_x(T)$, which are fairly well described by approximation curves that correspond to the exponential decay law. To simplify the perception, the fitting curves are not shown in

Figure 15b for the experimental points corresponding to the A50 wires. An analogue of Figure 15b with all fitting curves is given in SM Figure S3.

Since the XRD density ρ_x is inversely proportional to the cube of the parameter a ($\rho_x \sim 1/a^3$, see Formula (4)), it is satisfactory to consider the dependences $\rho_x(T)$ alone. For wires from cables of both types, the density ρ_x is maximum and practically does not change at a sufficient distance from the surface in the bulk of the wires. Near the surface, ρ_x is minimum, which is the result of the formation of defects of a void nature in the near-surface layers [10] (i.e., of the formation of NSDLs).

The main changes in the structure and density of the wire Al material occur in the near-surface layer with a thickness of T_{layer} = 36.4–39.1 μm for A50 wires or a bit smaller, T_{layer} = 35.9–38.2 μm, for AC50 wires, which were determined from the intersection of the approximation curves $\rho_x(T)$ referring to wires after operation with the fitting curves of unused wires of the corresponding type (see Figure 15b for AC50 samples and Ref. [10] for A50 wires). The approximation curves $\rho_x(T)$ of the unoperated A50 (N5-2) and AC50 (N5) wires are very close. At depths T from ~25 μm for both wires, the density defect is $\Delta\rho_x/\rho_x \approx -0.05\%$ and practically does not change. When approaching the surface, only a slight drop in density is observed in the wires of both types ($\Delta\rho_x/\rho_x \approx -0.2\%$ at a depth of $T \approx 12.5$ μm). Thus, NSDL with almost the same negative density defect (i.e., with a decrease in density due, apparently, to defects of a void nature), which does not exceed ~0.2% in absolute value at a depth of $T \approx 12.5$ μm, already exists in wires from unused cables of both A50 and AC50 types.

As follows from the definition of the T_{layer} value (see above), in A50 and AC50 wires after operation at a depth equal to T_{layer} ~35.9–39.1 μm from the surface, the density defect is the same as in non-operated wires ($\Delta\rho_x/\rho_x \approx -0.05\%$). However, after operation, the density of the near-surface layers noticeably decreases the greater the closer the layer is to the surface, and in A50 wires (from AAAC cables without a steel core), this decrease in density is significantly greater than in AC50 wires. For example, in A50 wires at depths of $T \approx 12.5$ μm, the density defect reaches $\Delta\rho_x/\rho_x \approx -0.68\%$ and -1.17% for wires after 10 (sample N8) and 18 years (N7) of operation in comparison with $\Delta\rho_x/\rho_x \approx -0.48\%$ and -1.04% for AC50 wires after service life durations of 8 (sample N2-2) and 20 years (N6), respectively, i.e., the density defect in AC50 wires is less in absolute value by ~30% and ~10% after ~10 and ~20 years of operation. Thus, the use of a steel core in AC50-type ACSR cables leads to less degradation (smaller decrease in the density) of near-surface layers with a thickness of at least ~25 μm (as can be seen from Figure 15b, the largest deviations of the approximation curves $\rho_x(T)$ for A50- and AC50-type wires start at depths from the surface of less than ~25 μm).

Extrapolation of the approximation curves $\rho_x(T)$ to large depths from the surface shows that, for wires of both types, the estimated density $\rho_x^{200\mu m}$ at depths of $\geq$200 μm does not change with an accuracy of a thousandth of a percent, and the dependence $\rho_x(T)$ almost reaches a plateau, which can be related to the mass density ρ_x^{bulk} of the wires in the bulk.

Estimates show that the density ρ_x^{Tlayer} at a depth of T_{layer} for A50 and AC50 wires after operation is ~99.6% of $\rho_x^{200\mu m}$ (i.e., of ρ_x^{bulk}). Most of the XRD mass density ρ_x drop is occurring in a layer with a thickness equal to T_{layer} from the surface ($\frac{\rho_x^{12.5\mu m} - \rho_x^{Tlayer}}{\rho_x^{12.5\mu m} - \rho_x^{200\mu m}} \cdot 100\%$ ~70% for most samples (N8 (A50 type, 10 years of service life), N7 (A50 type, 18 years), and N6 (AC50 type, 20 years)) and ~50% for AC50 sample N2-2 after 8 years of operation). Apparently, this significant drop of ρ_x evidences that most of the defects of a void nature after the operation of cables of both types are formed in a layer of wires with a thickness equal to the value of T_{layer} from the surface of the wires.

We have also made estimations of the depth T_{layer}^{sat} from the surface where the density ρ_x^{sat} is 99.99% of the value $\rho_x^{200\mu m}$ (which is taken as the value of ρ_x^{bulk}). In the inset of Figure 15b, these estimates are shown graphically for AC50 samples and were given in [10] for A50 wires. According to the $\frac{\rho_x^{12.5\mu m} - \rho_x^{sat}}{\rho_x^{12.5\mu m} - \rho_x^{200\mu m}} \cdot 100\% \approx 99\%$ criterion for all studied samples of both types, in the near-surface layer of thickness $T_{layer}{}^{sat}$, almost the entire observed decrease in the XRD mass density ρ_X occurs in comparison with the values of density $\rho_X{}^{bulk}$ in the bulk of the wires.

Figure 16 presents a comparison of the thickness estimates obtained for both characteristic near-surface layers, namely, T_{layer} (~50–70% reduction in density ρ_X) and $T_{layer}{}^{sat}$ (~99% reduction in density ρ_X). The thickness T_{layer} of NSDL, where most void defects are concentrated, increases to 39.2(1) μm in A50 wires after a service life of 10 years (sample N8), then remains practically unchanged (39.1(1) μm) in sample N7 after 18 years of operation. Perhaps due to the influence of the steel core, the thickness of the T_{layer} of a similar NSDL in AC50 wires is a bit less, reaching 35.9(1) μm in sample N2-2 after 8 years of operation and slightly increasing to 38.2(1) μm in sample N6 after 20 years of service. After the operation of wires in cables of overhead transmission power lines, the thickness $T_{layer}{}^{sat}$ of NSDL, where almost all void defects are concentrated, is significantly (3–4 times) greater than T_{layer} for wires from cables of both types. In A50 wires, the value of $T_{layer}{}^{sat}$ increases almost linearly with service life duration from 55.8(1) μm (N5-2, service life of 0 years) to 96.3(1) μm (N8, 10 years) and 114.7(1) μm (N7, 18 years). In AC50 wires (with steel core), the value of $T_{layer}{}^{sat}$ also increases, though non-linearly, from 21.6(1) μm (N5, 0 years) to 160.0(1) μm (N2-2, 8 years) and 119.1(1) μm (N6, 20 21.6(1) μm (N5, 0 years) to 160.0(1) μm (N2-2, 8 years) and 119.1(1) μm (N6, 20 years). As one can see, NSDL is already present in non-operated wires. The difference in its thickness by almost 2.5 times may be associated not with the type of wires but with the features of their manufacture and with experimental inaccuracies. Samples after 18 and 20 years of operation (respectively, A50 sample N7 and AC50 sample N6) are characterized by similar values of the $T_{layer}{}^{sat}$. At the same time, the AC50 sample N2-2 after 8 years of operation has a $T_{layer}{}^{sat}$ value that is ~1.5 times greater than that of the A50 sample N8 after a comparable 10 years of service. This difference may also be related to the features of operation and inaccuracies of the experiment.

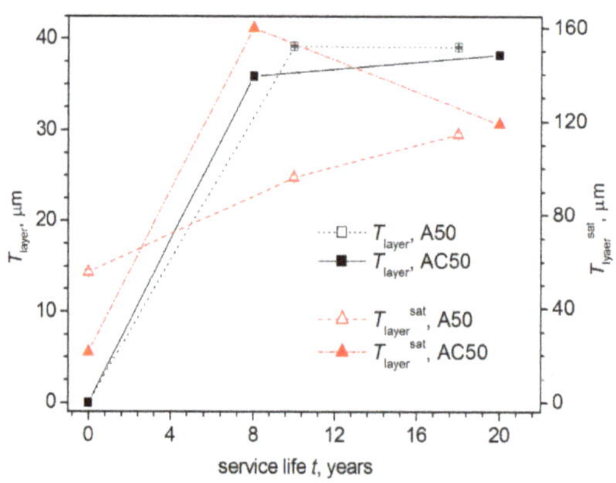

Figure 16. Comparison of dependences of thicknesses of the NSDLs, at which there is a ~50–70% (T_{layer}) and ~99% ($T_{layer}{}^{sat}$) drop in XRD mass density, on the service life t of the wires from the overhead power line cables of A50 and AC50 types (respectively, without and with steel core).

The unit cell parameter a and the calculated X-ray density ρ_X of the Al material are not exclusive characteristics which demonstrate smooth dependences on the depth T (= T^{hkl}_{pen} in the case of the hkl reflection) from the surface from which the diffracted X-rays come. There are the microstructural parameters too. As shown above, there is a microstrain in the Al crystallites. Average size D of crystallites and absolute values of average microstrain ε_s were estimated by WHP and SSP techniques using all the observed XRD reflections (see Table 4, Figure 14, and SM Figure S2).

At the same time, with an increase in the X-ray penetration depth T^{hkl}_{pen}, not only the values of the Bragg angles $2\theta_B$ of XRD reflections regularly change, which determine the value of the unit cell parameter a (i.e., the value ρ_X), but also their $FWHM_{obs}$, from which one can estimate the size of the crystallite D^{hkl} and the absolute value of microstrain ε_s^{hkl} after making the appropriate correction of $FWHM_{obs}$ for instrumental broadening (see [37,43]), corresponding to an hkl reflection.

Since it is impossible to determine two values, D^{hkl} and ε_s^{hkl}, from one reflection at once, in order to identify patterns of trends in changes in microstructural parameters, we first estimated the values $D = D^{hkl}_0$ within the model of the absence of microstrains ($\varepsilon_s = 0$) for the A50 and AC50 wires under study (Figure 17a,b).

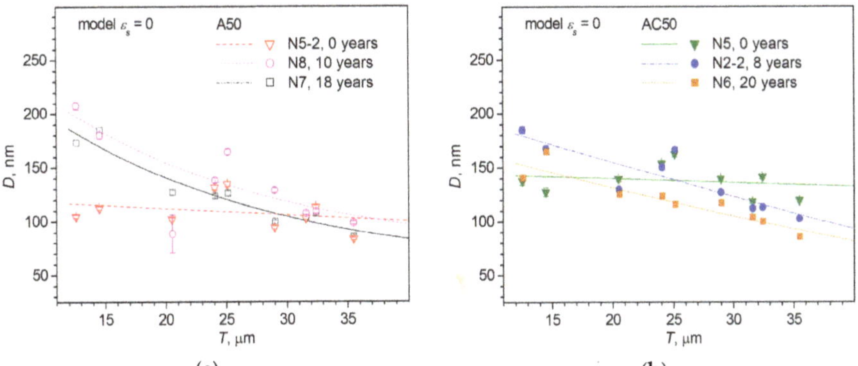

Figure 17. Distribution of the crystallite size $D = D^{hkl}_0$ estimated in the framework of zero microstrain assumption (model $\varepsilon_s = 0$) in the Al wires along the depth T from the surface of the (**a**) A50 and (**b**) AC50 wires. Sample numbers according to Table 2 and their service lifetimes are shown in (**a**,**b**). The approximation lines in (**a**,**b**) are drawn according to the exponential-decay law.

As one can see from Figure 17, wires from cables of both types (with and without a steel core) demonstrate identical trends of a smooth change in the crystallite size $D = D^{hkl}_0$, calculated in the zero microstrain approximation ($\varepsilon_s = 0$) for different reflections corresponding to different depths T from the surface. It is worth noting that, obviously due to the lower precision of determining the parameters of the microstructure compared to the parameters of the structure, the scatter of the experimental points D^{hkl}_0 around the approximation curves is rather large, which makes it difficult to obtain sufficiently accurate quantitative characteristics, although the trends in $D^{hkl}_0(T)$ are rather apparent.

The sizes of crystallites D^{hkl}_0 in non-exploited A50 (N5-2) and AC50 (N5) wires change slightly and almost linearly at different depths T from the surface (if one considers the approximation lines at experimentally achievable depths up to T ~35.5 μm; see Figure 17a,b). For the A50 N5-2 wire (service life of 0 years), the experimental values of D^{hkl}_0 corresponding to different reflections (i.e., to different depths T from the surface) give an average crystallite size of 109(16) nm, while for the AC50 wire N5 (0 years), the experimental points corresponding to different T are scattered around an almost straight horizontal line corresponding to a larger average value of 138(16) nm (model $\varepsilon_s = 0$ in Table 4). Approximation curves of operated samples for experimental values of D^{hkl}_0 obey the exponential decay law or the nearly linear drop law in the case of AC50 wires, i.e., in both cases, the farther from the surface (within ~35.5 μm thick NSDL), the smaller D^{hkl}_0. For wires of both types, the approximation curves for $D^{hkl}_0(T)$ in wires with a service life of 8–10 years go higher than for wires with a longer service life of 18–20 years. The presence of a steel core in AC50 type wires leads to a smaller crystallite size D^{hkl}_0 near the surface than in the case of A50 wires without a core, while the sizes D^{hkl}_0 of crystallites are almost the same far from the surface (compare 185 nm and 208 nm near surface at a depth T of ~12.5 μm for AC50 and A50 wires after 8 (sample N2-2) and 10 (N8) years of exploitation, respectively, and ≈100 nm for both wires at T ~35.5 μm with, correspondingly, 141 nm and 173 nm at T ~12.5 μm and

≈90 nm at T ~35.5 μm for AC50 and A50 wires after 20 (N6) and 18 (N7) years). As a result, the amplitude of variation of the $D^{hkl}{}_0(T)$ approximation curves in NSDL at depths from the surface T ~12.5 μm to ~35.5 μm for AC50 wires is noticeably smaller than for A50 wires (respectively, ≈74 nm and ≈62 nm in AC50 wires with service lives of 8 (N2-2) and 20 (N6) years compared with ≈94 nm (N8, 10 years) and ≈91 nm (N7, 18 years) in A50 wires).

As in [10] for wires of the A50 type, we have estimated the distribution of the absolute value of microstrain ε_s in crystallites at depths up to T ~12.5 μm for AC50 wires under the assumption that there are no microstrains near the surface at depths up to T ~12.5 μm ($\varepsilon_s = 0$), while the crystallite sizes at large depths $T \geq 15$ μm are fixed and equal to the crystallite size $D_{12.5\mu m}$ at a depth of T ~12.5 μm.

Figure 18 shows the $\varepsilon_s{}^{hkl}$ distributions obtained for AC50 wires with different service lifetimes under the above assumptions. Similar estimates of the $\varepsilon_s{}^{hkl}$ distribution for A50 wires with various service lifetimes can be found in [10].

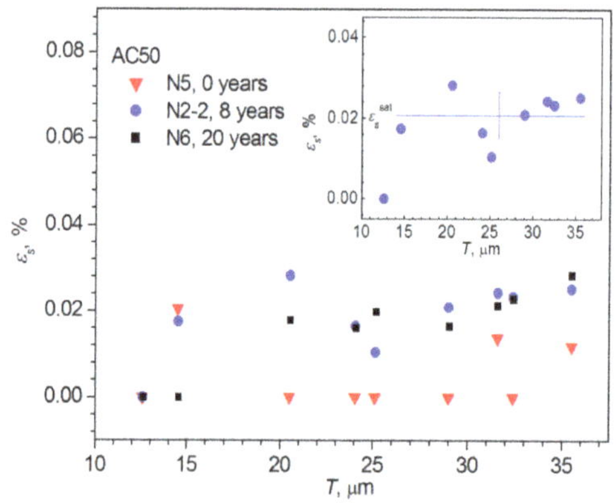

Figure 18. Distribution of microstrain $\varepsilon_s = \varepsilon_s{}^{hkl}$ along the depth T from the surface of the AC50 wire. The microstrain $\varepsilon_s{}^{hkl}$ is estimated under the assumption of a fixed crystallite size equal to the crystallite size $D_{12.5\mu m}$ near the surface (at a depth of T ~12.5 μm). Sample numbers according to Table 2 and their service lifetimes are shown. The inset shows an example of estimation of the average microstrain $\varepsilon_s{}^{sat}$ in the NSDL at depths $T \geq 15$ μm, if the crystallite size is fixed and equal to the crystallite size $D_{12.5\mu m}$.

An analysis of the obtained distributions of $\varepsilon_s{}^{hkl}$ over depth T from the surface for NSDLs of AC50 and A50 wires shows that the microstrains $\varepsilon_s{}^{hkl}$ estimated under the assumption of crystallite sizes fixed at $D_{12.5\mu m}$ at a depth T ~12.5 μm become constant in value ("saturated", $\varepsilon_s{}^{sat}$) already at depths of T ~15 μm. An example of evaluating the value of $\varepsilon_s{}^{sat}$ is given in the inset in Figure 18 for AC50 N2-2 wire with a service life of 8 years and in [10] for A50 wires. Upon comparing the distribution profiles $\varepsilon_s{}^{hkl}(T)$, one can conclude that the values of $\varepsilon_s{}^{sat}$ for AC50 wires are smaller than for A50 wires with a comparable service life. Figure 19 shows $\varepsilon_s{}^{sat}$ vs. service life t for both wire types. As can be seen, the form of the $\varepsilon_s{}^{sat}(t)$ dependences is very similar for both types of wires. Yet, for AC50 wires, the dependence $\varepsilon_s{}^{sat}(t)$ is shifted to smaller values, i.e., the presence of a steel core in AC50 type cables reduces the absolute value of microstrain in the wires of this type of cable compared to A50 wires without a core. The results obtained confirm the trends and even the numerical values of absolute values of average microstrain ε_s obtained by the WHP and SSP techniques for microstrains averaged over NSDL with a thickness of ~35.5 μm (see Figure 14b).

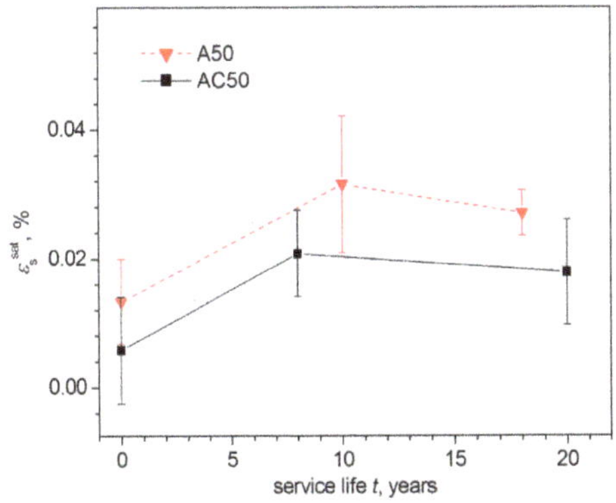

Figure 19. Microstrain $\varepsilon_s{}^{sat}$ calculated under the assumption of a fixed crystallite size equal to the crystallite size $D_{12.5\mu m}$ near the surface (at a depth of T ~12.5 μm) depending on the service life t of the Al wires of the overhead power lines of the A50 and AC50 types (correspondingly, AAAC and ACSR types, without and with steel cores).

It should be noted that both considered cases (the absence of microstrains ($\varepsilon_s = 0$, Figure 17a,b) and a fixed crystallite size equal to the size of crystallites near the surface of the samples (Figure 18)) are limiting. If, for example, the size D of crystallites from the surface to the depth first increases and then decreases, then, in accordance with the formulas for calculating the parameters of the microstructure (see [10,43]), the absolute value of microstrain ε_s also first increases and then decreases. As a result, in this case, the dependences $D^{hkl}{}_0(T)$ and $\varepsilon_s{}^{hkl}(T)$ in NSDL will acquire a form close to bell-shaped.

Thus, the XRD study of wires from A50 and AC50 cables of different service life lengths from 0 to 20 years revealed the features of changes in the structural and microstructural parameters of NSDL of wires (characteristics averaged over the NSDL with a thickness of ~35.5 μm in accordance with the values of the Bragg angles of the observed reflections and the profiles of these characteristics in depth from the surface of the wire) depending on the presence of a steel core in the cables of overhead power lines from which these wires are extracted.

3.5. Results of Acoustic Studies

To study the quite probable scatter of experimental data of the acoustic study along the wire segment, samples with a length of ~25 mm were cut from different sections of the wire (as a rule, there were three pieces for each service life). This scatter can be caused both by the different deformation prehistory of wire sections during cable manufacture and by its operation duration. As an example, the results of the measurements of the amplitude dependence (i.e., dependence on the amplitude of vibrational strain ε) of both the Young's modulus $E(\varepsilon)$ and the decrement of elastic vibrations $\delta(\varepsilon)$, as well as the dependence of oscillatory stress on the amplitude of the nonlinear inelastic deformation, $\sigma(\varepsilon_d)$, for three studied samples with the same service life are shown in Figures 20a and 21a (A50, samples N8, 10 years of service life) and Figures 20b and 21b (AC50, N6, 20 years of service life). It can be seen from the Figures that the data obtained from pieces cut from different sections of wire of the same service life noticeably differ from each other. The greatest difference (several times) is observed for the values of δ and σ. The Young's modulus E turns out to be structurally sensitive only in the third or fourth significant decimal place, and, apparently, this sensitivity is determined mainly by the presence of defects and the microplastic properties of a particular piece.

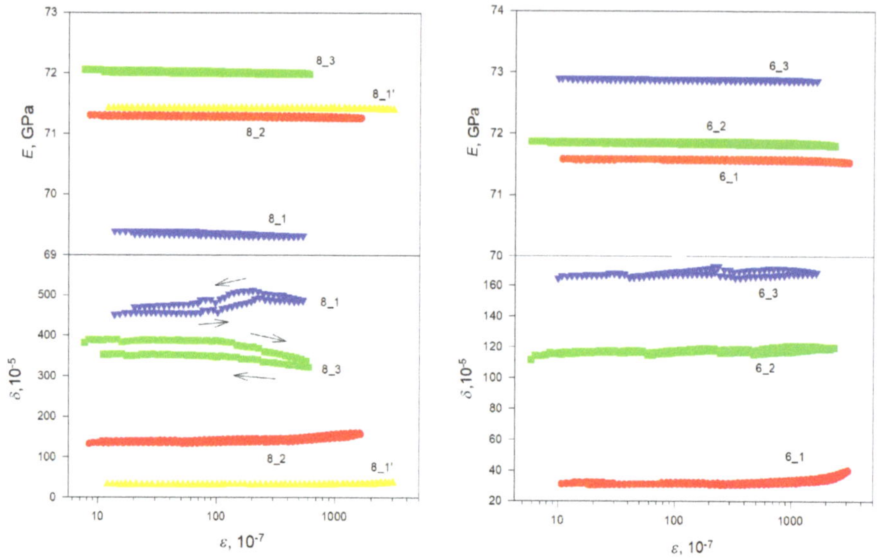

(a) (b)

Figure 20. Dependences of the Young's modulus E and decrement δ on the amplitude of vibrational deformation ε for Al wires of (**a**) A50 type after 10 years of service before (N8_1,2,3) and after (N8_1′) etching and (**b**) AC50 type after 20 years of service (N6_1,2,3). The measurements were taken at room temperature.

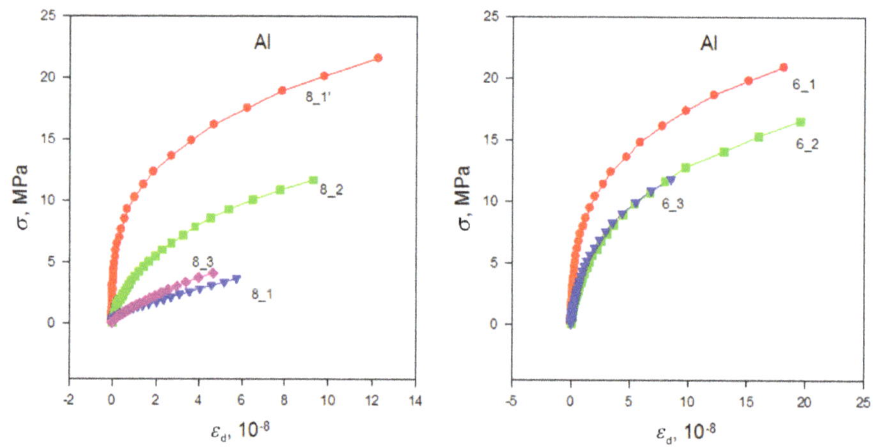

(a) (b)

Figure 21. Diagrams of microplastic deformation $\sigma(\varepsilon_d)$ of Al wires of (**a**) A50 type after 10 years of service before (N8_1,2,3) and after (N8_1′) etching and (**b**) AC50 type after 20 years of service (N6_1,2,3). The measurements were taken at room temperature.

We have made an attempt to reveal the effect of operating time on the studied parameters. To do that, one sample with the smallest value of the amplitude-independent decrement δ_i was selected for each material (A50 and AC50) and service life. In such a wire piece, the plastic deformation and, consequently, defect formation, which can occur during the manufacture of the cable, are apparently less than in the other wire areas. When these values are the minimum, the effect of unfavorable factors (wind, snow sticking, etc.) on the decrement, leading to the appearance of additional defects and their evolution during operation, should be more apparent. The obtained data are summarized in Table 5 and, for explicitness, are shown in Figure 22 in the form of dependences on the operating time, the Table and Figure show the values of the elastic modulus E, amplitude-independent decrement δ_i, and microplastic flow stress $\sigma_s = \sigma$ measured at inelasticstrain amplitude $\varepsilon_d = 4 \cdot 10^{-8}$ obtained for specimens with the lowest δ_i value for all investigated wires.

Table 5. Samples of cables of overhead power lines' Young's modulus E, amplitude-independent decrement of elastic vibrations δ_i, and microplastic flow stress σ_s of the aluminum samples prepared from wires of overhead power lines with different service life and selected according to the criterion of the smallest value of the amplitude-independent decrement δ_i. For each sample, the number of studied pieces cut from wires of the same service life is indicated.

Sample N	Type	Service Life t, Years	E, GPa	$\delta_i \cdot 10^5$	σ_s, MPa at $\varepsilon_d = 4 \cdot 10^{-8}$
5-2 (3)	A50	0	72.78	23	14
5 (1)	AC50	0	71.84	27	15
8 (3)	A50	10	71.30	133	8.0
2-2 (3)	AC50	8	72.87	17	16
7 (1)	A50	18	71.21	85	9.0
6 (3)	AC50	20	71.58	31	13

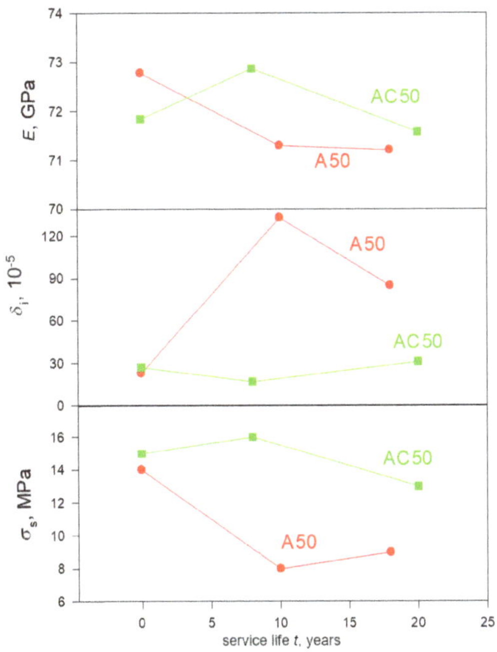

Figure 22. Dependences of Young's modulus E, amplitude-independent decrement δ_i, and microplastic stress σ_s for Al wires on operating time t. The measurements were taken at room temperature.

When looking at Figure 22, the first thing that catches the eye is the noticeable effect of the steel core. The use of this core in AC50 type samples maintains a low level of amplitude-independent decrement δ_i and a high level of stress σ_s of microplastic flow in comparison with the wires of the A50-type cables without a steel core. At the same time, as expected, the values of δ_i and σ_s differ insignificantly for A50 and AC50 unused samples (0 years of service life).

As for the dependences of δ_i and σ_s on the operating time (Figure 22), these parameters change much less for the wires of a steel core (ACSR) AC50 cable than for those of an all-aluminum (AAAC) A50 cable. One can also note the difference in the behavior of E, δ_i, and σ_s for wires from the former cable and the latter one, namely the fact that, at approximately the service life, where there is a maximum in the operating-time dependence of $E(t)$, $\delta_i(t)$, and $\sigma_s(t)$ for one, the minimum is observed for the other one. The presence of a steel core in an ACSR AC50 cable at the beginning of operation for a service life of up to 8 years is favorable for Al wires from the cable (δ_i decreases while both E and σ_s increase), then there are slight changes for the worse with an increase in service life up to 20 years. On the contrary, in Al wires of an all-aluminum (AAAC) A50 cable, at first, significant unhardening is observed after 10 years of service, then there is a slight improvement in deformation characteristics with an increase in service life up to 18 years. However, it should be noted that the results obtained should be treated with caution since the number of samples studied is small. Nevertheless, the positive role of the steel core for power transmission lines is not in doubt, as it has already been discussed in the review of the available literature in Section 1 (Introduction).

In this paper, apparently, we need to once again pay attention to a result discussed in detail in [10], namely, the fact that the defects are mainly located in the NSDL of the Al wires. Figures 20a and 21a show data for A50 sample N8_1 (service life of 10 years) in its original state and after etching (N8_1′), when a layer of about 35 μm thick was removed from the surface. In the initial state before etching, this sample had the largest amplitude-independent decrement δ_i, the lowest level of microplastic stress σ, and the smallest value of Young's modulus E. Upon etching, it turned out to have the smallest decrement δ_i and the largest value of σ among all the investigated samples with numbers N8_*i* ($i = 1 - 3$). Young's modulus of the sample N8_1′ also increased noticeably, i.e., an improvement in the deformation characteristics is observed after the removal of the near-surface layer by etching. Evidently, this indeed means that most of the defects are located in the near-surface layer. Undoubtedly, one can dare to attribute this result to other samples (N8_2 and N8_3, N6_1,2,3 and so on), for which etching was not carried out.

4. Discussion

First of all, let us discuss the difference of 2 years in the service life of the compared wires (A50 N8, 10 years and AC50, N2-2, 8 years; A50 N7, 18 years and AC50, N6, 20 years). The results of data analysis have shown that the dependences of the wire parameters on the service life (unit cell parameter $a(t)$ of Al, Figure 13a; XRD density $\rho_x(t)$, Figure 13b) are linear or close to linear (fraction of aluminum oxide $q(t)$, Figure 11). Thus, the difference of 2 years between the compared parameters for wires of different types does not play a role because the dependences for the wires of different types differ well from each other, and the values of these parameters for a wire of the same type with a difference of 2 years are close. The same can be said about the time dependences of other parameters (Figures 16 and 19), particularly about the dependence of the deformation characteristics of wires on their service life (Figure 22).

As for the profile dependences ($a(T)$ and $\rho_x(T)$, Figure 15a,b), then, taking into account that according to determined values of the rate of change of these quantities with time (see Section 3.4), in 2 years they will change only by ~$1 \cdot 10^{-4}$ Å and ~$2 \cdot 10^{-4}$ g/cm^3, respectively. Moreover, the parameter a of the Al unit cell will increase, and the density ρ_x will decrease. As a result, if we scale the values of a and ρ_x of A50 wires (service life of 10 years and 18 years) to the values for AC50 wires (service life of 8 years and 20 years) by substraction/addition of the corresponding quantity for 2 years, then the difference between the dependences $a(T)$ and $\rho_x(T)$ for different types of wire will change only slightly. All trends remain, and numerical values are virtually unchanged (up to the fourth digit after the decimal point). As an illustration, the SM Figure S3 shows an example of the dependencies of $\rho_x(T)$ Al wires on A50 cables, reduced to the same service life as the

AC50 cable wires. Obviously, this conclusion also applies to the profile dependence of the densitometric density ρ_{dL} of NSDL (density defect $\Delta\rho_{dL}/\rho_{dL}$, Figure 7).

Thus, the analysis of the data obtained indicates that when comparing the Al wires of A50 and AC50 type cables, the difference of 2 years in service life does not significantly affect the comparison results.

As stated in [37], the changes in structure, microstructure, and various physical properties, which are observed for wires in overhead power transmission cables, are probably related to processes occurring during operation. Among these processes, it is worth emphasizing the stretching of wires due to either vibrations or swaying of cables under the action of the variable wind and temperature of the surrounding air atmosphere, the phenomenon of fretting (vibration fatigue due to the friction of the wires in the cable against each other), and the formation of defects of a void nature, especially in the near-surface layers of the wires. AAAC (A50) type cables are characterized by approximately the same ~50 mm^2 cross-sectional area of their aluminum component as ACSR (AC50) cables (Table 1). The most significant difference in their design is the replacement of an additional central aluminum wire in AAAC (A50) cables with a steel core wire of approximately the same diameter in ACSR (AC50) cables. As can be expected, the presence of a steel core in ACSR (AC50) type cables, which is much stiffer than the Al wires of these cables, will reduce the influence of at least some of the above factors, such as stretching and abrasion of wires, in ACSR (AC50) type cable wires compared to AAAC (A50) cable wires. As a result of using the steel core, changes in the structure, microstructure, and, as a result, in the physical properties of AC50 wires will occur more slowly.

Indeed, a comparative study of wires from A50 (AAAC type) and AC50 (ACSR type) cables, similar in geometric dimensions and structure, with the exception of the material of the central wire (Al wire in A50 (AAAC) cable and steel core wire in AC50 (ACSR), by means of the XRD, EBSD, and densitometry methods in combination with acoustic studies, has shown a noticeable difference in the structure, microstructure, and elastic-plastic properties of these wires.

The methods of EBSD, densitometry, and XRD showed the presence of NSDL (near-surface defect layer, in which most of the defects of the void nature are concentrated) in the wires of cables of both types. A significant change in the Young's modulus, decrement, and stress of the microplastic flow after the removal of the surface layer also indirectly confirmed the existence of defects in the near-surface layer of the Al wire and their significant contribution to the physical properties of the wire.

Analysis of the EBSD maps (Figure 2 and Refs. [10,37]) taken in the center and at a distance of ~150 µm from the edge of the facets of the cross-sections of the wires, which, based on the constructed histograms of grain-boundary misorientation-angle distribution (Figure 6), has made it possible to gain a qualitative conclusion about the difference in the morphology of the near-surface layers and the bulk region in wires of both types after operation. The presence of a steel core in AC50 cables makes the grain misorientation change slightly both in the bulk and in the near-surface layers of AC50 wire over a service life, in contrast to the wire from A50 cables without the stabilizing effect of the steel core, where in the near-surface layers there is a more pronounced tendency to align the crystal lattice of grains along one common direction. The formation of such a textured structure in the near-surface layers of Al wires is probably associated with the combined effect for a long time of the ambient temperature during operation, electric current, and electric voltage, as well as the tensile load, which is noticeably higher in wires of A50 type cables than in AC50 ones due to the absence of a steel core in the A50 cables. Such low-temperature annealing under load contributes to the fact that the grain size in the direction of stretching increases more than in the transverse direction, that is, there is some "pulling" of the grains in the direction of the load [53], as it was observed on the EBSD maps obtained from the longitudinal sections of A50 wires [10].

The application of the densitometry method to the wires after the sequential removal of near-surface layers of a certain thickness by etching has made it possible to quantify the

NSDLs' characteristic thicknesses (Figure 7), amounting to ~10 μm and ~30 μm for wires of both types. In both cases, when approaching the surface, the integral density ρ_d measured by densitometric technique decreases according to a law close to the exponential-decay law. The greatest drop in ρ_d is observed in a narrow near-surface layer up to ~10 μm (~80–85% of the total reduction), the smallest value of ρ_d being near the surface. Taking into account the absence of light chemical elements in sufficiently large quantities according to the results of the EDX microanalysis (Table 3 and Section 3.1), obviously, it indicates the presence of defects of a void nature (nano and micropores, microcracks) in the narrow near-surface layer, the concentration of these defects increasing when approaching the surface. Deeper inside a wire, ρ_d grows weakly until stabilization at depths from ~10 μm to ~30 μm from the surface, apparently due to a decrease in the number of defects. Such a change in the density and in the number of defects is expected since it is the surface of a wire that is affected by the environment and neighboring wires in the first place.

As noted in Section 3.3, the presence of a steel core leads to the phenomenon that in a narrow (~5 μm thick) near-surface layer of an AC50 wire, the integral density defect $\Delta\rho_{dL}/\rho_{dL}$ (calculated by Formula (3)) is somewhat smaller (by ~0.1–0.2%) in absolute value than in an A50 wire (Figure 7), i.e., the integral density ρ_{dL} in this thin layer in the AC50 sample obtained by densitometric measurements of the wire after etching (Formula (2)) is slightly higher than that in the A50 wire. Moreover, the integral density ρ_d of the entire AC50 wire is ~0.05% higher than that of the A50 wire.

Thus, the density of NSDL in AC50 wire and the integral density of the entire AC50 wire are greater than in A50 wire, which can be caused by various reasons.

First, when the cable is stretched because of sagging and under the influence of the surrounding atmosphere, changes in ambient temperature, and vibrations caused by the wind, the parameter a of the unit cell of the wire Al-material increases, as was established by XRD studies (Figure 13a). As a result of stretching the crystal lattice of Al wires after operation, the X-ray density ρ_x of the wire Al material, calculated from structural data, decreases (see Formula (4) and Figure 13b). Obviously, this also leads to a decrease in the integral density ρ_{dL} of NSDL and the integral density ρ_d of the entire wire, measured by densitometry. The same causes lead to the formation of microvoids (microcracks) with a higher concentration of them near the surface due to the effect of fretting, which leads to a decrease in the ρ_{dL} in NSDL and the integral density ρ_d of the entire wire. In the presence of a steel core, the influence of many of the above listed factors (sagging, vibrations, etc.), leading to a reduction in the integral density of the NSDL and the whole wire, is reduced. As a result, the integral densities ρ_{dL} of NSDL in the AC50 wire and the ρ_d of the entire AC50 wire degrade (decrease) less than in the A50 wire.

Second, as is known, the maximum acceptable operating temperature of wires of the A50 and AC50 types does not exceed 90 °C, with the possible interval of air temperature being from −40 °C to +40 °C. It is worth recalling that the coefficients of linear thermal expansion are different for aluminum and steel. So, under certain conditions, despite the fact that steel is much stronger than aluminum, the tension of the aluminum part of the wire can noticeably weaken as the temperature rises, whereas the mechanical load on the steel elements of the wire increases slightly. It is possible that this is just a circumstance that provides a greater value of densities ρ_{dL} in NSDL and ρ_d of aluminum wires of an AC50 type cable after operation compared to an A50 cable without a steel core.

Furthermore, taking into account the fact that densities ρ_d of wire and ρ_{dL} of wire NSDL obtained by the densitometric method are integral quantities averaged, respectively, over the densities of all wire or wire NSDL components, the higher density ρ_{dL} in NSDL of the Al wires of the AC50 cable compared to the A50 cable, at least at depths up to ~5–10 μm (and, accordingly, a lower value in terms of the absolute magnitude of the negative density defect $\Delta\rho_{dL}/\rho_{dL}$, see Figure 7) can be due to a larger proportion of Al oxides formed in the NSDL of wires of the AC50 cable (Figure 11), probably as a result of the oxidizing action of the steel core. These aluminum oxides (δ- and/or δ*-Al_2O_3) are characterized by a higher nominal calculated XRD density (~3.7 g/cm^3) [51,52] compared to Al (~2.7 g/cm^3) [50].

Hence, the observed integral density ρ_{dL} of NSDL and, consequently, the total integral density ρ_d of the Al wires of an AC50 type cable will be higher than that of an A50 cable. However, it should be noted that alumina crystallites formed in NSDLs of wires are much denser and harder than aluminum crystallites. It enhances the influence of the fretting effect and, as noted above, leads to the opposite effect of a decrease in density due to the formation of void defects. Since in AC50 wires, probably due to the oxidizing effect of the steel core, the proportion of aluminum oxides is higher compared to A50 wires, the influence of both effects in AC50 wires will arise competitively. On the one hand, due to the greater proportion of aluminum oxides with a higher XRD density than the XRD density of aluminum, integral densities ρ_d and ρ_{dL} increase. On the other hand, due to the greater influence of fretting, more voids are formed and the integral densities ρ_d and ρ_{dL} decrease.

Profiling the near-surface layer by analyzing XRD reflections with different Bragg angles $2\theta_B$ corresponding to different X-ray-penetration depths has confirmed the presence of NSDLs in Al wires of both types and made it possible to obtain their quantitative characteristics. All trends indicated by the results of density measurements discussed above are also confirmed. As a result of the smooth expansion of the Al lattice (an increase in the unit cell parameter a of the wire Al material) while approaching the sample surface, the X-ray density ρ_x calculated from the structural data decreases (see Formula (4)). This decrease in ρ_x with decreasing depth T from the surface occurs according to an exponential decay law similar to the dependence of the densitometric density ρ_{dL} of the near-surface layer on the thickness T_{etch} of the layer removed by etching, although flatter (cf. Figures 7 and 15b).

Based on the dependences $\rho_x(T)$ obtained from the analysis of XRD reflections corresponding to different X-ray penetration depths (i.e., different depths from the wire surface), two characteristic thicknesses of NSDLs of the wires were estimated. One of the characteristic thicknesses obtained from XRD studies (Section 3.4) is close to the characteristic NSDL thickness of ~30 μm obtained from densitometric measurements. In particular, the characteristic thickness T_{layer}, corresponding to the density $\rho_x{}^{Tlayer}$ ~99.6% of the density $\rho_x{}^{bulk}$ in the bulk of wire, is T_{layer} = 36.4–39.1 μm for A50 wires and T_{layer} = 35.9–38.2 μm for AC50 wires of different service life lengths (Figure 16).

Thus, the presence of a steel core (in ACSR (AC50) cable) results in a lower T_{layer}, i.e., in a smaller thickness of that part of the NSDL where the main drop of the X-ray density ρ_x occurs (~50–70% of the total decrease from $\rho_x{}^{bulk}$) when approaching the surface. Although the second characteristic thickness $T_{layer}{}^{sat}$, which corresponds to the density $\rho_x{}^{sat}$ ~99.99% of $\rho_x{}^{bulk}$ and thus to the entire thickness of the layer from the surface, where almost the entire observed decrease in the X-ray mass density ρ_x (~99%) occurs, is practically the same for A50 and AC50 type wires after 18–20 years of operation (115 μm–119 μm) and ~1.5 times more after 8–10 years of service for AC50 wire (with steel core) than for A50 wire without steel core (160 μm vs. 96 μm, respectively). At the same time, in wires without operation, the value of $T_{layer}{}^{sat}$ of the total thickness of NSDL in AC50 wire was, on the contrary, ~2.5 times less compared to A50 wire (respectively, ~56 μm vs. ~22 μm, see Figure 16). It is possible, however, that such a non-smooth (irregular) dependence of the total thickness of $T_{layer}{}^{sat}$ of NSDL in AC50 wires on the cable-operation duration t, in contrast to the almost linear increase in $T_{layer}{}^{sat}$ for A50 wires, is not due to the presence of a steel core in AC50 wires, but to the peculiarities of manufacturing the cables under study.

A noticeable difference in the amplitude of changes in the unit cell parameter a of the wire Al material and, thus, in both the calculated X-ray density ρ_x and density defect $\Delta\rho_x/\rho_x$ depending on the depth T from the surface (Figure 15a,b) is another peculiarity obtained from the analysis of XRD reflections from wires, which indicates a notable effect of the steel core in cables AC50 and A50 of the same cross section of ~50 mm^2 of the Al component of the cables. For unused wires and at depths $T \approx 25$ μm–36 μm from the wire surface after operation for both types of cables, the approximation dependences $\rho_x(T)$ practically coincide. On approaching the surface, the approximation curves $\rho_x(T)$ begin to diverge for operated wires of different types, and the closer to the surface they get, the greated. Moreover, in the presence of a steel core, the decrease in $\rho_x(T)$ and in $\Delta\rho_x/\rho_x$

is smaller. For example, near the surface ($T \approx 12$ μm) of wires from cables of the AC50 and A50 types, respectively, after 8 years–10 years of operation, the density defect $\Delta\rho_x/\rho_x$ decreases in absolute value by ≈9 and ≈14 times compared to the value established for the depth from the surface T_{layer}. After 18 years–20 years of operation, this difference is ≈21 and ≈23 times for AC50 and A50 wires, respectively.

Thus, in the presence of a steel core in the cable after operation in overhead power lines, the decrease in the X-ray density ρ_x in the NSDLs of the cable wires when approaching the surface from the depth of the wire occurs more gently. This flatter course of $\rho_x(T)$ in NSDLs of AC50 wires indicates that the degradation ("aging") of the AC50 wires from cables with steel cores is slower. The quantitative characteristics of wire degradation were obtained by analyzing the wire parameters averaged over all observed reflections, i.e., over the NSDL with a thickness of ~35.5 μm (Figure 13a,b). In the presence of a steel core (ACSR (AC50) cable), the rate of expansion of the crystal lattice and, thus, that of decrease in the X-ray density ρ_x of the Al material of wires in NSDL with a thickness of ~35.5 μm is ~1.2 times lower ($-2.13(7) \cdot 10^{-4}$ g/cm^3/year for AC50 wires in comparison to $-2.52(8) \cdot 10^{-4}$ g/cm^3/year for A50 ones, see Section 3.4). As a result, the delay in the degradation of the lattice and ρ_x of the Al material of the AC50-type wires ranges from ~1 year after a service life of ~10 years up to ~3 years after ~20 years of operation.

The method of calculating ρ_x (namely, from the unit cell volume of the Al material in NSDLs of wires (Formula (4)) directly indicates that the cables are less stretched because of vibrations due to wind and, possibly, temperature fluctuations of the surrounding atmosphere, which are the main reasons for the higher density ρ_x in NSDL of Al wires from AC50 cables compared to A50 ones. Due to the stabilizing effect of the steel core, the wires of an AC50 cable are less affected by vibrations and stretches. As a result, the Al lattice of the wire material of AC50 cable expands less, and accordingly, the X-ray density ρ_x estimated from XRD structural data decreases less compared to A50 wires.

The ~10% lower rate of amplification of the effects of the preferential orientation of XRD reflections attributed to the Al material of the wires of the AC50-type cable during operation with an increase in service life from 8 to 20 years compared to those of the A50 one with an increase in service life from 10 to 18 years (Table 4 and Figure 10) can probably also be associated with the stabilizing effect of the steel core in AC50 wires and, thus, limitation of the cable stretch during vibrations because of the wind influence. This observation is also consistent with the result of the EBSD analysis (Section 3.2), that in the near-surface layer of AC50 wires there is a less pronounced tendency to align the crystal lattice of grains along one common direction than for A50 wires.

The lower stretch of the AC50 cable during service due to the stabilizing effect of a steel core in comparison to the A50 cable without a steel core is probably also a reason for the lower value of average microstrain ε_s formed in the Al crystallites of NSDL of wires of AC50 cables after service in comparison with that of A50 cables (Figure 14b). Moreover, apparently, the same reason is associated with a lower value of microstrains $\varepsilon_s{}^{sat}$ observed at depths T below the wire surface in AC50 wires after operation compared to A50 ones (Figure 19). In this case, directly at a depth from the surface down to $T \approx 12.5$ μm, the wires of both types are apparently relaxed ($\varepsilon_s = 0$), which seems to be the natural state of the surface of the wires.

One can see from Figure 15b for AC50 wires and from [10] for A50 samples that the value of the X-ray density $\rho_x{}^{bulk}$ in the bulk of the wire at depths T of 200 μm and more (and, accordingly, the value $\rho_x{}^{sat} = 0.9999 \cdot \rho_x{}^{bulk}$), which are estimated by extrapolation from the approximation curves $\rho_x(T)$ for unexploited samples A50 and AC50, are close to each other. Quantitative estimation by approximating curves $\rho_x(T)$ gives $\rho_x{}^{bulk} = 2.6987(2)$g/cm^3 and 2.6970(2) g/cm^3 for A50 and AC50, respectively. These $\rho_x{}^{bulk}$ values also agree satisfactorily with the average density values in NSDL of wires $\rho_x = 2.6973(2)$ g/cm^3 and 2.6972(2) g/cm^3 (Table 4). Therefore, the X-ray densities ρ_x estimated for NSDLs of wires and their bulk agree satisfactorily for the two types of zero-service-life wires. These estimated values of ρ_x are larger than the tabular calculated X-ray density of Al material at a temperature of 312.3 K, which is close to the temperature of XRD measurements in this work (ρ_x = 2.6964 g/cm^3 according to PDF-2 card 01-071-4008). As discussed in Section 3.4, this discrepancy may be because of the inclusion of a few Si and Fe atoms in the Al structure, which are present in the composition of wires according to EDX (Table 3, Figure 1a,b and Ref. [37]).

For wires of both types, when service life increases from 0 to 18–20 years, a decrease in both the X-ray density ρ_x of the Al material averaged over NSDL of 35.5 μm thick (i.e., averaged over the near-surface layer with a thickness of about 1st characteristic thickness T_{layer} of NSDL from the wire surface, see Figure 13b) and the integral density ρ_d of entire wire (Figure 8) obtained from densitometry measurements is observed. Moreover, for A50 wire from the AAAC type cable, the average X-ray density defect $\Delta\rho_x/\rho_x$ (NSDL characteristic) and the defect of the integral (densitometric) density of the entire wire increase in absolute value from ~0% for unused wire to practically the same value $\approx -0.17\%$ after 18 years of operation. In the case of AC50 wires from the ACSR cable, the average defect of the X-ray density of NSDL wire is also ~0% for unused wire and $\Delta\rho_x/\rho_x = -0.162(1)\%$ after 20 years of operation, i.e., somewhat less in absolute value than for A50 wire with a comparable service life. Thus, although the tendency to decrease the density ρ_x of the NSDL with a thickness of T_{layer} ~35.5 μm is the same for both types of cables, for AC50 wire the decrease is less than for A50 wire, which once again emphasizes the stabilizing effect of the steel core in AC50 cables.

However, as one can see from the inset to Figure 15b for AC50 wires and Ref. [10] for A50 wires, unlike the average ρ_x in NSDL of a thickness from the surface of about T_{layer} ~35.5 μm, the XRD density $\rho_x{}^{sat} = 0.9999 \cdot \rho_x{}^{bulk}$ at depths $T = T_{layer}{}^{sat}$ (~100 μm–160 μm for wires after operation ~10–20 years) does not decrease but increases with an increase in service life from 0 to ~20 years. Figure 23 shows the obtained values of the density $\rho_x{}^{bulk}$ and the unit cell parameter a^{bulk} of the Al material in the bulk, recalculated from $\rho_x{}^{bulk}$ using Formula (4) depending on the service life t. For both kinds of wires, the initial value (t = 0 years of operation) of $\rho_x{}^{bulk}$ is close to the tabular value ρ_x of pure Al powder at a temperature approximately equal to the temperature of XRD measurements in this work. As can be seen, the dependences $\rho_x{}^{bulk}(t)$ for wires of both types qualitatively resemble the dependences of the fraction $q(t)$ of the Al_2O_3 phases in the wires (Figure 11). For wires of the A50 cables, the dependences $\rho_x{}^{bulk}(t)$ and $q(t)$ are close to linear in the interval from 0 to 18 years, with the density $\rho_x{}^{bulk}$ increasing from the initial value by $\approx$0.4% to $\rho_x{}^{bulk}$ = 2.7092(2) g/cm^3 after 18 years of service. In the presence of a steel core (the AC50 cables), $\rho_x{}^{bulk}$ increases from the initial value to almost the same value $\rho_x{}^{bulk}$ = 2.7091(2) g/cm^3 after 8 years of service, slightly decreasing to $\rho_x{}^{bulk}$ = 2.7085(2) g/cm^3 after 20 years of operation.

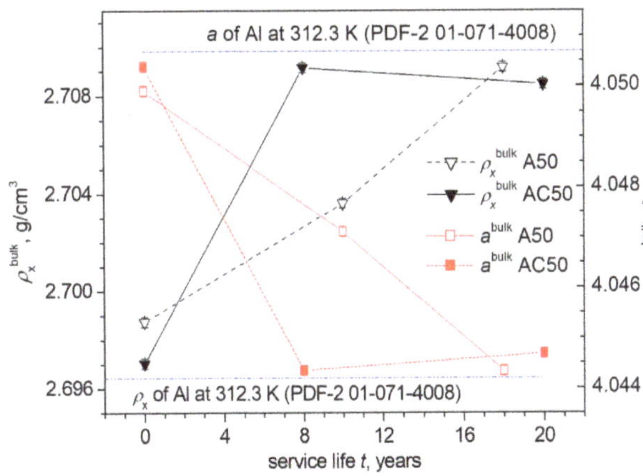

Figure 23. Dependences of the densities $\rho_x{}^{bulk}$ and unit cell parameters a^{bulk} in the bulk of the wires from the cables of AAAC (A50) and ACSR (AC50) types on the service life t of the cables.

This contradiction can probably be qualitatively explained as follows. The integral density ρ_d obtained in densitometric investigations is a characteristic of the entire wire volume (NSDL and bulk, including all crystalline and non-crystalline phases), which is measured experimentally. On the contrary, the X-ray densities ρ_x^{bulk} in the bulk of the wires are estimated from the approximation dependences of the X-ray densities $\rho_x(T)$ down to large depths T ~200 μm from the surface, where the experimental values of ρ_x are calculated from the experimental values of the unit cell parameter a of the Al material in the NSDL of wires at depths T from ≈12.5 μm to ≈35.5 μm. Along with a possible partial overestimation of the value of ρ_x^{bulk} due to experimental errors in determining the Bragg angles $2\theta_B$ of the observed Al reflections and, accordingly, the individual values of the X-ray density ρ_x corresponding to these reflections, another physical reason for the increase in the X-ray density of ρ_x^{bulk} in the volume (in the bulk) of wires can be considered. The increase in the density ρ_x^{bulk} in the bulk of the wires corresponds to the compression of the lattice in the bulk and correlates with the increase in the proportion of aluminum oxides δ- and/or δ*-Al_2O_3 (Figure 11), with higher mass densities ~3.7 g/cm^3 than ~2.7 g/cm^3 of aluminum. Presumably, the formed aluminum oxide crystallites compress the Al lattice of the wire material, the values of ρ_x^{bulk} increasing consequently.

Structural and microstructural data obtained by the methods of EBSD, XRD, and densitometry fully explain the changes in elastic and microplastic properties during the operation of wires (Figure 22). The decrease in the elastic modulus E for the A50 for the entire 18-year period of operation and after 8 years of service for the AC50 is most likely due to a decrease in the integral (ρ_d) and XRD (ρ_x) density of samples detected by densitometry and XRD methods. A slight increase in the E modulus for the AC50 from 0 to 8 years of operation can be explained by taking into account the data on the proportion of aluminum oxides shown in Figure 11. For AC50, after 8 years of service, the volume fraction of Al_2O_3 oxides in the wire NSDL is almost three times greater than for A50. According to the literature data, the elastic modulus for Al_2O_3 is E = 247–380 GPa compared to E ~70 GPa for Al, therefore, the formation of such a layer contributes to the increase in the elastic modulus of the AC50 wire. With an increase in the service life of up to 20 years, it is likely that the formation of defects of a hollow nature (nano and micropores, etc.) in Al wires is decisive for changing the E modulus and leads to its decrease in AC50 wires. In addition, the alignment of the crystal structure of grains in the near-surface layer along one direction and their elongation in the direction of stretching [10] and the crystallographic texture of crystallites along [011], which is enhanced for both types of wires during their operation, can also affect the modulus of elasticity [54].

In the behavior of the amplitude-independent decrement δ_i, depending on the operating time t (Figure 22), it can be noted that for AC50, the decrement δ_i changes slightly, while for A50, its changes are substantial. The observed change for the A50 is most likely due to more intensive plastic deformation processes caused by a greater load resulting from the absence of a steel core. This is also indirectly confirmed by the results of EBSD for longitudinal sections of A50 wires, according to which the process of changing the shape of the grains to a more elongated one in the direction of the load action takes place in the surface layer [10]. At the same time, the non-monotonic nature of the change in $\delta_i(t)$ for A50 correlates with the dependence of the sizes of the crystallites $D(t)$ (Table 4 and Figure 14a). These microstructural changes are associated, among other things, with a change in the dislocation structure of the material, which determines the changes in the decrement δ_i. The same structural changes also affect the change in the microplastic stress $\sigma_s(t)$ (Figure 22), which demonstrates the positive effect of the steel core in overhead power line cables, leading to the preservation of a high level of σ_s in AC50 wires with a service life of up to 20 years for the studied samples.

Thus, EBSD, densitometry, and XRD methods have shown that the steel core in AC50 type cables leads to a slower change in NSDL of Al wires from cables of this type compared to wires from A50-type cables (without steel core). Remarkably, acoustic studies demonstrated less change and even improvement in the deformation characteristics of

AC50 wires compared to A50 wires, which correlates with the results of EBSD, densitometry, and XRD.

5. Conclusions

The studies carried out using EDX, EBSD, densitometry, XRD, and acoustic techniques have made it possible to quantitatively estimate the structural, microstructural, and deformation characteristics of Al wires from AAAC (A50) and ACSR (AC50, or AC50/8 in full designation) overhead power line cables after different service life periods from 0 to 20 years.

It has been found that NSDLs with defects of a void nature are formed in Al wires of both types. The integral density ρ_d, which is measured using the densitometric method, and the X-ray density ρ_x, which is calculated from the unit cell parameter a of the lattice of the Al material in NSDL of wires, decrease according to a law close to the exponential-decay law as the depth T from the surface decreases, although more gentle in the case of $\rho_x(T)$ than $\rho_d(T)$. With an increase in the operating time, the ρ_x and ρ_d values at depth ~12.5 μm near the wire surface decrease significantly (density defect $\Delta\rho_x/\rho_x \approx -1.1\%$ after a service life of ~20 years in comparison to $\Delta\rho_x/\rho_x \approx -0.2\%$ in non-exploited samples and $\approx 0\%$ in the bulk).

The characteristic thicknesses of NSDLs of ~10 μm, ~30 μm–40 μm, and ~100 μm–160 μm for the two types of wires after operation were established, corresponding to ~(80–85)% of the total drop of ρ_d, ~100% of the total decrease in the ρ_d and ~50%–70% of the total decrease in ρ_x, and a ~99% decrease in the X-ray mass density ρ_x, respectively. The difference in the results obtained by the methods of densitometric and XRD profiling is associated with the different sensitivity of the methods to different effects on the wires during operation. In the case of X-ray density ρ_x, which is calculated from structural data, the structural state of the Al material of the NSDL of wire will play a major role. The Al lattice of the NSDL of wire expands when approaching the surface, and, accordingly, the X-ray density ρ_x decreases, which is mainly due to the formation of defects of a void nature in the NSDL of wires under the action of vibrations due to wind, temperature fluctuations of the surrounding atmosphere, and fretting. In the case of densitometric density ρ_d, which is an integral value, an important role is also played by the contribution of other phases, aluminum oxides in particular, the crystallites of which are formed when the service life increases, which, in turn, leads to the formation of a larger number of void defects near the surface because of the fretting amplification.

The presence of a steel core plays a stabilizing role in changing the structural, microstructural, and deformation characteristics.

In the presence of a steel core in the AC50 type cables, the wires of this cable show a ~0.2% lower drop in both integral and X-ray densities in comparison with the wires of AAAC A50 cables with comparable service life durations. Furthermore, in the presence of a steel core (ACSR-type AC50 cable), the thickness T_{layer} of that part of the NSDL where the main decrease in X-ray density ρ_x (~50–70% of the total drop) occurs when approaching the surface is 1–2% less than that in wires from AAAC-type A50 cables without a steel core. For wires from AC50 cables, a smaller value of microstrains is observed, which are formed at depths $T \geq 15$ μm from the wire surface in NSDL after operation, compared to A50 cables, and it is associated with the steel core effect limiting cable vibrations because of wind.

The degradation rate of the average parameter a of the Al cubic unit cell of wires with a service life of 0–20 years and their density ρ_x, which is calculated from XRD data in NSDL with a thickness of ~35.5 μm, with the service life in the presence of a steel core is noticeably lower (respectively, $1.07(3) \cdot 10^{-4}$ Å/year and $-2.13(7) \cdot 10^{-4}$ g/cm^3/year for AC50 wires compared to $1.26(4) \cdot 10^{-4}$ Å/year and $-2.52(8) \cdot 10^{-4}$ g/cm^3/year for A50 wires). As a result, the expansion of the lattice in the NSDL and, accordingly, the decrease in the X-ray density ρ_x of the NSDL Al wire material from A50 type cables without a steel core occurs faster by ~1 year after a service life of ~10 years and by ~3 years after ~20 years of operation.

At the same time, in comparison with A50 cables, the presence of a steel core in AC50-type cables leads to a significantly smaller change in the deformation characteristics of AC50 wires and, even, to some improvement in their early stages of operation, lasting at least up to 8 years. This smaller change and improvement in the deformation characteristics of AC50 wires is associated with less development of defects in the near-surface layer of up to ~30 μm thick for wires made of AC50 type cables compared to wires made of all-aluminum A50 cables.

Supplementary Materials: The following supporting information can be downloaded at https://www.mdpi.com/article/10.3390/cryst12091267/s1, Section S1. X-ray diffraction details: Figure S1. Comparison of dependences of the average crystallite size *D*, calculated within the framework of the zero microstrain model ($\varepsilon_s = 0$), on the service life duration *t* for Al-wires of A50 and AC50 type cables of the overhead power lines; Figure S2. WHP and SSP of AC50 wires, without exploitation and after 8 (and 20 years of exploitation.; Figure S3. Distribution of the mass X-ray density $\rho_x(T)$ of the wire Al material along the depth *T* from the surface of the A50 and AC50 wires.

Author Contributions: A.A.L., Conceptualization, Investigation (XRD), Methodology, Validation, Project administration, Software, Data curation, Formal analysis, Visualization, Writing—original draft preparation, Writing—review and editing; M.V.N., Conceptualization, Investigation (densitometry, acoustic), Methodology, Validation, Software, Visualization, Writing—original draft preparation; A.I.L., Investigation (SEM, EDX, EBSD), Methodology, Software, Visualization, Writing—original draft preparation; B.K.K., Investigation (acoustic), Methodology, Software, Visualization; A.G.K., Conceptualization, Investigation (acoustic), Methodology, Software; N.D.P., Data curation; Validation, Project administration; A.G.P., Writing—original draft preparation, Conceptualization; R.V.S., Conceptualization, Methodology, Project administration; P.N.B., Supervision, Conceptualization, Methodology, Funding acquisition; M.M.S., Conceptualization, Methodology; Resources; Funding acquisition; A.V.S., Conceptualization, Methodology; Resources; I.A.B., Conceptualization, Methodology; Resources. All authors have read and agreed to the published version of the manuscript.

Funding: This research was funded by Russian Federation public contract No. FSWF-2020-0025 "Technique development and method analysis for ensuring power system object security and competitiveness based on the digital technologies".

Institutional Review Board Statement: Not applicable.

Informed Consent Statement: Not applicable.

Data Availability Statement: The data presented in this study are available on request from the corresponding author.

Acknowledgments: The work was carried out using the equipment and software of the Center of Joint Use "Materials Science and Diagnostics in Advanced Technologies" (Ioffe Institute, St. Petersburg, Russia).

Conflicts of Interest: The authors declare no conflict of interest.

Abbreviations

The following abbreviations and nomenclature are used in this manuscript:

A50	AAAC cable with an Al conductor cross-section of 49.5 mm^2
A7E	grade of cold-drawn aluminum according to GOST
AAAC	all aluminum alloy aonductor
AC50 (AC50/8)	ACSR cable with Al conductor cross-section of ~50 mm^2 and ~8 mm^2 cross-section of steel core
ACSR	aluminum conductor steel reinforced
a.m.u.	atomic mass unit
EBSD	electron backscattering diffraction
EDX	energy dispersive X-ray microanalysis
e.s.d., e.s.d.s.	estimated standard deviation(s)
FWHM	full width at half-maximum of XRD reflection
GOST	interstate technical standard
NSDL, NSDLs	near-surface defect layer(s)

PDF-2	Powder Diffraction File-2
pV	pseudo-Voigt
Ref., Refs.	Reference(s)
SEM	scanning electron microscopy
SSP	size-strain plot
SM	Supplementary Materials
WHP	Williamson–Hall plot
XRD	X-ray diffraction
e.g.,	Latin" exempli gratia" (for example)
i.e.,	Latin "id est" (that is)
etc.	Latin "et cetera" (and so on)
cf.	Latin "confer" (compare)
2θ	diffraction angle
$2\theta_B$	Bragg angle of XRD reflection corrected to angular corrections (zero shift $\Delta 2\theta_{zero}$ and displacement $\Delta 2\theta_{displ}$)
$2\theta_{obs}$	observed Bragg angle of XRD reflection
A_r	tabular value of the atomic mass of aluminium
a	cubic unit cell parameter of Al
a_0	parameter of the Al cubic unit cell in the bulk of a new (0 years of service) wire
a^{bulk}	parameter of the Al cubic unit cell in the bulk of wire of non-zero service life
B_{int}	integral width of XRD reflection
$C_{a.m.u.}$	conversion factor of a.m.u. into gram
D	average size of crystallites
D_0	mean square root value of crystallite size after averaging the $D^{hkl}{}_0$ values over all reflections in framework of model with absence of microstrains ($\varepsilon_s = 0$)
D^{hkl}	size of crystallite corresponding to XRD reflection with Miller indices *hkl*
t	size of crystallite corresponding to XRD reflection with Miller indices *hkl* in framework of model with absence of microstrains ($\varepsilon_s = 0$)
E	Young's modulus (modulus of elasticity, elastic modulus)
E_i	amplitude-independent Young's modulus
f	oscillation frequency of wire-samples
$FWHM_{corr}$	$FWHM_{obs}$ corrected for instrumental broadening
$FWHM_{obs}$	observed full width at half-maximum of XRD reflection
hkl	Miller indices of XRD reflection
[*hkl*]	Miller indices of crystallographic direction
$K_{Scherrer}$	coefficient in Scherrer equation
K_{strain}	coefficient in Wilson–Stokes equation
I_{max}	maximum intensity of XRD reflection
$I_{max}{}^{022}$, $I_{max}{}^{111}$, $I_{max}{}^{002}$	maximum intensitie of XRD reflection with Miller indices 022, 111, and 002, respectively
I_{int}	integral intensity of XRD reflection
$I_{int}{}^{Al}$	integral intensity of the strongest Al reflection
$I_{int}{}^{Al_2O_3}$	integral intensity of the 121 δ*-(212 δ-) Al_2O_3 reflection
l	wire-sample length
m_0	mass of the sample before polishing (etching)
m_i	mass of the sample after polishing of the *i*th layer
N5, N5-1, etc.	Number 5, Number 5-1, etc.
p	a parameter of wire Al material (unit cell parameter *a* or X-ray mass density ρ_x)
p_0	a bulk parameter of Al material of a new (0 years of service) wire (unit cell parameter a_0 or X-ray mass density ρ_x)
q	fraction of the Al_2O_3 crystalline phases in wires
R_0	radius of the wire-sample in the form of a cylindrical rod (before polishing (etching))
t	service life duration

T	depth from the wire surface or or penetratiun depth or temperature (in dependence on context)
T_{etch}	thickness of the layer removed by polishing (etching)
T_{layer}	layer thickness corresponding to depth where ~70% of the decrease in the mass density ρ_x occurs in comparison to the $\rho_x^{200\mu m}$ value
T_{layer}^{sat}	layer thickness corresponding to depth where ~99% of the decrease in the mass density ρ_x occurs in comparison to the $\rho_x^{200\mu m}$ value
T_{pen}^{hkl}	penetration depth of X-ray radiation, corresponding to XRD reflection with Miller indices *hkl*
V_{cell}	volume of the Al cubic unit cell
Z	number of formula units in the Al unit cell
$\Delta 2\theta_{displ}$	displacement correction
$\Delta 2\theta_{step}$	2θ step of XRD scanning
$\Delta 2\theta_{zero}$	correcting shift of the zero of the counter
$\Delta a/a$	defect (relative change) of the unit cell parameter a
$(\Delta E/E)_h$	amplitude-dependent Young's modulus defect (amplitude-dependent variation of E)
$\Delta\rho_d/\rho_d$	defect (relative change) of the mass density ρ_d
$\Delta\rho_{dL}/\rho_{dL}$	density defect value of the etched layer
$\Delta\rho_x/\rho_x$	defect (relative change) of the mass density ρ_x
δ	(logarithmic) decrement of the sample material ((logarithmic) decrement of elastic vibrations)
$\delta 2\theta_B$	e.s.d. of Bragg angle $2\theta_B$
$\delta\rho_d/\rho_d$	relative error of integral density ρ_d
δa	e.s.d. of the Al cubic unit cell parameter a
δ_i	amplitude-independent part of the logarithmic decrement δ (amplitude-independent decrement of elastic vibrations)
ε	vibrational strain amplitude (vibrational deformation)
ε_d	non-linear inelastic strain (non-linear inelastic deformation)
ε_s	absolute value of average magnitude of microstrains
ε_s^{hkl}	absolute value of microstrain corresponding to XRD reflection with Miller inices *hkl*
ε_s^{sat}	average microstrain ε_s^{sat} in the NSDL at depths $T \geq 15$ μm if the crystallite size is fixed and equal to the crystallite size at a depth of T ~12.5 μm
θ	half a diffraction angle
θ_B	half a Bragg angle of XRD reflection $2\theta_B$
λ	wavelength of Cu-$K_{\alpha 1}$ radiation (after correction of Cu-$K_{\alpha 2}$ contribution)
μ_l	the linear mass absorption coefficient of material
ρ	mass density
ρ_d	mass density of material measured by densitometry technique ('densitometric density', 'integral density')
ρ_{d0}	density of a sample before polishing (etching) measured by densitometry technique
ρ_{di}	density of a sample after polishing (etching) of the *i*th layer measured by densitometry technique
ρ_{dL}	mass density of the etched layer obtained by densitometry technique
ρ_x	mass density of material calculated according to XRD structural data ('XRD density', 'X-ray density')
$\rho_x^{12.5\mu m}$	XRD mass density at depth of 12.5 μm from the surface
$\rho_x^{200\mu m}$	XRD mass density at depth of 200 μm from the surface
ρ_x^{bulk}	XRD mass density in the bulk of the wire-sample
ρ_x^{Tlayer}	XRD mass density of Al material at a depth of T_{layer}
ρ_x^{sat}	XRD mass density of Al material at a depth of T_{layer}^{sat}
ρ_{x0}	XRD density in the bulk of a new (0 years of service) Al wire
$\rho_{x\,0\,years}$	mean density of the A50 wire N5-2 (service life of 0 years)
σ	amplitude of vibrational stress (microplastic deformation, microplastic stress)
σ_s	microplastic flow stress (microplastic stress, micro-flow stress) equal to σ at inelastic strain $\varepsilon_d = 5.0 \times 10^{-8}$
Φ	Euler angle (angle of nutation)
φ_1	Euler angle (angle of intrinsic rotation)
φ_2	Euler angle (angle of precession)

References

1. Nakhimova, L.I. (Ed.) *GOST 839-2019. Non-Insulated Conductors for Overhead Power Lines. Specifications*; Strojizdat: Moscow, Russia, 2019. (In Russian)
2. Karabay, S.; Ertürk, A.T.; Zeren, M.; Yamanoğlu, R.; Karakulak, E. Failure analysis of wire-breaks in aluminum conductor production and investigation of early failure reasons for transmission lines. *Eng. Fail. Anal.* **2018**, *83*, 47–56. [CrossRef]
3. Karabay, S.; Feyzullahoğlu, E. Determination of early failure sources and mechanisms for Al 99.7% and Al–Mg–Si alloy bare conductors used in aerial transmission line. *Eng. Fail. Anal.* **2014**, *38*, 1–15. [CrossRef]
4. Betekhtin, V.I.; Sklenicka, V.; Kadomtsev, A.G.; Kolobov, Y.R.; Narykova, M.V. Defect Structure and Thermomechanical Stability of Nano- and Microcrystalline Titanium Obtained by Different Methods of Intense Plastic Deformation. *Phys. Solid State* **2017**, *59*, 960–966. [CrossRef]
5. Betekhtin, V.I.; Kadomtsev, A.G.; Narykova, M.V. Evolution of a Defect Structure during Creep Tests of Ultrafine-Grained Metals and Alloys Produced by Severe Plastic Deformation. *Phys. Solid State* **2020**, *62*, 318–324. [CrossRef]
6. Brettschneider, S.; Fofana, I. Evolution of Countermeasures against Atmospheric Icing of Power Lines over the Past Four Decades and Their Applications into Field Operations. *Energies* **2021**, *14*, 6291. [CrossRef]
7. Lu, J.; Hu, J.; Fang, Z.; Qiao, X.; Zhang, Z. Electric Field Distribution and AC Breakdown Characteristics of Polluted Novel Lightning Protection Insulator under Icing Conditions. *Energies* **2021**, *14*, 7493. [CrossRef]
8. Zhou, Z.R.; Cardou, A.; Goudreau, S.; Fiset, M. Fundamental investigations of electrical conductor fretting fatigue. *Tribol. Int.* **1996**, *29*, 221–232. [CrossRef]
9. Azevedo, C.R.F.; Cescon, T. Failure analysis of aluminum cable steel reinforced (ACSR) conductor of the transmission line 1232 crossing the Parana River. *Eng. Fail. Anal.* **2002**, *9*, 645–664. [CrossRef]
10. Narykova, M.V.; Levin, A.A.; Prasolov, N.D.; Lihachev, A.I.; Kardashev, B.K.; Kadomtsev, A.G.; Panfilov, A.G.; Sokolov, R.V.; Brunkov, P.N.; Sultanov, M.M.; et al. The structure of the near-surface layer of the AAAC overhead power line wires after operation and its effect on their elastic, microplastic, and electroresistance properties. *Crystals* **2022**, *12*, 166. [CrossRef]
11. Kalombo, R.B.; Araujo, J.A.; Ferreira, J.L.A.; da Silva, C.R.M.; Alencar, R.; Capra, A.R. Assessment of the fatigue failure of an All Aluminium Alloy Cable (AAAC) for a 230 kV transmission line in the Center-West of Brazil. *Eng. Fail. Anal.* **2015**, *61*, 77–87. [CrossRef]
12. Matos, I.M.; Rocha, P.H.C.; Kalombo, R.B.; Veloso, L.A.C.M.; Araujo, J.A.; Castro, F.C. Fretting fatigue of 6201 aluminum alloy wires of overhead conductors. *Int. J. Fatique* **2020**, *141*, 105884. [CrossRef]
13. Kalombo, R.B.; Martínez, J.M.G.; Ferreira, J.L.A.; da Silva, C.R.M.; Araújo, J.A. Comparative fatigue resistance of overhead 1222 conductors made of aluminium and aluminium alloy: Tests and analysis. *Proc. Eng.* **2015**, *133*, 223–232. [CrossRef]
14. Kalombo, R.B.; Pestana, M.S.; Freire, R.C.S.; Ferreira, J.L.A.; da Silva, C.R.M.; Veloso, L.A.C.M.; Camara, E.C.B.; Araujo, J.A. Fatigue life estimation of an all aluminium alloy 1055 MCM conductor for different mean stresses using an artificial neural network. *Int. J. Fatique* **2020**, *140*, 105814. [CrossRef]
15. Lévesque, F.; Goudreau, S.; Cardou, A.; Cloutier, L. Strain Measurements on ACSR Conductors During Fatigue Tests I—Experimental Method and Data. *IEEE Trans. Power Deliv.* **2010**, *25*, 2825–2834. [CrossRef]
16. Rocha, P.H.C.; Díaz, J.I.M.; Silva, C.R.M.; Araújo, J.A.; Castro. F., C. Fatigue of two contacting wires of the ACSR Ibis 397.5 MCM conductor: Experiments and life prediction. *Int. J. Fatique* **2019**, *127*, 25–35. [CrossRef]
17. Kreislova, K.; Jaglova, M.; Turek, L.; Koukalova, A. Evaluation of corrosion of long-term exposed aluminium conductor. *Koroze a Ochrana Materiálu* **2013**, *57*, 25–34. [CrossRef]
18. De Araujo, M.M.; Manheimmer, W.A.; Serra, E. Corrosion aspects in aluminum electrical conductors. *Mater. Sci.* **1988**, *8*, 175–197. [CrossRef]
19. De la Fuente, D.; Otero-Huerta, E.; Morcillo, M. Studies of long-term weathering of aluminium in the atmosphere. *Corros. Sci.* **2007**, *49*, 3134–3148. [CrossRef]
20. Vargel, C. *Corrosion of Aluminium*, 2nd ed.; Elsevier: Amsterdam, The Netherlands, 2020. [CrossRef]
21. Sawada, S. Corrosion and overhead aluminum conductors and the prevention of it. *Smitomo Electr. Tech. Rev.* **1968**, *11*, 22–32.
22. Graedel, E. Corrosion mechanism for aluminum exposed to the atmosphere. *J. Electrochem. Soc.* **1989**, *136*, 204C–212C. [CrossRef]
23. Isozaki, M.; Adachi, K.; Hita, T.; Asano, Y. Study of corrosion resistance improvement by metallic coating for overhead transmission line conductor. *Electr. Eng. Jpn.* **2008**, *163*, 41–47. [CrossRef]
24. Harvard, D.G.; Bellamy, G.; Buchan, P.G.; Ewing, H.A.; Horrocks, D.J.; Krishnasamy, S.G.; Motlis, J.; Yoshiki-Gravelsins, K.S. Aged ACSR conductors. I. Testing procedures for conductors and line items. *IEEE Trans. Power Deliv.* **1992**, *7*, 581–587. [CrossRef]
25. Deng, Y.J.; Yu, J.C.; Xia, K.Q.; Yang, L. Corrosion Conditions Analysis of In-service ACSR Overhead Lines. *Appl. Mech. Mater.* **2013**, *446–447*, 753–758. [CrossRef]
26. Havard, D.G.; Bissada, M.K.; Fajardo, C.G.; Horrocks, D.J.; Meale, J.R.; Motlis, J.Y.; Tabatabai, M.; Yoshiki-Gravelsins K., S. Aged ACSR conductors. II. Prediction of remaining life. *IEEE Trans. Power Deliv.* **1992**, *7*, 588–595. [CrossRef]
27. Chen, G.H.; Wang, X.; Wang, J.Q.; Liu, J.J.; Zhang, T.; Tang, W.M. Damage investigation of the aged aluminium cable steel reinforced (ACSR) conductors in a high-voltage transmission line. *Eng. Fail. Anal.* **2012**, *19*, 13–21. [CrossRef]
28. Lequien, F.; Auzoux, Q.; Moine, G.; Rousseau, M.; Pasquier-Tilliette, S.; Holande, A.; Ammi, S.; Heurtault, S.; Prieur, P. Characterization of an aluminum conductor steel reinforced (ACSR) after 60 years of operation. *Eng. Fail. Anal.* **2021**, *120*, 105039. [CrossRef]

29. Kim, S.D.; Morcos, M.M. Mechanical deterioration of ACSR conductors due to forest fires. *IEEE Trans. Power Deliv.* **2003**, *18*, 271–276. [CrossRef]
30. Hakansson, E.; Predecki, P.; Kumosa, M.S. Galvanic Corrosion of High Temperature Low Sag Aluminum Conductor Composite Core and Conventional Aluminum Conductor Steel Reinforced Overhead High Voltage Conductors. *IEEE Trans. Realib.* **2015**, *64*, 928–934. [CrossRef]
31. Achiriloaiei, D.; Medeleanu, M. Studies on the Effects of Environmental Pollution on ACSR Conductors. *Rev. Chim.* **2019**, *70*, 3984–3986. [CrossRef]
32. Jaffrey, N.A.; Hettiwatte, S. Corrosion Detection in Steel Reinforced Aluminium Conductor Cables. In Proceedings of the Australasian Universities Power Engineering Conference (AUPEC), Perth, Australia, 28 September–1 October 2014. [CrossRef]
33. Karabay, S.; Önder, F.K. An approach for analysis in refurbishment of existing conventional HV–ACSR transmission lines with AAAC. *Electr. Power Syst. Res.* **2004**, *72*, 179–185. [CrossRef]
34. Kopsidas, K.; Rowland, S.M. A Performance Analysis of Reconductoring an Overhead Line Structure. *IEEE Trans. Power Deliv.* **2009**, *24*, 2248–2256. [CrossRef]
35. Kalombo, R.B.; Pestana, M.S.; Ferreira, J.L.A.; da Silva, C.R.M.; Araujo, J.A. Influence of the catenary parameter (H/w) on the fatigue life of overhead conductors. *Tribol. Int.* **2017**, *108*, 141–149. [CrossRef]
36. Ma, X.C.; Gao, L.; Zhang, J.X.; Zhang, L.C. Fretting Wear Behaviors of Aluminum Cable Steel Reinforced (ACSR) Conductors in High-Voltage Transmission Line. *Metals* **2017**, *7*, 373. [CrossRef]
37. Levin, A.A.; Narykova, M.V.; Lihachev, A.I.; Kardashev, B.K.; Kadomtsev, A.G.; Brunkov, P.N.; Panfilov, A.G.; Prasolov, N.D.; Sultanov, M.M.; Kuryanov, V.N.; et al. Modification of the structural, microstructural, and elastoplastic properties of aluminum wires after operation. *Metals* **2021**, *11*, 1955. [CrossRef]
38. Kul'kov, V.G.; Tyshkevich, V.N.; Kuryanov, V.N.; Sultanov, M.M.; Norov, D.S.; Narykova, M.V.; Kadomtsev, A.G.; Prasolov, N.D.; Brunkov, P.N.; Likhachev, A.I.; et al. Experimental studies of fatigue strength and surface electrical resistance of aluminum wire of overhead power transmission lines. *Saf. Reliab. Power Ind.* **2021**, *14*, 189–195. (In Russian) [CrossRef]
39. Nakhimova, L.I. (Ed.) *GOST 11069-2001. Primary Aluminium. Grades*; IPK Izdatel. Standartov: Moscow, Russia, 2004. (In Russian)
40. Bruker AXS. *Karlsruhe, Diffrac. Suite Eva*, Version 5.1.0.5. DIFFRAC.SUITE User Manual. Bruker AXS GmbH: Karlsruhe, Germany, 2019.
41. International Centre for Diffraction Data (ICDD). *Powder Diffraction File-2 Release 2014*; ICDD: Newton Square, PA, USA, 2014.
42. Maunders, C.; Etheridge, J.; Wright, N.; Whitfield, H.J. Structure and microstructure of hexagonal $Ba_3Ti_2RuO_9$ by electron diffraction and microscopy. *Acta Crystallogr. Sect. B Struct. Sci.* **2005**, *61*, 154–159. [CrossRef]
43. Levin, A.A. Program SizeCr for Calculation of the Microstructure Parameters from X-ray Diffraction Data. Preprint. 2022. Available online: https://www.researchgate.net/profile/Alexander-Levin-6/research (accessed on 5 June 2022).
44. Terlan, B.; Levin, A.A.; Börrnert, F.; Simon, F.; Oschatz, M.; Schmidt, M.; Cardoso-Gil, R.; Lorenz, T.; Baburin, I.A.; Joswig, J.-O.; et al. Effect of Surface Properties on the Microstructure, Thermal, and Colloidal Stability of VB_2 Nanoparticles. *Chem. Mater.* **2015**, *27*, 5106–5115. [CrossRef]
45. Terlan, B.; Levin, A.A.; Börrnert, F.; Zeisner, J.; Kataev, V.; Schmidt, M.; Eychmüller, A. A Size-Dependent Analysis of the Structural, Surface, Colloidal, and Thermal Properties of $Ti_{1-x}B_2$ (x = 0.03–0.08) Nanoparticles. *Eur. J. Inorg. Chem.* **2016**, *2016*, 3460–3468. [CrossRef]
46. Welzel, U.; Ligot, J.; Lamparter, P.; Vermeulen, A.C.; Mittemeijer, E.J. Stress analysis of polycrystalline thin films and surface regions by X-ray diffraction. *J. Appl. Crystallogr.* **2005**, *38*, 1–29. [CrossRef]
47. Nikanorov, S.P.; Kardashev, B.K. *Elasticity and Dislocation Inelasticity of Crystals*; Moscow Izdatel. Nauka: Moscow, Russia, 1985. (In Russian)
48. Mason, J.K.; Schuh, C.A. Representation of Texture. In *Electron Backscatter Diffraction in Material Science*, 2nd ed.; Schwartz, A.I., Kumar, M., Adams, B.L., Field, D.P., Eds.; Springer: Boston, MA, USA, 2009; pp. 45–46. [CrossRef]
49. Miller Jr, P.H.; DuMond, J.W.M. Tests for the Validity of the X-Ray Crystal Method for Determining N and e with Aluminum, Silver and Quartz. *Phys. Rev.* **1940**, *57*, 198–206. [CrossRef]
50. Otte, H.M.; Montague, W.G.; Welch, D.O. X-ray diffractometer determination of the thermal expansion coefficient of aluminum near room temperature. *J. Appl. Phys.* **1963**, *34*, 3149–3150. [CrossRef]
51. Tsybulya, S.; Kryukova, G. New X-ray powder diffraction data on δ-Al_2O_3. *Powder Diffr.* **2003**, *18*, 309–311. [CrossRef]
52. Fargeot, D.; Mercurio, D.; Dauger, A. Structural characterization of alumina metastable phases in plasma sprayed deposits. *Mater. Chem. Phys.* **1990**, *24*, 299. [CrossRef]
53. Kaibyshev, O.A.; Valiev, R.Z. *Grain Boundaries and Properties of Metals*; Metallurgiya: Moscow, Russia, 1987; pp. 156–157. (In Russian)
54. Tromans, D. Elastic anisotropy of HCP metal crystals and polycrystals. *Int. J. Res. Rev. Appl. Sci.* **2011**, *6*, 462–483.

crystals

MDPI

Article

The Griffith Crack and the Interaction between Screw Dislocation and Semi-Infinite Crack in Cubic Quasicrystal Piezoelectric Materials

Jiandong Pi [1,2] and Lianhe Li [1,2,*]

1 College of Mathematics Science, Inner Mongolia Normal University, Huhhot 010022, China
2 Inner Mongolia Center for Applied Mathematics, Inner Mongolia Normal University, Huhhot 010022, China
* Correspondence: nmglilianhe@163.com

Abstract: The Griffith crack problem and the interaction between screw dislocation and semi-infinite crack in cubic quasicrystal piezoelectric materials are studied by using the complex variable function method. The stress intensity factors and electric displacement intensity factors are obtained. The effects of the linear force and coupling elastic coefficient on the stress intensity factor of phonon field and phason fields are discussed in detail. By numerical examples, it is found that the linear force and the coupling elastic constant have a significant effect on the stress intensity factor.

Keywords: cubic quasicrystal piezoelectric materials; crack; screw dislocation; complex variable function method

Citation: Pi, J.; Li, L. The Griffith Crack and the Interaction between Screw Dislocation and Semi-Infinite Crack in Cubic Quasicrystal Piezoelectric Materials. *Crystals* **2022**, *12*, 1250. https://doi.org/10.3390/cryst12091250

Academic Editors: Ulrich Prahl, Sergey Guk and Faisal Qayyum

Received: 14 August 2022
Accepted: 30 August 2022
Published: 2 September 2022

1. Introduction

Since solid quasicrystals were reported in 1984, their physical properties have attracted extensive attention. In terms of the piezoelectric effect of quasicrystals, researchers have also done a lot of research. Hu et al. [1] derived the piezoelectric constants of two-dimensional (2D) and three-dimensional (3D) quasicrystals using thermodynamics and group representation theory. Altay and Dökmeci [2] studied the basic equation of quasicrystal piezoelectricity and gave the basic equation of the quasi-static electric effect of 3D quasicrystal elastic materials through differential and variational forms. Li et al. [3] studied the basic solution of the one-dimensional (1D) hexagonal quasicrystal piezoelectric problem, and expressed the basic solution by five quasi-harmonic functions using operator theory and generalized Almansi theorem. Yu et al. [4] derived the governing equations of each crystal system of 1D quasicrystal piezoelectric materials, and gave the general solutions of the governing equations by using the operator method and complex method.

Since the establishment of the elastic theory of quasicrystal piezoelectric materials, although many achievements have been made, most of the research has been focused on 1D and 2D quasicrystal piezoelectric materials [5–13]; there have been few studies on the fracture of 3D quasicrystal piezoelectric materials. Because the elasticity of quasicrystal piezoelectric materials is similar to that of quasicrystal materials, the defect problem of 3D cubic quasicrystal piezoelectric materials can be studied by means of the defect problem of 3D cubic quasicrystal materials. However, due to the piezoelectric effect, the defects of quasicrystal piezoelectric materials are more complicated. For cubic quasicrystal materials, some research results have focused on the defects problem. Zhou et al. [14] studied the anti-plane problem and mode III crack problem of cubic quasicrystals, resulting in new developments in the elastic theory of cubic quasicrystals. Zhang [15] studied the anti-plane conjugate crack problem; general solutions for the stress and strain of conjugate cracks in cubic quasicrystal were obtained. Gao et al. [16] studied the elliptical hole or crack problem, and obtained analytical expressions for both entire and asymptotic fields. Suo et al. [17] studied the effect of T stress on the cross-type cracks in cubic quasicrystals.

These studies focused on cubic quasicrystal materials, and the methods involved can be used to solve the defect problems of cubic quasicrystal piezoelectric materials. For cubic quasicrystal piezoelectric materials, Zhang et al. [18] studied the free vibration of plates, producing one of the few research studies focusing on the defect problem of cubic quasicrystal piezoelectric materials. In this paper, the Griffith crack problem and the interaction between screw dislocation and semi-infinite crack in 3D cubic quasicrystal piezoelectric materials are studied. The main factors affecting the stress intensity factor will be discussed.

2. Basic Equations of Cubic Quasicrystal Piezoelectric Material

In the spatial rectangular coordinate system $x_i(i = 1, 2, 3)$, the basic equations of the anti-plane elasticity problem for cubic quasicrystal piezoelectric materials can be expressed as follows:

The constitutive equation [19]

$$\begin{bmatrix} \sigma_{yz} \\ \sigma_{zx} \\ H_{yz} \\ H_{zx} \\ D_x \\ D_y \end{bmatrix} = \begin{bmatrix} C_{44} & 0 & R_3 & 0 & -e_{14} & 0 \\ 0 & C_{44} & 0 & R_3 & 0 & -e_{14} \\ R_3 & 0 & K_{44} & 0 & -d_{123} & 0 \\ 0 & R_3 & 0 & K_{44} & 0 & -d_{123} \\ e_{14} & 0 & d_{123} & 0 & \lambda_{11} & 0 \\ 0 & e_{14} & 0 & d_{123} & 0 & \lambda_{11} \end{bmatrix} \begin{bmatrix} 2\varepsilon_{yz} \\ 2\varepsilon_{zx} \\ 2\omega_{yz} \\ 2\omega_{zx} \\ E_x \\ E_y \end{bmatrix}, \tag{1}$$

the equilibrium equation (regardless of body force) [20]

$$\sigma_{iz,i} = 0, H_{iz,i} = 0, D_{i,i} = 0 \ (i = x, y, z), \tag{2}$$

and the geometric equation

$$\begin{aligned} \varepsilon_{yz} &= \frac{1}{2}\frac{\partial u_z}{\partial y}, \varepsilon_{zx} = \frac{1}{2}\frac{\partial u_z}{\partial x}, \omega_{yz} = \frac{1}{2}\frac{\partial w_z}{\partial y}, \\ \omega_{zx} &= \frac{1}{2}\frac{\partial w_z}{\partial x}, E_x = -\frac{\partial \Phi}{\partial x}, E_y = -\frac{\partial \Phi}{\partial y}, \end{aligned} \tag{3}$$

where u_z represents phonon field displacement, w_z represents phason field displacement, and Φ represents electric potential. The σ_{ij}, ε_{ij} represent the stress and strain of phonon field, respectively; H_{ij}, ω_{ij} represent the stress and strain of phason field, respectively; D_{ij} represents electric displacement; E_i represents electric field strength; C_{44} is the elastic coefficient of the phonon field; K_{44} is the elastic coefficient of the phason field; R_3 is the coupling coefficient between the phonon field and phason field; e_{14} and d_{123} are the piezoelectric constant; and λ_{11} is the dielectric constant.

By Equations (1)–(3), the governing equations expressed by displacement and electric potential can be obtained as follows:

$$\begin{aligned} C_{44}\nabla^2 u_z + R_3\nabla^2 w_z + 2e_{14}\frac{\partial^2 \Phi}{\partial x \partial y} &= 0, \\ R_3\nabla^2 u_z + K_{44}\nabla^2 w_z + 2d_{123}\frac{\partial^2 \Phi}{\partial x \partial y} &= 0, \\ 2e_{14}\frac{\partial^2 u_z}{\partial x \partial y} + 2d_{123}\frac{\partial^2 w_z}{\partial x \partial y} - \lambda_{11}\nabla^2 \Phi &= 0, \end{aligned} \tag{4}$$

where $\nabla^2 = \frac{\partial^2}{\partial x^2} + \frac{\partial^2}{\partial y^2}$ is the Laplace operator.

Thus, the anti-plane elasticity problem of cubic quasicrystal piezoelectric materials is reduced to solving the system of partial differential Equation (4) under appropriate boundary conditions.

3. Basic Solution for the Fracture Problem of Cubic Quasicrystal Piezoelectric Materials

Equation (4) can be rewritten [4] as follows:

$$AU = 0 \tag{5}$$

where $U = (u_z, w_z, \Phi)^T$ and A is a matrix of differential operators

$$A = \begin{bmatrix} C_{44}\nabla^2 & R_3\nabla^2 & 2e_{14}\frac{\partial^2}{\partial x\partial y} \\ R_3\nabla^2 & K_{44}\nabla^2 & 2d_{123}\frac{\partial^2}{\partial x\partial y} \\ 2e_{14}\frac{\partial^2}{\partial x\partial y} & 2d_{123}\frac{\partial^2}{\partial x\partial y} & -\lambda_{11}\nabla^2 \end{bmatrix}. \tag{6}$$

From Equation (6), the determinant of A can be given as follows:

$$|A| = a\frac{\partial^6}{\partial y^6} + b\frac{\partial^2}{\partial x^2}\frac{\partial^4}{\partial y^4} + b\frac{\partial^4}{\partial x^4}\frac{\partial^2}{\partial y^2} + a\frac{\partial^6}{\partial x^6}, \tag{7}$$

in which

$$a = \lambda_{11}\left(R_3^2 - C_{44}K_{44}\right), b = 4\left(2e_{14}d_{123}R_3 - d_{123}^2C_{44} - e_{14}^2K_{44}\right) + 3\lambda_{11}\left(R_3^2 - C_{44}K_{44}\right).$$

Now, we introduce a function F that satisfies the equation

$$\nabla_1^2\nabla_2^2\nabla_3^2F = 0 \tag{8}$$

where $\nabla^2 = \frac{\partial^2}{\partial x^2} + \frac{1}{t_i^2}\frac{\partial^2}{\partial y^2}(i = 1,2,3)$ and t_i^2 are the three characteristic roots of equation $at^6 - bt^4 + bt^2 - a = 0$.

According to the operator theory, the general solution of Equation (5) can be expressed as follows:

$$u_z = A_{i1}F, w_z = A_{i2}F, \Phi = A_{i3}F, (i = 1,2,3). \tag{9}$$

Let $i = 2$, then the cofactor of the matrix A can be expressed as follows:

$$\begin{aligned} A_{21} &= \lambda_{11}R_3\tfrac{\partial^4}{\partial x^4} + 2(\lambda_{11}R_3 + 2e_{14}d_{123})\tfrac{\partial^4}{\partial x^2\partial y^2} + \lambda_{11}R_3\tfrac{\partial^4}{\partial y^4}, \\ A_{22} &= -\lambda_{11}C_{44}\tfrac{\partial^4}{\partial x^4} - 2(\lambda_{11}C_{44} + 2e_{14}^2)\tfrac{\partial^4}{\partial x^2\partial y^2} - \lambda_{11}C_{44}\tfrac{\partial^4}{\partial y^4}, \\ A_{23} &= 2(e_{14}R_3 - d_{123}C_{44})\left(\tfrac{\partial^4}{\partial x^3\partial y} + \tfrac{\partial^4}{\partial x\partial y^3}\right). \end{aligned} \tag{10}$$

Substituting Equation (10) into Equation (9), one has the following:

$$u_z = \lambda_{11}R_3\frac{\partial^4F}{\partial x^4} + 2(\lambda_{11}R_3 + 2e_{14}d_{123})\frac{\partial^4F}{\partial x^2\partial y^2} + \lambda_{11}R_3\frac{\partial^4F}{\partial y^4}, \tag{11}$$

$$w_z = -\lambda_{11}C_{44}\frac{\partial^4F}{\partial x^4} - 2\left(\lambda_{11}C_{44} + 2e_{14}^2\right)\frac{\partial^4F}{\partial x^2\partial y^2} - \lambda_{11}C_{44}\frac{\partial^4F}{\partial y^4}, \tag{12}$$

$$\Phi = 2(e_{14}R_3 - d_{123}C_{44})\left(\frac{\partial^4F}{\partial x^3\partial y} + \frac{\partial^4F}{\partial x\partial y^3}\right). \tag{13}$$

Assuming that the form of the complex function $F(x,y)$ is $F(x,\mu y)$, then μ must satisfy the following characteristic equation [7]:

$$a\mu^6 + b\mu^4 + b\mu^2 + a = 0 \tag{14}$$

Equation (14) has six pure imaginary roots

$$\mu_1 = is_1 = i, \mu_2 = is_2 = i\frac{\sqrt{b+a}+\sqrt{b-3a}}{2\sqrt{a}}, \mu_3 = is_3 = i\frac{\sqrt{b+a}-\sqrt{b-3a}}{2\sqrt{a}},$$

$$\mu_4 = -\mu_1, \mu_5 = -\mu_2, \mu_6 = -\mu_3, \left(i = \sqrt{-1}, s_1 = 1, \mu_2\mu_3 = -1(or\ s_2s_3 = 1)\right)$$

According to the properties of the analytic function and Equation (8), the complex representation of the displacement function F can be given as follows:

$$F = 2Re\sum_{k=1}^{3} F_k(z_k), \tag{15}$$

where Re represents the real part and $F_k(z_k)$ represents three arbitrary analytic functions with the argument $z_k = x + \mu_k y$ $(k = 1, 2, 3)$.

By Equations (1), (3), (11)–(13), and (15), the complex representations of the stresses and electric potential can be obtained as follows:

$$\sigma_{yz} = -4e_{14}(d_{123}C_{44} - e_{14}R_3)Re\sum_{k=1}^{3}\left(\mu_k - \mu_k^3\right)F_k^{(5)}(z_k), \tag{16}$$

$$\sigma_{zx} = 4e_{14}(d_{123}C_{44} - e_{14}R_3)Re\sum_{k=1}^{3}\left(\mu_k^2 - \mu_k^4\right)F_k^{(5)}(z_k), \tag{17}$$

$$H_{yz} = Re\sum_{k=1}^{3}\left(l\mu_k + m\mu_k^3 + n\mu_k^5\right)F_k^{(5)}(z_k), \tag{18}$$

$$H_{zx} = Re\sum_{k=1}^{3}\left(n + m\mu_k^2 + l\mu_k^4\right)F_k^{(5)}(z_k), \tag{19}$$

$$D_x = 2\lambda_{11}(d_{123}C_{44} - e_{14}R_3)Re\sum_{k=1}^{3}\left(\mu_k - \mu_k^5\right)F_k^{(5)}(z_k), \tag{20}$$

$$D_y = -2\lambda_{11}(d_{123}C_{44} - e_{14}R_3)Re\sum_{k=1}^{3}\left(1 - \mu_k^4\right)F_k^{(5)}(z_k), \tag{21}$$

in which

$$\begin{aligned} l &= 2\lambda_{11}\left(R_3^2 - C_{44}K_{44}\right) - 4d_{123}^2C_{44} + 4e_{14}d_{123}R_3, \\ m &= 4\lambda_{11}\left(R_3^2 - C_{44}K_{44}\right) - 4d_{123}^2C_{44} - 8e_{14}^2K_{44} + 12e_{14}d_{123}R_3, \\ n &= 2\lambda_{11}\left(R_3^2 - C_{44}K_{44}\right). \end{aligned}$$

4. Griffith Crack Problem

Now, we study the Griffith crack problem with a crack length of $2a$, which is subjected to a pair of linear forces and charges at the distancefrom the origin of the coordinates. The linear force intensity of the phonon field and the phason field per unit length are Q_1 and Q_2, respectively, and the linear charge density is q, as shown in Figure 1.

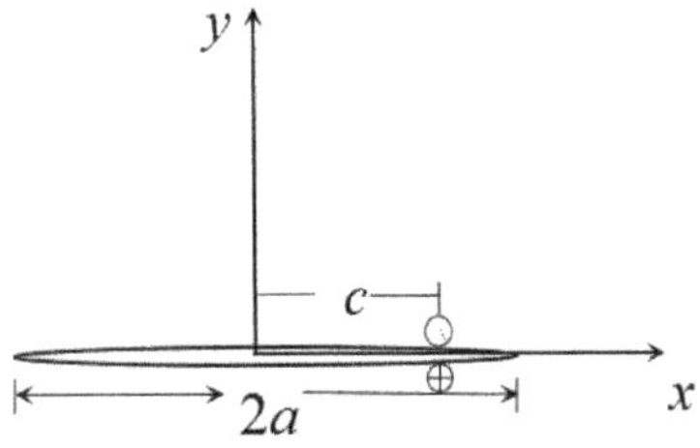

Figure 1. Griffith crack in cubic quasicrystal piezoelectric materials.

The complex function $F_k{}^{(5)}(z_k)$ is assumed as follows:

$$F_k{}^{(5)}(z_k) = \frac{A_k}{(z_k - c)\sqrt{z_k^2 - a^2}}, \tag{22}$$

the constants $A_k = \alpha_k + \beta_k i \ (k = 1, 2, 3)$ can be determined by the boundary conditions.

When $y = 0$, $|x| > a$, it is required to $u_z = 0, w_z = 0, \Phi = 0$, then

$$2\lambda_{11}R_3 Re\sum_{k=1}^{3} A_k + 4(\lambda_{11}R_3 + 2e_{14}d_{123})Re\sum_{k=1}^{3}\mu_k^2 A_k + 2\lambda_{11}R_3 Re\sum_{k=1}^{3}\mu_k^4 A_k = 0, \tag{23}$$

$$-2\lambda_{11}C_{44}Re\sum_{k=1}^{3} A_k - 4\left(\lambda_{11}C_{44} + 2e_{14}^2\right)Re\sum_{k=1}^{3}\mu_k^2 A_k - 2\lambda_{11}C_{44}Re\sum_{k=1}^{3}\mu_k^4 A_k = 0, \tag{24}$$

$$-4(d_{123}C_{44} - e_{14}R_3)\left(Re\sum_{k=1}^{3}\mu_k A_k + Re\sum_{k=1}^{3}\mu_k^3 A_k\right) = 0. \tag{25}$$

and then we can deduce the following:

$$\begin{aligned}
&\sum_{k=1}^{3}\left(2\lambda_{11}R_3 - 4(\lambda_{11}R_3 + 2e_{14}d_{123})s_k^2 + 2\lambda_{11}R_3 s_k^4\right)\alpha_k = 0,\\
&\sum_{k=1}^{3}\left(-2\lambda_{11}C_{44} + 4(\lambda_{11}C_{44} + 2e_{14}^2)s_k^2 - 2\lambda_{11}C_{44}s_k^4\right)\alpha_k = 0,\\
&\sum_{k=1}^{3}\left(s_k^3 - s_k\right)\beta_k = 0.
\end{aligned} \tag{26}$$

The integration along a small semi-circle around the point of action of a linear force and charge on $x = c$, one has

$$\begin{aligned}
\textstyle\sum_{k=1}^{3}\left(s_k + s_k^3\right)\beta_k &= \tfrac{Q_1\sqrt{a^2-c^2}}{4\pi e_{14}(C_{44}d_{123}-e_{14}R_3)},\\
\pi\lambda_{11}\left(R_3^2 - C_{44}K_{44}\right)&\sum_{k=1}^{3}\left(-\tfrac{1}{s_k} + s_k + s_k^3 - s_k^5\right)\beta_k\\
+4\pi e_{14}(d_{123}R_3 - e_{14}K_{44})&\sum_{k=1}^{3}\left(s_k + s_k^3\right)\beta_k = Q_2\sqrt{a^2 - c^2},\\
\sum_{k=1}^{3}\left(s_k^4 - 1\right)\alpha_k &= \tfrac{q\sqrt{a^2-c^2}}{2\pi\lambda_{11}(C_{44}d_{123}-e_{14}R_3)}.
\end{aligned} \tag{27}$$

Combining Equations (26) and (27), the expressions of the complex constant A_k can be obtained as follows:

$$\begin{aligned}
A_1 = i&\frac{Q_1\sqrt{a^2-c^2}\left(4e_{14}(e_{14}K_{44}-d_{123}R_3)s_3^2-\left(C_{44}K_{44}-R_3^2\right)\left(-1+s_3^2\right)^2\lambda_{11}\right)}{8\pi e_{14}(-C_{44}d_{123}+e_{14}R_3)\left(C_{44}K_{44}-R_3^2\right)\left(-1+s_3^2\right)^2\lambda_{11}}\\
&-i\frac{Q_2\sqrt{a^2-c^2}s_3^2}{2\pi\left(C_{44}K_{44}-R_3^2\right)\left(-1+s_3^2\right)^2\lambda_{11}},
\end{aligned}$$

$$
\begin{aligned}
A_2 &= \frac{q\sqrt{a^2-c^2}s_3^4}{4\pi(C_{44}d_{123}-e_{14}R_3)\left(-1+s_3^2\right)\left(1+s_3^2\right)\lambda_{11}} \\
&+i\frac{\sqrt{a^2-c^2}(Q_1(e_{14}K_{44}-d_{123}R_3)+Q_2(C_{44}d_{123}-e_{14}R_3))s_3^5}{2\pi(C_{44}d_{123}-e_{14}R_3)\left(C_{44}K_{44}-R_3^2\right)\left(-1+s_3^2\right)^2\left(1+s_3^2\right)\lambda_{11}}, \\
A_3 &= \frac{q\sqrt{a^2-c^2}}{4\pi(C_{44}d_{123}-e_{14}R_3)\left(-1+s_3^4\right)\lambda_{11}} \\
&+i\frac{\sqrt{a^2-c^2}(Q_1(e_{14}K_{44}-d_{123}R_3)+Q_2(C_{44}d_{123}-e_{14}R_3))s_3}{2\pi(C_{44}d_{123}-e_{14}R_3)\left(C_{44}K_{44}-R_3^2\right)\left(-1+s_3^2\right)^2\left(1+s_3^2\right)\lambda_{11}}.
\end{aligned} \tag{28}
$$

Thus, the complete elastic field can be obtained. After derivation, the corresponding stress and electrical displacement intensity factors can be expressed as follows:

$$K_\sigma = \frac{Q_1\sqrt{a+c}}{\sqrt{a-c}\sqrt{\pi a}}, \tag{29}$$

$$K_H = \frac{Q_2\sqrt{a+c}}{\sqrt{a-c}\sqrt{\pi a}}, \tag{30}$$

$$K_{D_y} = \frac{q\sqrt{a+c}}{\sqrt{a-c}\sqrt{\pi a}}. \tag{31}$$

Now take $a = 5 \times 10^{-3}$ m to analyze the influence of linear force Q_1 on the phonon field stress intensity factor K_σ (both the phason field stress intensity factor and the electric displacement intensity factor have similar variation rules with the phonon field stress intensity factor).

It can be seen from Figure 2 that the stress intensity factor of the phonon field increases with the increase of c, indicating that the farther the action point is from the origin, the greater the stress intensity factor. Furthermore, the stress intensity factor increases with the increase of linear force Q_1, and the farther the action point is from the origin, the greater the influence of the linear force.

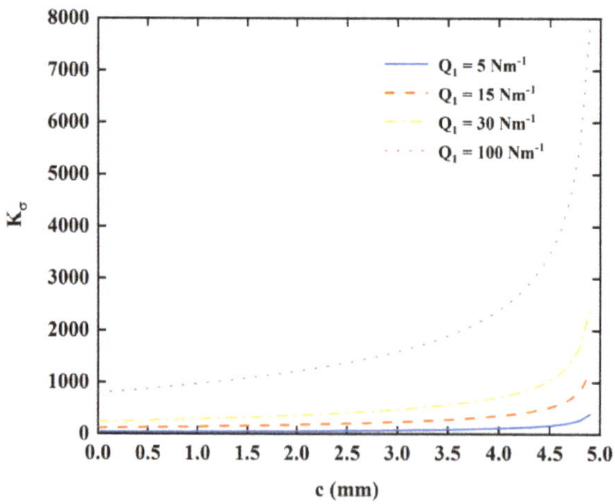

Figure 2. Influence of linear force Q_1 on the phonon field stress intensity factor K_σ.

5. Interaction between Screw Dislocation and Semi-Infinite Crack

In this section, the interaction between a semi-infinite crack and a screw dislocation with the Burgers vector $B = (b_1, b_2)$ is studied. Suppose that the dislocationi s located at the distance c from the semi-infinite crack tip, as shown in Figure 3.

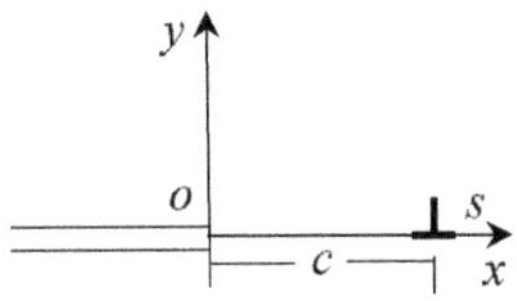

Figure 3. A screw dislocation near a semi-infinite crack.

Suppose the analytical formula F_k has the following form:

$$F_k^{(5)}(z_k) = \frac{B_k}{(z_k - c)\sqrt{z_k}}, (k = 1,2,3). \tag{32}$$

the constants $B_k = \gamma_k + \delta_k i$ $(k = 1,2,3)$.

When $x < 0$, make $\sigma_{yz} = 0, H_{yz} = 0, D_y = 0$, we have the following:

$$\begin{gathered} \sum_{k=1}^{3}(s_k + s_k^3)\gamma_k = 0, \\ \sum_{k=1}^{3}(ls_k - ms_k^3 + ns_k^5)\gamma_k = 0, \\ \sum_{k=1}^{3}(1 - s_k^4)\delta_k = 0. \end{gathered} \tag{33}$$

when $x = c$, we have the following:

$$\begin{gathered} 2\pi\sum_{k=1}^{3}[-\lambda_{11}R_3 + 2(\lambda_{11}R_3 + 2e_{14}d_{123})s_k^2 - \lambda_{11}R_3 s_k^4]\delta_k = b_1\sqrt{c}, \\ 2\pi\sum_{k=1}^{3}[\lambda_{11}C_{44} - 2(\lambda_{11}C_{44} + 2e_{14}^2)s_k^2 + \lambda_{11}C_{44}s_k^4]\delta_k = b_2\sqrt{c}, \\ \sum_{k=1}^{3}(s_k^3 - s_k)\gamma_k = 0. \end{gathered} \tag{34}$$

The following can be obtained from Equations (33) and (34):

$$\gamma_1 = \gamma_2 = \gamma_3 = 0, \tag{35}$$

$$\delta_1 = \frac{\sqrt{c}\left(b_1\left(4e_{14}^2 s_3^2 - C_{44}(-1 + s_3^2)^2\lambda_{11}\right) + b_2\left(4d_{123}e_{14}s_3^2 - R_3(-1 + s_3^2)^2\lambda_{11}\right)\right)}{8\pi e_{14}(-C_{44}d_{123} + e_{14}R_3)(-1 + s_3^2)^2\lambda_{11}}, \tag{36}$$

$$\delta_2 = \frac{\sqrt{c}(b_2 d_{123} + b_1 e_{14})s_3^4}{4\pi(C_{44}d_{123} - e_{14}R_3)(-1 + s_3^2)^2\lambda_{11}}, \tag{37}$$

$$\delta_3 = \frac{\sqrt{c}(b_2 d_{123} + b_1 e_{14})}{4\pi(C_{44}d_{123} - e_{14}R_3)(-1 + s_3^2)^2\lambda_{11}}. \tag{38}$$

Substituting Equations (35)–(38) into Equations (16) and (18), the analytical expressions of the stresses were obtained, as follows:

$$\begin{gathered} \sigma_{yz} = 4e_{14}(d_{123}C_{44} - e_{14}R_3)\sum_{k=1}^{3}\frac{(s_k + s_k^3)\delta_k}{(x-c)\sqrt{x}}, \\ H_{yz} = \sum_{k=1}^{3}\frac{(-ls_k + ms_k^3 - ns_k^5)\delta_k}{(x-c)\sqrt{x}}. \end{gathered} \tag{39}$$

Substituting Equation (39) into the expressions of the stress intensity factors for phonon field and phason field, respectively, and introducing

$$K_\sigma = \lim_{x\to 0}\sqrt{2\pi x}\sigma_{yz},$$
$$K_H = \lim_{x\to 0}\sqrt{2\pi x}H_{yz}.$$

then one has the following:

$$K_\sigma = -\frac{\sqrt{2}\left(b_1\left(2e_{14}^2 s_3 + C_{44}(1+s_3)^2\lambda_{11}\right) + b_2\left(2d_{123}e_{14}s_3 + R_3(1+s_3)^2\lambda_{11}\right)\right)}{\sqrt{\pi c}(1+s_3)^2\lambda_{11}}, \tag{40}$$

$$\begin{aligned} K_H = b_1 & \left(\frac{2\left(2e_{14}^2(e_{14}K_{44}-d_{123}R_3)s_3-C_{44}d_{123}R_3(1+s_3)^2\lambda_{11}\right)}{\sqrt{2\pi c}(C_{44}d_{123}-e_{14}R_3)(1+s_3)^2\lambda_{11}} \right. \\ & \left. -\frac{e_{14}\left(C_{44}K_{44}(-1+s_3)^2-R_3^2\left(1+s_3^2\right)\right)}{\sqrt{2\pi c}(C_{44}d_{123}-e_{14}R_3)s_3} \right) \\ & +b_2\left(\frac{2e_{14}R_3\left(-2d_{123}^2 s_3+K_{44}(1+s_3)^2\lambda_{11}\right)}{\sqrt{2\pi c}(C_{44}d_{123}-e_{14}R_3)(1+s_3)^2\lambda_{11}} \right. \\ & \left. +\frac{\left(d_{123}\left(4e_{14}^2K_{44}s_3^2+(1+s_3)^2\left(R_3^2(-1+s_3)^2-C_{44}K_{44}\left(1+s_3^2\right)\right)\lambda_{11}\right)\right)}{\sqrt{2\pi c}(C_{44}d_{123}-e_{14}R_3)s_3(1+s_3)^2\lambda_{11}} \right). \end{aligned} \tag{41}$$

The influence of coupled elastic coefficient on stress intensity factor will be discussed in detail by numerical examples. To this end, the material properties have been chosen by referring to the previous work [3], such as follows:

$$C_{44} = 50\times10^9\ \mathrm{Nm}^{-2},\ R_3 = 0.5\times10^9\ \mathrm{Nm}^{-2},\ K_{44} = 0.3\times10^9\ \mathrm{Nm}^{-2},$$
$$e_{14} = -0.138\ \mathrm{Cm}^{-2}, d_{123} = -0.16\ \mathrm{Cm}^{-2}, \lambda_{11} = 82.6\times10^{-12}\ \mathrm{C}^2\mathrm{N}^{-1}\mathrm{m}^{-2}.$$

Other parameters are $x = 1\times10^{-3}$ m, $b_1 = 1.6\times10^{-9}$ m, $b_2 = 10.7\times10^{-9}$ m.

It can be seen from Figures 4 and 5 that the coupling elastic coefficient R_3 has a significant impact on both the phonon and phason filed stress intensity factors. When the distance between the dislocation and the crack tip is determined, it can be found that the stress intensity factor decreases with the increase of R_3.

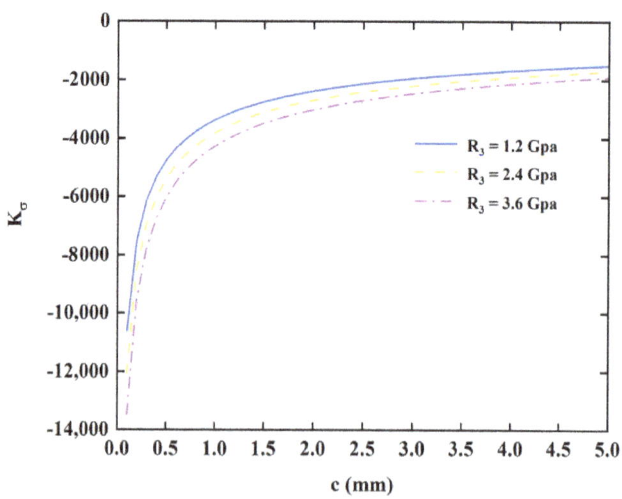

Figure 4. Influence of coupling elastic coefficient R_3 on the phonon filed stress intensity factor.

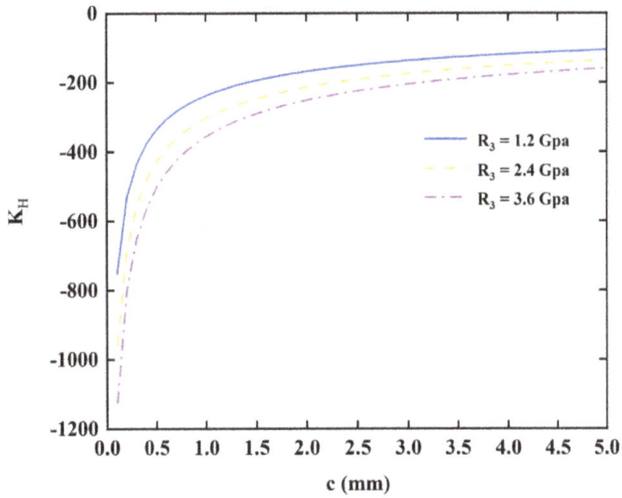

Figure 5. Influence of coupling elastic coefficient R_3 on the phason field stress intensity factor.

6. Conclusions

The Griffith crack problem and the interaction between screw dislocation and semi-infinite crack in cubic quasicrystal piezoelectric materials were studied. With the help of complex function theory, the analytical expression of stress intensity factor was obtained. Numerical examples showed that for the Griffith crack problem, the magnitude of the linear force and the distance from the crack tip have obvious effects on stress intensity factor, while for the interaction between screw dislocation and semi-infinite crack problem, the coupling elastic constant of the phonon and phason fields has a significant effect on the stress intensity factor.

Author Contributions: Project administration, Writing—review & editing, J.P.; Conceptualization, Methodology, L.L. All authors have read and agreed to the published version of the manuscript.

Funding: This study was supported by the National Natural Science Foundation of China (Nos. 11962026, 12002175, 12162027 and 62161045), the Natural Science Foundation of Inner Mongolia (Nos. 2020MS01018, 2021MS01013) and the Fundamental Research Funds for Inner Mongolia Normal University (Nos. 2022JBYJ020, 2022JBZD010).

Institutional Review Board Statement: Not applicable.

Informed Consent Statement: Not applicable.

Data Availability Statement: Not applicable.

Conflicts of Interest: The authors declare that they have no conflict of interest with this work. We declare that we do not have any commercial or associative interest that represents a conflict of interest in connection with the work submitted.

References

1. Hu, C.Z.; Wang, R.H.; Ding, D.H.; Yang, W.G. Piezoelectric effects in quasicrystals. *Phys. Rev. B* **1997**, *56*, 2463–2468. [CrossRef]
2. Altay, G.; Dökmeci, M.C. On the fundamental equations of piezoelasticity of quasicrystal media. *Int. J. Solids Struct.* **2012**, *49*, 3255–3262. [CrossRef]
3. Li, X.Y.; Li, P.D.; Wu, T.H.; Shi, M.X.; Zhu, Z.W. Three-dimensional fundamental solutions for one-dimensional hexagonal quasicrystal with piezoelectric effect. *Phys. Lett. A* **2014**, *378*, 826–834. [CrossRef]
4. Yu, J.; Guo, J.H.; Pan, E.N.; Xing, Y.M. General solutions of plane problem in one-dimensional quasicrystal piezoelectric materials and its application on fracture mechanics. *Appl. Math. Mech. (Engl. Ed.)* **2015**, *36*, 793–814. [CrossRef]
5. Zhou, Y.B.; Liu, G.T. Electroplastic analysis of anti-plane type III crack in one-dimensional hexagonal quasicrystal piezoelectric materials. *Chin. J. Solid Mech.* **2015**, *36*, 63–68.

6. Guo, J.h.; Zhang, Z.Y.; Xing, Y.M. Antiplane analysis for an elliptical inclusion in 1D hexagonal piezoelectric quasicrystal composites. *Philos. Mag.* **2016**, *96*, 349–369. [CrossRef]
7. Cui, X.W.; Li, L.H. Anti-plane problem of finite large one-dimensional hexagonal quasicrystal piezoelectric wedge with screw dislocation. *Chin. J. Appl. Mech.* **2019**, *36*, 1058–1062.
8. Zhou, Y.B.; Liu, G.T.; Li, L.H. Effect of T-stress on the fracture in an infinite one-dimensional hexagonal piezoelectric quasicrystal with a Griffith crack. *Eur. J. Mech.—A/Solids* **2021**, *86*, 104184. [CrossRef]
9. Loboda, V.; Sheveleva, A.; Komarov, O.; Lapusta, Y. An interface crack with mixed electrical conditions at it faces in 1D quasicrystal with piezoelectric effect. *Mech. Adv. Mater. Struct.* **2022**, *29*, 3334–3344. [CrossRef]
10. Jiang, L.J.; Liu, G.T. The interaction between a screw dislocation and a wedge-shaped crack in one-dimensional hexagonal piezoelectric quasicrystals. *Chin. Phys. B* **2017**, *26*, 249–255. [CrossRef]
11. Zhang, F.; Li, X. The Analytic Solutions of a Circular Hole with Four Cracks of One-Dimensional Hexagonal Piezoelectric Quasicrystals. *J. Jiangxi Norm. Univ. (Natural Sci. Ed.)* **2015**, *39*, 50–54.
12. Bai, Q.M.; Ding, S.H. Anti-plane problem of hexagonal hole edge crack in one-dimensional hexagonal quasicrystal piezoelectric. *Appl. Math. Mech.* **2019**, *40*, 1071–1080.
13. Pi, J.D.; Zhao, Y.; Li, L.H. Interaction between a ScrewDislocation and Two Unequal Interface racks Emanating from an Elliptical Hole in One Dimensional Hexagonal Piezoelectric Quasicrystal Bi-Material. *Crystals* **2022**, *12*, 314. [CrossRef]
14. Zhou, W.M.; Fan, T.Y.; Yin, S.Y.; Wang, N.P. Anti-Plane Elasticity Problem and Mode III Crack Problem of Cubic Quasicrystal. *J. Beijing Inst. Technol. (Engl. Ed.)* **2001**, *10*, 250–254.
15. Zhang, L. Stress Intensity of Antiplane Conjugate Cracks in Cubic Quasicrystal. *J. Southwest JT. Univ. (Engl. Ed.)* **2008**, *16*, 285–289.
16. Gao, Y.; Ricoeur, A.; Zhang, L. Plane problems of cubic quasicrystal media with an elliptic hole or a crack. *Phys. Lett. A* **2011**, *375*, 2775–2781. [CrossRef]
17. Suo, Y.R.; Zhou, Y.B.; Liu, G.T. Effect of T stress on the presence of cross-shaped cracks in an optically shaped quasicrystal. *Acta Mech. Solida Sin.* **2022**, *43*, 95–110.
18. Zhang, J.M.; Mao, Z.H.; Feng, X.; Zhang, L.L.; Gao, Y. Free Vibration of Three-Dimensional Piezoelectric Cubic Quasicrystal Plates. In Proceedings of the 2020 15th Symposium on Piezoelectrcity, Acoustic Waves and Device Applications (SPAWDA), Zhengzhou, China, 16–19 April 2021.
19. Wang, R.H.; Hu, C.Z.; Gui, J.N. *Quasicrystal Physics*; Science Press: Beijing, China, 2004.
20. Gao, Y. Governing equations and general solutions of plane elasticity of cubic quasicrystals. *Phys. Lett. A* **2009**, *8*, 885–889. [CrossRef]

crystals

MDPI

Article

Intergranular Cracking in Mg-Gd-Y Alloy during Tension Test

Jianhua Liu [1,2], Jie Sun [1,2,*], Qingqiang Chen [1,2] and Laixiao Lu [1,2]

1 School of Mechanical and Electronic Engineering, Shandong Jianzhu University, Jinan 250101, China; ljh@sdjzu.edu.cn (J.L.); 13770@sdjzu.edu.cn (Q.C.); lulaixiao@sdjzu.edu.cn (L.L.)
2 Research Center of High Performance Materials Forming, Shandong Jianzhu University, Jinan 250101, China
* Correspondence: sunjie20@sdjzu.edu.cn; Tel.: +86-0531-86367021; Fax: +86-0531-86367021

Abstract: The intergranular cracking in the Mg-Gd-Y alloy was investigated by an in situ tension test combined with an electron backscattered diffraction (EBSD) measurement and digital image correlation (DIC). During the tension test, the crack was found at the triangle point of the grain boundary area with profuse slip traces. DIC results show that inhomogeneous strain distribution can be found at the triangle point area, leads to serious deformation incompatibility at this point. This also leads to a weak point for crack initiation, and as tension strain increases, the crack would extend from the weak point along the grain boundary.

Keywords: magnesium alloy; ductility; grain boundary; characteristics; electron microscopy

Citation: Liu, J.; Sun, J.; Chen, Q.; Lu, L. Intergranular Cracking in Mg-Gd-Y Alloy during Tension Test. *Crystals* **2022**, *12*, 1040. https://doi.org/10.3390/cryst12081040

Academic Editor: Sergio Brutti

Received: 10 July 2022
Accepted: 25 July 2022
Published: 27 July 2022

1. Introduction

The elongation of wrought magnesium at ambient temperature can achieve 20% or even more, which is larger than that of traditional aluminum alloys [1–4]. However, the formability of wrought magnesium still cannot be compared with aluminum alloys at ambient temperature, which impedes the development of wrought magnesium.

It is common to think that damage nucleation during the deformation process occurs at locations with large strain concentrations, where substantial heterogeneous deformation occurs. If large local strains are effective in accommodating the required geometry changes, they may prevent damage nucleation, whereas it is conceivable that damage may nucleate where insufficient strain or shape accommodation occurs. Such variability in shape accommodation is connected to crystal orientations and crystallographic deformation mechanisms. In previous studies, cracking was thought to be related to {10–12} tension twinning [5], {10–11}–{10–12} double twinning [6,7], persistent slip bands [8] or grain boundaries [9]. J. Koike et al. [7] thought {10–11}–{10–12} double twinning would lead to the formation of large surface steps, cracks and final failure. An Luo et al. [10] thought that fatigue cracks mainly exist as transgranular cracks, and the crack propagation path is related to the orientation of target grains and the loading direction. Small cracks usually initiate on basal planes in grains with a larger Schmid factor.

This paper focused on the microstructure evolution and fracture behavior in the Mg-Gd-Y alloy, which is critical to its ductility. In situ tension tests were used in combination with electron backscattered diffraction (EBSD) and digital image correlation (DIC) to correlate the activation of deformation modes and fracture behavior. Strain accommodation in the grain boundary area was carefully analyzed, and the effects of grain orientation and grain boundary compatibility were discussed.

2. Experimental Procedure

The alloy used had a chemical composition of Mg-8.0Gd-3.0Y (wt%) (GW83). It was prepared by extrusion at 450 °C with an extrusion ratio of 25:1.

The in situ tensile test was performed using a micro-test system provided by Deben UK Ltd, Suffolk UK. with a load cell of 5 KN capacity, installed inside an FEI Quanta

250 FEG SEM. Before tensile tests, the sample surfaces were mechanically polished using a sequence of ethanol-based diamond suspensions of 6, 3 and 1 μm. This was followed by fine-polishing using a colloidal silica suspension (OPS), and a final 2–4 s etching using a solution of 5% HNO_3, 15% acetic acid, 20% H_2O and 60% ethanol before SEM and EBSD observations. The microstructure and grain orientation analysis was performed via EBSD (TSL^{TM} OIM) measurements. Micro grids, illustrated in this study that had a 1 μm step size and 20 μm width were deposited on polished surfaces of the tension samples by using an FEI Nova 600 I dual-beam focused ion beam miller. During the tensile test, the samples were deformed at room temperature at a displacement rate of 0.5 mm/min along the loading direction. The loading test was carried out in steps. SEM pictures from the same area of the sample surface were collected during interrupted loading with the displacement held.

3. Results and Discussion

3.1. Initial Microstructure

Figure 1a shows the microstructure from the longitudinal section of the extruded bar. The specimen exhibits a full recrystallization microstructure with obvious extrusion streaks along the extrusion direction. Figure 1b shows the (0001) PF (pole figure) of the extruded tube. The texture intensity is only 2.752, which is much lower than that of an extruded AZ31 alloy. Figure 1c shows the grain size distribution of this alloy, and the average grain size is 15 μm. Figure 1d shows the stress–strain curve of this alloy. Yield strength and ultimate strengths are 180 MPa and 295 MPa, respectively. The elongation of this alloy reached 24% due to the randomized texture.

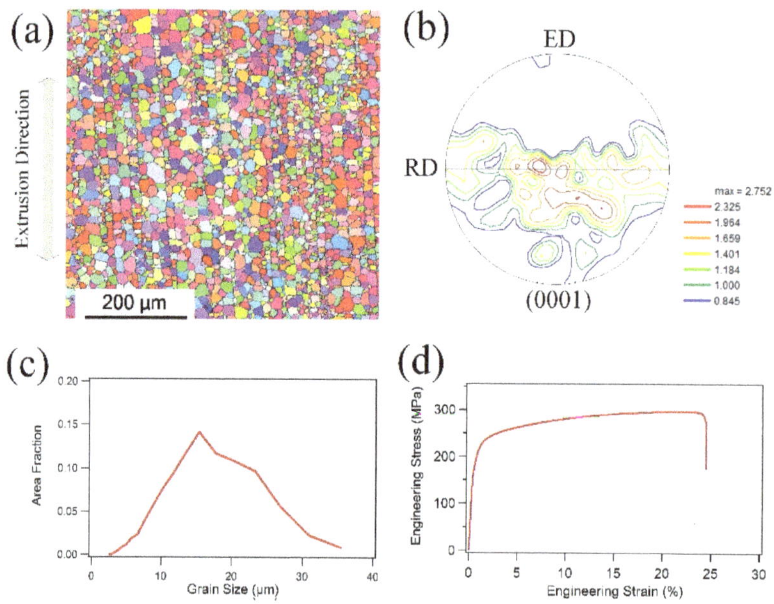

Figure 1. (**a**) Initial microstructure, (**b**) (0001) pole figure (**c**) grain size distribution and (**d**) stress-strain curve of this alloy.

3.2. Microstructure Evolution during Tension Test

Figure 2 shows the microstructure during the in situ tension test. At the strain of 9.5% (Figure 2a), a small amount of slip trace can be observed in Grain 1. This kind of straight slip trace was thought to be a basal slip trace. However, no obvious slip trace can be found in Grain 2. At a strain of 11.5% (Figure 2b), the slip trace in Grain 1 becomes denser at the boundary close to Grain 2. As the strain increases to 12.5%, the slip trace grows much denser than before and an obvious change in the shape of the grain boundary can be found between Grain 1 and Grain 2. At the strain of 13.5%, a crack initiated at the boundary between Grain 1 and Grain 2 with an obvious slip trace in Grain 1 and negligible shape change in Grain 2. As the strain increases to 15.5%, the crack starts to propagate. Due to the in situ observation at the initiation of the grain boundary crack, it can be deduced that the slip close to the boundary and strain concentration had a significant effect on the initiation of the grain boundary crack. Moreover, noticeable fluctuation at the grain boundaries can be found as the strain increased. This indicates that strain accommodation is incompatible at the grain boundary area, which causes strain concentration during the deformation process.

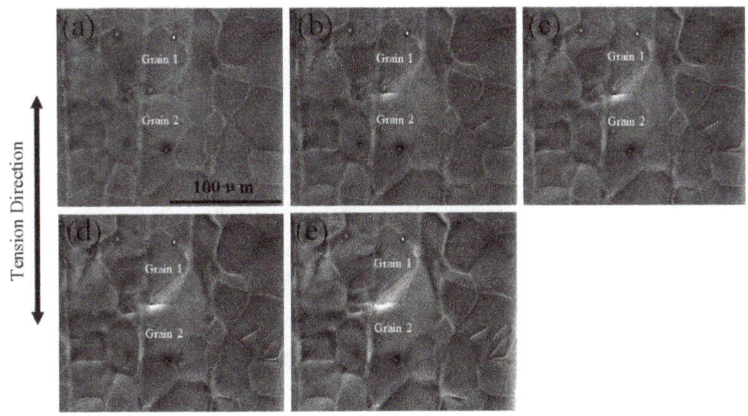

Figure 2. Microstructure during in situ tension test at strain of (**a**) 9.5%, (**b**) 11.5%, (**c**) 12.5%, (**d**) 13.5%, (**e**) 15.5%.

In order to quantitatively analyze the effect of the slip and strain concentration of the initiation of the grain boundary crack, the grid was marked on the surface of the polished tension sample to calculate the strain distribution after deformation. Figure 3 shows the deformed microstructure during tension at the strain of 0.15. Profuse slip traces can be found in grains and transferred slip traces can be found in grain boundary areas. The red lines in Figure 3 indicate basal slip traces, which were drawn by OIM software, and the blue lines indicate prismatic slip traces. An obvious crack can be found between Grain 1 and Grain 2. Furthermore, this crack was linked to the triangle point among Grain 1, Grain 2 and Grain 3.

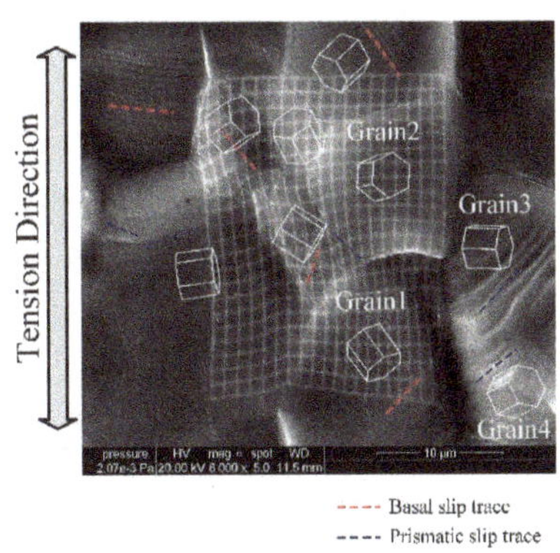

Figure 3. Microstructure after 0.15 tension strain.

The strain tensor caused by the activation of a single slip (x-axis is the a-axis in the crystallography coordinates and z-axis is the c-axis in the crystallography coordinates) is assumed as:

$$\varepsilon = \begin{bmatrix} 0 & 0 & 1 \\ 0 & 0 & 0 \\ 1 & 0 & 0 \end{bmatrix} \quad (1)$$

The direction of the crack is set as the y-axis and the direction perpendicular to the crack is set as the x-axis (the crack coordinate); the y-axis is approximately perpendicular to the tension direction. The strain caused by the basal slip can be calculated and the results are:

$$\varepsilon_1 = \begin{bmatrix} 0.7294 & -0.6368 & 0.0481 \\ -0.6368 & -0.5651 & 0.3912 \\ 0.0481 & 0.3912 & -0.1643 \end{bmatrix} \quad \varepsilon_2 = \begin{bmatrix} 0.0903 & -0.3818 & -0.8464 \\ -0.3818 & -0.2953 & 0.2630 \\ -0.8464 & 0.2630 & 0.2051 \end{bmatrix} \quad (2)$$

ε_1 is the strain tensor caused by the basal slip in Grain 1 and ε_2 is the strain tensor caused by the basal slip in Grain 2. ε_{yy} in Grain 1 is -0.5651 and the absolute value is much higher than that in Grain 2. The negative value indicates a compressive strain, which is induced by the activity of the basal slip. Thus, a large compressive strain is induced by the activity of the basal slip in Grain 1. The DIC analysis shown in Figure 4 confirmed the calculation above, that a large compressive strain exists at Grain 1 close to the boundary. The macro engineering strain is 0.15 and the local strain achieves -0.2 along the x-axis, which shows a huge strain concentration.

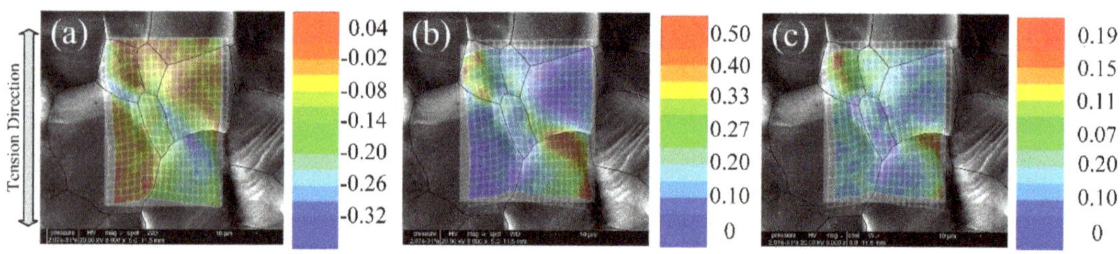

Figure 4. DIC analysis of (**a**) ε_{yy}, (**b**) ε_{xx} and (**c**) γ_{yx} at the strain of 0.15 (x-axis is parallel to the tension direction).

3.3. Analysis of Crack Initiation

Figure 5a,b show microstructure evolution during tension at the strain of 0.1 and 0.15 superimposed with grain maps. Strain concentration can be found at the grain boundary area between Grain 1 and Grain 2, especially at the strain of 0.15. This is because the crack extended along the grain boundary as the strain increased. Strain amplitude at the GB1 area is also high compared to that at the GB2 area. Moreover, both cracks at these two strain conditions are linked to the triangle point among Grain 1, Grain 2 and Grain 3.

It is a common thought that crack initiation is related to the activation of deformation modes in the surrounding area. Therefore, firstly, the Schmid factors of basal slip systems and prismatic systems in Grains 1–4 were calculated and shown in Table 1. The Schmid factor of the basal slip system in Grain 1 is 0.486, which is close to 0.5. Additionally, a large amount of basal slip traces can be found in Grain 1, especially in the grain boundary area near Grain 3 and Grain 4 in Figure 3. Conversely, in Grain 2, the Schmid factor of the prismatic slip system shows the highest value at 0.483. However, slip traces cannot be found obviously in Grain 2, especially at the grain boundary area near Grain 3 in Figure 3. This is because activation of the slip in the magnesium alloy needs not only its preferred grain orientation, but also needs good deformation compatibility in the adjacent grains.

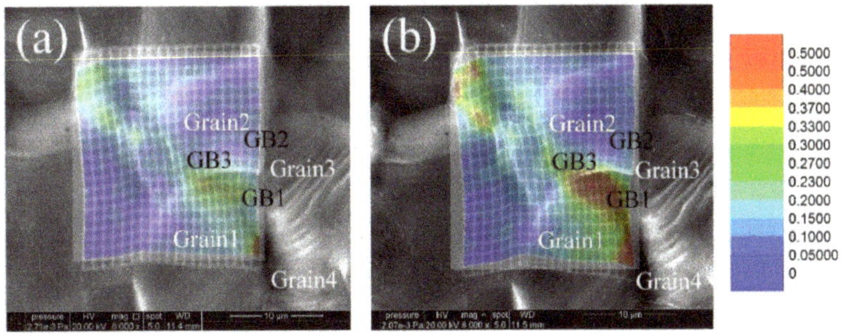

Figure 5. (**a**) Strain map after 0.1 tension strain and (**b**) strain map after 0.15 tension strain.

Table 1. Schmid factors of basal slip systems and prismatic slip systems in Grains 1–4.

		Grain 1			
Basal {11−20}	Basal {1−210}	Basal {2−1−10}	Prismatic {11−20}	Prismatic {1−210}	Prismatic {2−1−10}
0.486	0.330	0.156	0.220	0.088	0.132
		Grain 2			
Basal {11−20}	Basal {1−210}	Basal {2−1−10}	Prismatic {11−20}	Prismatic {1−210}	Prismatic {2−1−10}
0.161	0.125	0.035	0.483	0.284	0.199
		Grain 3			
Basal {11−20}	Basal {1−210}	Basal {2−1−10}	Prismatic {11−20}	Prismatic {1−210}	Prismatic {2−1−10}
0.020	0.019	0.002	0.467	0.388	0.079
		Grain 4			
Basal {11−20}	Basal {1−210}	Basal {2−1−10}	Prismatic {11−20}	Prismatic {1−210}	Prismatic {2−1−10}
0.216	0.266	0.050	0.290	0.450	0.160

In our previous study [11,12], strain accommodation at the grain boundary area was found to be related to both grain orientation and grain boundary misorientation. Grain orientation is quantified by the Schmid factor and grain boundary misorientation is quantified by deformation compatibility parameter m' [13,14]. Therefore, the m' values of two slip systems were calculated and shown in Table 2. Because three different prismatic slip systems own different slip planes, the activated prismatic slip system was identified as the (1–100) {11−20} slip system by slip trace analysis. This method was fully introduced in our previous paper [15]. The m' value between the (0001) {1−210} slip system in Grain 1 and the (1–100) {11−20} slip system in Grain 3 is 0.425, which shows good deformation compatibility between these two slip systems. The Schmid factors for these two slip systems are 0.330 and 0.467, which are also high numbers. Both high values of the Schmid factor and m' value indicated high strain amplitudes at the grain boundary area. Because the activation of basal slip systems is not favored in Grain 2 and Grain 3, and because of the relatively high CRSS of the prismatic slip, strain amplitudes are low at the grain boundary area.

It appears that the GB1 area shows high strain amplitudes and the GB2 area shows low strain amplitudes. This leads to serious deformation incompatibility at the junction point, which is a weak point for crack initiation. Thereafter, as the tension strain-increased crack extends from the weak point along GB3, it leads to intergranular cracking as seen in the results shown in Figure 5a,b. This kind of crack can also be found in the Mg-2% Gd (wt%) alloy used in a previous study [16] that used a predictive fracture initiation parameter to evaluate the probability of microcrack nucleation at the grain boundary.

Table 2. m' value between different slip systems from Grain 1 and Grain 3.

m' Value		Grain 3 Basal {11−20}	Basal {1−210}	Basal {2−1−10}	Prismatic {11−20}	Prismatic {1−210}	Prismatic {2−1−10}
Grain 1	Basal {11−20}	0.427	0.341	0.086	0.258	0.135	0.092
	Basal {1−210}	0.037	0.562	0.526	0.425	0.012	0.563
	Basal {2−1−10}	0.391	0.221	0.612	0.167	0.123	0.655
	Prismatic {11−20}	0.035	0.545	0.509	0.629	0.016	0.364
	Prismatic {1−210}	0.013	0.011	0.003	0.376	0.605	0.027
	Prismatic {2−1−10}	0.367	0.207	0.574	0.004	0.387	0.617

There are other researchers who have studied grain boundary cracking behavior. Ashmawi and Zikry [17,18] used the slip transfer criteria to correlate dislocation density with damage based upon two assumptions: (1) damage occurs due to the dislocation accumulated and (2) the slip transfer mitigates dislocation accumulation. Because there is no reason to correlate concentrated dislocation activity with damage (concentrated dislocation activity could either mitigate or cause damage), this approach may not be effective. Ma et al. [19,20] also developed grain boundary elements that increase deformation resistance in boundaries and allow some slip transfer, but this has not been used to examine damage nucleation. There is clearly a need for an effective but simple way to identify how damage nucleates that takes into account the active slip system history in the vicinity of the boundary. In this study, intergranular cracking was found to be related to deformation incompatibility at the triangle grain boundary. If inhomogeneous strain distribution at the triangle grain boundary can be quantified, the prediction of intergranular cracking would be more accurate. Furthermore, this would provide a new perspective for preventing cracking and achieve higher ductility in magnesium alloys.

4. Conclusions

In situ tension in SEM combined with EBSD and DIC techniques was applied to investigate fracture behavior in a Mg-Gd-Y alloy. Intergranular cracking was found at the triangle grain boundary area with serious inhomogeneous strain distribution conditions. The difference in strain amplitude between linked grain boundaries along the tension direction led to severe deformation incompatibility at the triangle point, which makes it a weak point for crack initiation. As the tension strain increases, the crack extends from the weak point along the grain boundary and causes intergranular cracking.

Author Contributions: Formal analysis, J.S.; Methodology, J.S., J.L., Q.C. and L.L.; Project administration, J.S.; Writing—original draft, J.S. All authors have read and agreed to the published version of the manuscript.

Funding: This work was supported by the National Natural Science Foundation of China (Grant No. 52001188) and the Natural Science Foundation of Shandong Province (ZR2020QE161).

Conflicts of Interest: The authors declare no conflict of interest.

References

1. Jin, L.; Mishra, R.K.; Sachdev, A.K. Texture modification during extrusion of some Mg alloys. *Metall. Mater. Trans. A* **2012**, *43*, 2148–2157. [CrossRef]
2. Zhou, N.; Zhang, Z.; Jin, L.; Dong, J.; Chen, B.; Ding, W. Ductility improvement by twinning and twin–slip interaction in a Mg-Y alloy. *Mater. Des.* **2014**, *56*, 966–974. [CrossRef]

3. Sandlöbes, S.; Pei, Z.; Friák, M.; Zhu, L.F.; Wang, F.; Zaefferer, S.; Raabe, D.; Neugebauer, J. Ductility improvement of Mg alloys by solid solution: Ab initio modeling, synthesis and mechanical properties. *Acta Mater.* **2014**, *70*, 92–104. [CrossRef]
4. Luo, A.; Wu, W.; Mishra, R.; Jin, L.; Sachdev, A.; Ding, W. Microstructure and Mechanical Properties of Extruded Magnesium-Aluminum-Cerium Alloy Tubes. *Metall. Mater. Trans. A* **2010**, *41*, 2662–2674. [CrossRef]
5. Prasad, N.S.; Kumar, N.N.; Narasimhan, R.; Suwas, S. Fracture behavior of magnesium alloys—Role of tensile twinning. *Acta Mater.* **2015**, *94*, 281–293. [CrossRef]
6. Niknejad, S.; Esmaeili, S.; Zhou, N.Y. The role of double twinning on transgranular fracture in magnesium AZ61 in a localized stress field. *Acta Mater.* **2016**, *102*, 1–16. [CrossRef]
7. Koike, J.; Fujiyama, N.; Ando, D.; Sutou, Y. Roles of deformation twinning and dislocation slip in the fatigue failure mechanism of AZ31 Mg alloys. *Scr. Mater.* **2010**, *63*, 747–750. [CrossRef]
8. Wang, F.; Dong, J.; Jiang, Y.; Ding, W. Cyclic deformation and fatigue of extruded Mg–Gd–Y magnesium alloy. *Mater. Sci. Eng. A* **2013**, *561*, 403–410. [CrossRef]
9. Tian, J.; Deng, J.-F.; Chang, Y.; Shi, Q.-X.; Liang, W.; Ma, J. A study of unstable fracture of a magnesium alloy caused by uneven microstructure. *Mater. Lett.* **2022**, *314*, 131799. [CrossRef]
10. Luo, A.; Deng, Y.; Guan, L.; Guo, X. Role of twinning and texture on fatigue resistance enhancement of Mg-6Gd-3Y-1Nd-0.5Zr alloys. *Int. J. Fatigue* **2021**, *153*, 106471. [CrossRef]
11. Sun, J.; Jin, L.; Dong, J.; Ding, W.; Luo, A.A. Microscopic deformation compatibility during monotonic loading in a Mg-Gd-Y alloy. *Mater. Charact.* **2016**, *119*, 195–199. [CrossRef]
12. Sun, J.; Jin, L.; Dong, J.; Wang, F.; Dong, S.; Ding, W.; Luo, A.A. Towards high ductility in magnesium alloys—The role of intergranular deformation. *Int. J. Plast.* **2019**, *123*, 121–132. [CrossRef]
13. Luster, J.; Morris, M.A. Compatibility of deformation in two-phase Ti-Al alloys: Dependence on microstructure and orientation relationships. *Metall. Mater. Trans. A* **1995**, *26*, 1745–1756. [CrossRef]
14. Xin, R.L.; Liang, Y.C.; Ding, C.H.; Guo, C.F.; Wang, B.S.; Liu, Q. Geometrical compatibility factor analysis of paired extension twins in extruded Mg-3Al-1Zn alloys. *Mater. Des.* **2015**, *86*, 656–663. [CrossRef]
15. Wang, F.; Liu, M.; Sun, J.; Feng, M.; Jin, L.; Dong, J.; Jiang, Y. Effects of initial {10–12} twins on cyclic deformation and fatigue of magnesium alloy at low strain amplitudes. *Mater. Charact.* **2019**, *149*, 118–123. [CrossRef]
16. Liu, C.; Jin, L.; Dong, J.; Wang, F. The use of the fracture initiation parameter F1 to predict microcrack nucleation at grain boundaries in Mg-2%Gd alloy. *Mater. Des.* **2016**, *111*, 369–374. [CrossRef]
17. Ashmawi, W.M.; Zikry, M.A. Single void morphological and grain-boundary effects on overall failure in F.C.C. polycrystalline systems. *Mater. Sci. Eng. A* **2003**, *343*, 126–142. [CrossRef]
18. Ashmawi, W.M.; Zikry, M.A. Grain boundary effects and void porosity evolution. *Mech. Mater.* **2003**, *35*, 537–552. [CrossRef]
19. Ma, A.; Roters, F.; Raabe, D. A dislocation density based constitutive model for crystal plasticity FEM including geometrically necessary dislocations. *Acta Mater.* **2006**, *54*, 2169–2179. [CrossRef]
20. Ma, A.; Roters, F.; Raabe, D. On the consideration of interactions between dislocations and grain boundaries in crystal plasticity finite element modeling—Theory, experiments, and simulations. *Acta Mater.* **2006**, *54*, 2181–2194. [CrossRef]

crystals

MDPI

Article

Low Temperature and High-Pressure Study of Bending L-Leucinium Hydrogen Maleate Crystals

Kseniya D. Skakunova [1,2] and Denis A. Rychkov [1,2,*]

1 Institute of Solid State Chemistry and Mechanochemistry, Siberian Branch of Russian Academy of Sciences, 630090 Novosibirsk, Russia; skakunova.kseniya@mail.ru

2 Laboratory of Physicochemical Fundamentals of Pharmaceutical Materials, Faculty of Natural Sciences, Novosibirsk State University, 630090 Novosibirsk, Russia

* Correspondence: rychkov.dennis@gmail.com; Tel.: +7-383-233-24-10 (ext. 1130)

Abstract: The polymorphism of molecular crystals is a well-known phenomenon, resulting in modifications of physicochemical properties of solid phases. Low temperatures and high pressures are widely used to find phase transitions and quench new solid forms. In this study, L-Leucinium hydrogen maleate (LLHM), the first molecular crystal that preserves its anomalous plasticity at cryogenic temperatures, is studied at extreme conditions using Raman spectroscopy and optical microscopy. LLHM was cooled down to 11 K without any phase transition, while high pressure impact leads to perceptible changes in crystal structure in the interval of 0.0–1.35 GPa using pentane-isopentane media. Surprisingly, pressure transmitting media (PTM) play a significant role in the behavior of the LLHM system at extreme conditions—we did not find any phase change up to 3.05 GPa using paraffin as PTM. A phase transition of LLHM to amorphous form or solid–solid phase transition(s) that results in crystal fracture is reported at high pressures. LLHM stability at low temperatures suggests an alluring idea to prove LLHM preserves plasticity below 77 K.

Keywords: plastic crystals; Raman spectroscopy; low temperature; high-pressure; L-Leucinium hydrogen maleate; plasticity; bending crystal

Citation: Skakunova, K.D.; Rychkov, D.A. Low Temperature and High-Pressure Study of Bending L-Leucinium Hydrogen Maleate Crystals. *Crystals* **2021**, *11*, 1575. https://doi.org/10.3390/cryst11121575

Academic Editors: Ulrich Prahl, Sergey Guk and Faisal Qayyum

Received: 23 November 2021
Accepted: 14 December 2021
Published: 16 December 2021
Corrected: 21 April 2022

1. Introduction

Studying molecular crystals and their phase transitions is of great importance for many scientific fields such as crystallography [1,2], thermodynamics [3,4], computational [5–7] and solid state chemistry [8–10], etc. Solid forms of many organic molecules are being developed, studied, and produced in the pharmaceutical industry [11–15] and in arising subfields of materials science [16–18]. The last one, among others, focuses the attention on the mechanical properties of molecular crystals [19–25] and metal-organic complexes [26–28]. Several dozens of organic crystals show anomalous plasticity and elasticity at ambient and extreme conditions under mechanical stress [29–33]. L-Leucinium hydrogen maleate (LLHM) is a unique example of organic crystals that preserves plasticity at a cryogenic temperature [34]. This phenomenon was studied and explained recently using a mainly crystallographic approach [34,35].

One of the methods to understand the nature of an important property of molecular crystal is to apply significant outside impact—low temperature, hydrostatic pressure, mechanochemical stress, etc. [36–39]. These methods help to follow the behavior of the systems at changing environments at macroscopic (thermodynamics) and microscopic (molecular contacts) levels, depending on the availability of experimental and theoretical techniques [40–47]. Moreover, low temperatures and high pressures often trigger phase transitions, resulting in new solid forms, including polymorphs. The application of extreme conditions is a powerful tool to find and sometimes stabilize new forms of organic molecules.

Thus, studies of new forms of molecular crystals using extreme conditions [11,37,48] as well as a search of new bending crystals [30,49,50] are of great interest for modern science. Nevertheless, the interrelation of these two areas is just an emerging field. There are very few works that search for new forms of plastic crystals at high pressures or low temperatures [34,35,51]. To fulfill this gap, in this work we applied low temperatures and high pressures to crystals of LLHM, chasing new forms of these bending crystals. In this work we continue examination of the LLHM system at different conditions, reporting experimental behavior of this crystal at extreme conditions.

L-Leucinium hydrogen maleate crystals were grown for the first time by Arkhipov et al. [52] as an individual system in a series of amino acid maleates. Providing single crystal X-ray diffraction (SCXRD) experiments, the plasticity of LLHM was noted and formulated as "interesting mechanical behavior: mechanical action on crystals of (LLHM) results in elastic, and then plastic bending". Further, a detailed study showed that crystals preserve their plasticity at temperatures down to 77 K. It was proven using video and photo recording of the bending process [34]. Using SCXRD, authors showed that no significant changes in crystal structure of LLHM occurred during the cooling down to 100 K, preserving the layered structure [34]. The system behavior on bending was also investigated using optical and scanning electron microscopy, as well as SCXRD [35]. Scrupulous analysis of crystal structure allows authors to provide a simple model for bending LLHM crystals—layers of L-Leucinium cations and maleic acid anions connected via H-bonds (forming *bc* planes) were stacked over *a* direction (interacting with weak VdW interactions). This results in the possible slipping of layers along b direction (Figure 1).

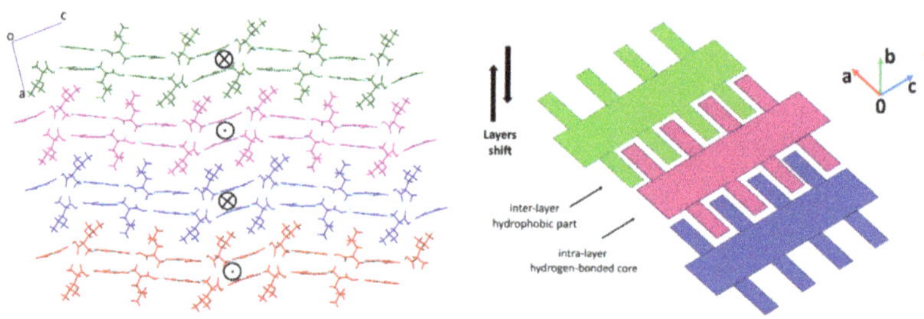

Figure 1. Crystallographic (**left**) and schematic (**right**) representation of the layered structure of LLHM, showing layers that shift resulting in plasticity under mechanical stress (adapted from [34,35]). Layers' shifts are marked with an arrow sign depending on the direction of displacement.

Nevertheless, no experiments below 100 K and high pressures were provided before, leaving an opportunity for a combined study of possible phase transitions at extreme conditions using powerful methods of XRD and Raman spectroscopy.

2. Materials and Methods

2.1. Crystal Growth

Crystals of LLHM were obtained by slow evaporation of an equimolar aqueous solution of L-leucine and maleic acid using the 'sitting-drop' approach [53] as described in previous work [34]. L-leucine (>98%, HPLC) and maleic acid (>99%, HPLC) were purchased from Sigma-Aldrich (St. Louis, MO, USA) and used without further purification. The crystallization procedure leads to 'hedgehog' polycrystalline material as presented in [35]. This bulk material was used to find and cut single crystals for low-temperature and high-pressure experiments. Crystals were operated with care at every stage of isolation and setting to the diamond anvil cell (DAC) to avoid accidental bending.

2.2. High-Pressure Generation and Measurement

Hydrostatic pressure was generated in diamond-anvil cells (DAC) of 'Almax–Boehler' type without beryllium backing plates [54] with natural diamonds suitable both for X-ray diffraction and for Raman experiments. Paraffin (ROTH GmbH, Karlsruhe, Germany, hydrostatic limit ~3 GPa) and a 1:1 stoichiometric pentane–isopentane mixture (PIP) (hydrostatic limit 7 GPa) were used as a pressure-transmitting medium (PTM) in two separate experiments [55]. A special chamber was used to facilitate the DAC loading [56]. Pressure measurement was done with a precision of 0.05 GPa using the ruby fluorescence method [57,58].

2.3. Optical Microscopy

Optical microscopy was provided using the OLYMPUS BX41 microscope (Olympus, Tokyo, Japan) with a MPlan N 10×/0.25 FN22 objective. Euromex fiber optic light source EK-1 was used for illumination.

2.4. Single-Crystal X-ray Diffraction (SCXRD)

Data were collected using an Oxford Diffraction Gemini R Ultra X-ray diffractometer (Crawley, Australia) with a CCD area detector and Mo Kα radiation. The quality of data was not high enough to refine the atomic coordinates or determine unit cell parameters, mainly because of the DAC usage. It was mentioned before that the diffraction of LLHM crystals is not very high for accessible laboratory instruments either. [34]

2.5. Raman Spectroscopy

Raman experiments were performed for low-temperature and high-pressure samples. Raman spectra were recorded using a LabRam HR 300 spectrometer from HORIBA Jobin Yvon (Edison, NJ, USA) with a CCD detector. For spectral excitation, a 488 nm line of an Ar^+ laser was used with a beam size of ~1 µm at the surface of the sample and a power of ~8 mW. All data were collected using a Raman microscope in backscattering geometry. The spectral resolution was ~2 cm^{-1} providing seven scans 30 s each for every spectrum.

Raman spectra of LLHM were recorded in the temperature range of 300–11 K during cooling without repeating on heating.

Two distinct experiments were performed for samples at high pressures. Raman spectra for LLHM in PIP were recorded at pressures of 1.35, 2.03, 2.48, 3.06, 3.63, 4.05, 4.48, 5.03, 5.50, and 6.15 GPa on loading and 3.05, 1.82, 1.15, 0.41, and 0 GPa on pressure release. Spectra for LLHM in Paraffin were recorded at pressures of 0, 0.38, 2.55, and 2.90 GPa.

2.6. Computational Methods

Gas-phase calculations of vibrational spectra (both IR and Raman) were done for L-leucine cation, Maleic acid anion, and LLHM dimer to provide a more reasonable assignment of experimental modes. Ions of L-Leucine, maleic acid, and their dimer were extracted from the LLHM crystal structure from [34] and were further freely optimized at B3LYP/6-311+G(d,p) level of theory, providing vibrational calculations subsequently. None of the atoms or groups were fixed for gas-phase optimization, making possible the ion formation of L-Leucine and Maleic acid in the gas phase calculations. Gaussian09 package was used for all calculations [59]

Solid state calculations, as suggested in literature, were attempted to perform for the simulation of high-pressure behavior [3,7,46] and the vibrational band assignment. Nevertheless, even the usage of supercomputers (80 cpu, 384 Gb RAM, max time for task without interruption—240 h) did not allow for the performance of such calculations in reasonable time, providing the 'simple' optimization of one full unit cell in several months. Thus, only gas-phase calculations were used for this work.

3. Results

Crystals of LLHM were grown as described in the' Section 2. One crystal was used for the low temperature experiment, and another two were used in high pressure experiments in PIP and paraffin PTM. All crystals were selected using optical polarized microscopy.

3.1. Room Temperature Raman Spectra and Vibrational Mode Assignment

The crystal structure of LLHM contains six molecules in the asymmetric unit cell, resulting in 24 molecules in the full unit cell (Figure 2). Multiple H-bonds are located in directions not coinciding with cell axes or crystal faces, which limits the utility of polarized Raman spectroscopy.

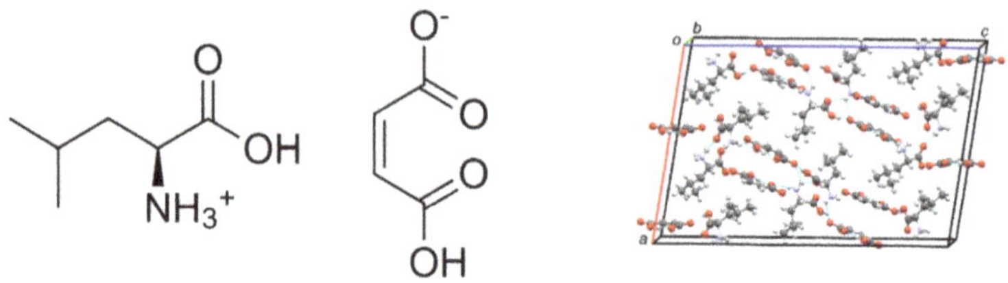

Figure 2. Molecular (**left**) and crystal structure fragment of LLHM at 298 K (**right**), showing 12 molecules of L-leucine and maleic acid each in the unit cell (CCDC# 1889564). Hydrogen bonds are shown by dashed lines.

The complicated structure of LLHM significantly constrains precise band assignment in Raman spectra. Nevertheless, literature data and calculated gas-phase vibrational spectra of L-Leucine and maleic acid ions and their dimer helps to assign main band regions of obtained experimental spectra (Figure 3) [60–66]. All literature data except ref. [67], which contains multiple inaccuracies, confirm the suggested band assignment.

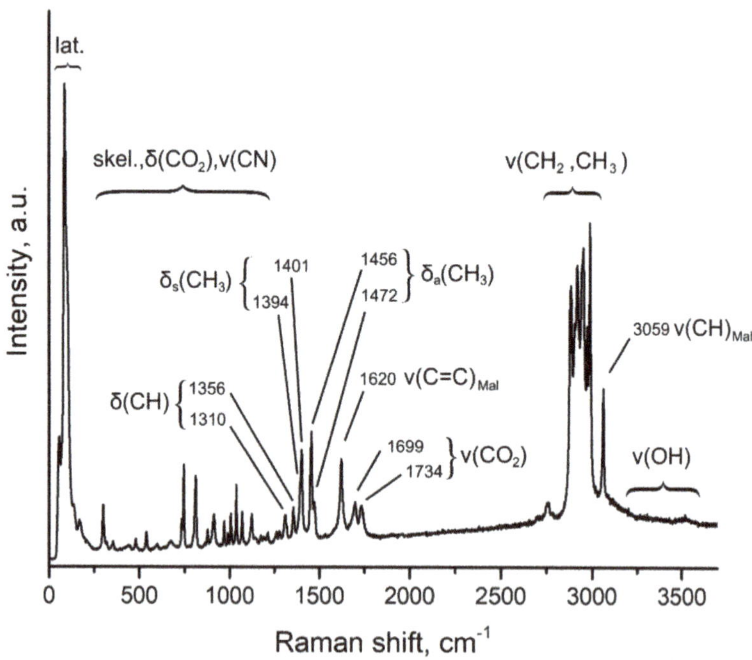

Figure 3. Experimental Raman spectra bands assignment in LLHM crystal structure at 298 K. Calculated gas-phase vibrational spectra are presented in Figure S1 (please see Supplementary Materials).

Main changes are expected in the region before 200 cm^{-1} (lattice vibrations) and 3000–3400 cm^{-1} (valence -OH and -NH_3 vibrations). A significant change in mode position or appearance of new modes is key evidence of phase transition, while slight changes in peak positions or their intensities are traditional for low temperature or high-pressure behavior of the original phase.

3.2. Low Temperature Study

Low temperature Raman spectra of LLHM crystal were recorded in the temperature range of 300–11 K, cooling down the sample. No additional peaks of a new phase were found in spectra (Figure 4). No crystal changes were also found using optical microscopy on cooling.

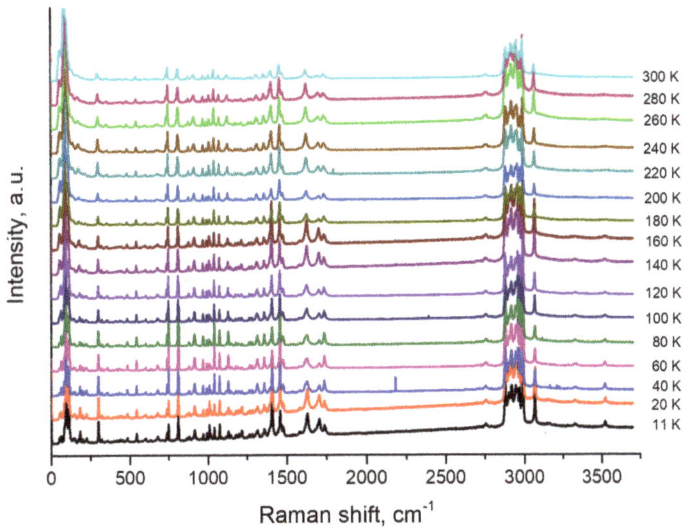

Figure 4. LLHM Raman spectra at low temperatures, showing no phase transition on cooling. Lone spurious peaks in the region of 2000–2500 cm^{-1} at temperatures 40 and 100 K were not vanished from spectra to preserve original data. Same is true for spurious peaks at 40 K in the region between 3100–3250 cm^{-1}.

Wide bands split on cooling due to the thermal motion decrease. At low temperatures, the motions of functional groups in the molecule (e.g., different CH_3 groups motion in L-Leucine ion) become distinguishable in Raman spectra. Moreover, vibrations of the same groups in symmetry unequal molecules are close in spectra, but they have slightly different frequencies (2–10 cm^{-1}) because of the crystalline environment and become distinguishable at low temperatures as well. This is a typical behavior of molecular crystals, especially with a high Z′ number [68–70]. Low and high-frequency regions are shown in detail in Figure S2. Stretching vibration of CO_2^- group (1699 cm^{-1} and 1734 cm^{-1} at 298 K, see Figure 3) has a different intensity ratio, which changes at 160 K and 80 K, and may be evidence of a phase transition, e.g., a doubling of the unit cell. Nevertheless, we assume change of this vibration intensity is the result of structure shrinking and corresponding intermolecular interaction changes, but not phase transition. SCXRD in ref. [34] showed no phase transition at 160 K or at the 0–200 cm^{-1} region in Raman spectra. Based on scrupulous analysis of Raman spectra at low temperatures, one can suggest no phase transition on cooling. This coincides well with our previous X-Ray study in the temperature range of 100–300 K [34]. Summing up these two experiments, LLHM plasticity preservation can be proposed below 77 K down to liquid helium temperatures.

3.3. High Pressure Study

High pressure experiments were provided in different PTMs using DAC as reported in the Section 2. Usage of SCXRD was limited due to the poor diffraction data, which is a result of the nature of the LLHM crystal structure (defining plasticity) and DAC construction. Thus, only Raman spectra were recorded for all high-pressure experiments.

Surprisingly, the behavior of LLHM crystals differs significantly in paraffin and PIP. An effect of PTM, as well as experiment protocol on molecular crystals phase transition at high pressure, is a documented phenomenon [43,71,72]. The low-frequency region in Raman spectra shows no phase transition in paraffin up to 2.9 GPa (close to pressure limit for paraffin) and some changes in LLHM crystal in PIP at 1.35 GPa (Figure 5).

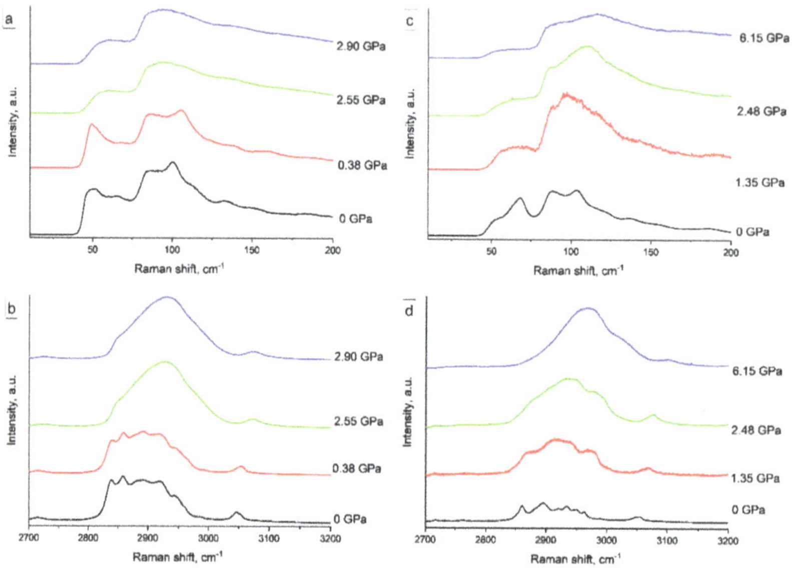

Figure 5. Raman spectra of LLHM at multiple pressures at (**a**,**c**) low-frequency and (**b**,**d**) high-frequency regions, showing different behavior in paraffin (**left**) and PIP (**right**) PTM.

The high-frequency Raman region is not very informative, lacking -OH and -NH_3 vibrational modes information at high pressures [73,74]. Nevertheless, some changes in -CH_x vibration modes confirm crystal changes in PIP in contrast to paraffin (full Raman spectra at all pressures are shown in Figure S3). We also report spectra of LLHM crystal in PIP at 0 GPa before pressure impact and after relaxation from 6.15 Gpa (Figure 6). Significant background level (halo) may be explained with possible amorphization of plastic LLHM crystal at high pressure or luminescence, which was not observed before pressure impact.

Possible phase transition in PIP was additionally confirmed by optical microscopy, which recorded crystal destruction at 1.35 GPa (Figure 7). No obvious changes of LLHM crystal in paraffin occurred in the whole pressure range according to optical microscopy. This confirms different behavior of LLHM crystals under pressure in different media.

An absence of SCXRD did not allow for the report of phase transition in LLHM crystal at high pressure in PIP unequivocally. Nevertheless, relevant changes in optical microscopy and Raman spectra allowed us to speculate about crystal destruction because of one or a cascade of phase transitions in the crystal structure or amorphization of the LLHM sample.

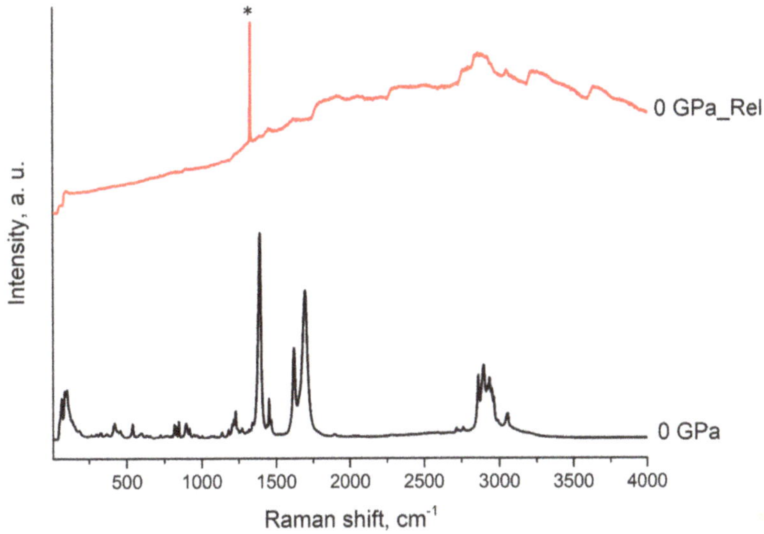

Figure 6. Raman spectra of LLHM at PIP PTM at 0 GPa before (bottom black line) and after (upper red line) pressure impact. * Asterisk mark the signal from DAC.

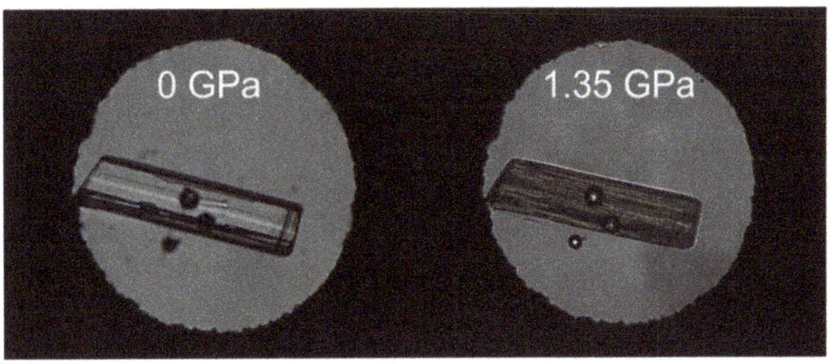

Figure 7. Photographs showing LLHM crystal destruction in PIP at pressure 1.35 GPa. LLHM crystal at 0 GPa (**left**) and 1.35 GPa (**right**). Crystal size is 0.3 mm × 0.05 mm × 0.02 mm.

4. Discussion

A search of new phases of molecular crystals at extreme conditions is an important area of modern chemistry. In this work, a unique bending crystal of LLHM, which preserves its plasticity at 77 K, was studied at low temperatures and high pressure. Based on literature data and computational vibrational spectra, major band regions were assigned to follow their changes at extreme conditions. Raman spectra of LLHM were recorded down to 11 K, showing no phase transition. Based on this experimental fact, we suggest the possibility to preserve LLHM plasticity below liquid nitrogen temperature. High pressure experiments showed different behavior of LLHM crystal in paraffin and PIP used as PTM. We found no phase transition using paraffin up to 2.95 GPa. Raman spectra and optical microscopy gave reasons to postulate the LLHM phase transition, phase transition cascade, or amorphization at high pressure, which results in crystal destruction at PIP PTM at 1.35 GPa. Additional XRPD or SCXRD using synchrotron radiation may be applied to highlight additional details of LLHM high pressure behavior in future studies.

Supplementary Materials: The following are available online at https://www.mdpi.com/article/10.3390/cryst11121575/s1, Figure S1: Gas phase calculated (a) Raman and (b) IR spectra of L-Leucine (Black), Maleic acid (Red), and L-Leucinium hydrogen maleate (Blue), Figure S2: Experimental LLHM Raman spectra at low temperatures, showing no phase transition on cooling. Detailed three different regions of Raman spectra, Figure S3: Raman spectra of LLHM at multiple pressures in PIP (left) and paraffin (right) PTM. Diamond peak from anvil cell is marked with *.

Author Contributions: Conceptualization, D.A.R.; methodology, D.A.R.; validation, D.A.R. and K.D.S.; formal analysis, K.D.S.; investigation, D.A.R.; resources, D.A.R.; data curation, D.A.R. and K.D.S.; writing—original draft preparation, D.A.R.; writing—review and editing, D.A.R.; visualization, K.D.S.; supervision, D.A.R.; project administration, D.A.R.; funding acquisition, D.A.R. All authors have read and agreed to the published version of the manuscript.

Funding: Computational part of this research was funded by the Russian Science Foundation, grant number 21-73-00094.

Data Availability Statement: The data presented in this study are available in the Supplementary Information section.

Acknowledgments: Denis Rychkov acknowledge Elena Boldyreva for basic idea to study the crystals of amino acids and their salts, including L-Leucinium Hydrogen Maleate system, at low temperatures and high pressures. All the experiments were performed at REC-008 "Molecular design and ecologically safe technologies" of the Novosibirsk State University using its instruments: The crystals of L-Leucinium Hydrogen Maleate were grown by T. Nguyen (under joint guidance of S. Arkhipov and D. Rychkov), S. Arkhipov has collected and preliminary processed Raman spectroscopy data at low temperature. E. Losev performed high-pressure X-ray diffraction and Raman spectroscopy experiments and provided the optical microphotographs. S. Arkhipov and E. Losev have performed preliminary data analysis. The authors thank S. Arkhipov and E. Losev for fruitful discussions. The authors are also grateful to A. A. Iurchenkova for assistance in preparing figures. The Siberian Branch of the Russian Academy of Sciences (SB RAS) Siberian Supercomputer Center is gratefully acknowledged for providing supercomputer facilities. The authors also acknowledge the Supercomputing Center of the Novosibirsk State University (http://nusc.nsu.ru) for providing computational resources.

Conflicts of Interest: The authors declare no conflict of interest. The funders had no role in the design of the study; in the collection, analyses, or interpretation of data; in the writing of the manuscript, or in the decision to publish the results.

References

1. Montis, R.; Fusaro, L.; Falqui, A.; Hursthouse, M.B.; Tumanov, N.; Coles, S.J.; Threlfall, T.L.; Horton, P.N.; Sougrat, R.; Lafontaine, A.; et al. Complex structures arising from the self-assembly of a simple organic salt. *Nature* **2021**, *590*, 275–278. [CrossRef] [PubMed]
2. Sobczak, S.; Ratajczyk, P.; Katrusiak, A. High-pressure Nucleation of Low-Density Polymorphs. *Chemistry* **2021**, *27*, 7069–7073. [CrossRef] [PubMed]
3. Rychkov, D.A.; Stare, J.; Boldyreva, E.V. Pressure-driven phase transition mechanisms revealed by quantum chemistry: L-serine polymorphs. *Phys. Chem. Chem. Phys.* **2017**, *19*, 6671–6676. [CrossRef] [PubMed]
4. Ghosh, S.; Mondal, A.; Kiran, M.S.R.N.; Ramamurty, U.; Reddy, C.M. The Role of Weak Interactions in the Phase Transition and Distinct Mechanical Behavior of Two Structurally Similar Caffeine Co-Crystal Polymorphs Studied by Nanoindentation. *Cryst. Growth Des.* **2013**, *13*, 4435–4441. [CrossRef]
5. Korabel'nikov, D.V.; Zhuravlev, Y.N. Semi-empirical and ab initio calculations for crystals under pressure at fixed temperatures: The case of guanidinium perchlorate. *RSC Adv.* **2020**, *10*, 42204–42211. [CrossRef]
6. Fedorov, A.Y.; Rychkov, D.A. Comparison of Different Computational Approaches for Unveiling the High-Pressure Behavior of Organic Crystals at a Molecular Level. Case Study of Tolazamide Polymorphs. *J. Struct. Chem.* **2020**, *61*, 1356–1366. [CrossRef]
7. Rychkov, D.A. A Short Review of Current Computational Concepts for High-Pressure Phase Transition Studies in Molecular Crystals. *Crystals* **2020**, *10*, 81. [CrossRef]
8. Errandonea, D. Pressure-Induced Phase Transformations. *Crystals* **2020**, *10*, 595. [CrossRef]
9. Casati, N.; Macchi, P.; Sironi, A. Molecular Crystals Under High Pressure: Theoretical and Experimental Investigations of the Solid-Solid Phase Transitions in [$Co_2(CO)_6(XPh_3)_2$] (X=P, As). *Chemistry* **2009**, *15*, 4446–4457. [CrossRef]
10. Zakharov, B.A.; Boldyreva, E.V. High pressure: A complementary tool for probing solid-state processes. *CrystEngComm* **2019**, *21*, 10–22. [CrossRef]

11. Mazurek, A.H.; Szeleszczuk, Ł.; Pisklak, D.M. Periodic DFT Calculations—Review of Applications in the Pharmaceutical Sciences. *Pharmaceutics* **2020**, *12*, 415. [CrossRef]
12. McGregor, L.; Rychkov, D.A.; Coster, P.L.; Day, S.; Drebushchak, V.A.; Achkasov, A.F.; Nichol, G.S.; Pulham, C.R.; Boldyreva, E.V. A new polymorph of metacetamol. *CrystEngComm* **2015**, *17*, 6183–6192. [CrossRef]
13. Bauer, J.; Spanton, S.; Henry, R.; Quick, J.; Dziki, W.; Porter, W.; Morris, J. Ritonavir: An extraordinary example of conformational polymorphism. *Pharm. Res.* **2001**, *18*, 859–866. [CrossRef]
14. Zhang, G. Phase transformation considerations during process development and manufacture of solid oral dosage forms. *Adv. Drug Deliv. Rev.* **2004**, *56*, 371–390. [CrossRef]
15. Llinàs, A.; Goodman, J.M. Polymorph control: Past, present and future. *Drug Discov. Today* **2008**, *13*, 198–210. [CrossRef]
16. Chen, Y.; Chen, C.; Rehman, H.U.; Zheng, X.; Li, H.; Liu, H.; Hedenqvist, M.S. Shape-Memory Polymeric Artificial Muscles: Mechanisms, Applications and Challenges. *Molecules* **2020**, *25*, 4246. [CrossRef]
17. Liu, H.; Lu, Z.; Tang, B.; Qu, C.; Zhang, Z.; Zhang, H. A Flexible Organic Single Crystal with Plastic-Twisting and Elastic-Bending Capabilities and Polarization-Rotation Function. *Angew. Chem. Int. Ed.* **2020**, *59*, 12944–12950. [CrossRef]
18. Liu, H.; Bian, Z.; Cheng, Q.; Lan, L.; Wang, Y.; Zhang, H. Controllably realizing elastic/plastic bending based on a room-temperature phosphorescent waveguiding organic crystal. *Chem. Sci.* **2019**, *10*, 227–232. [CrossRef]
19. Ghosh, S.; Reddy, C.M. Elastic and Bendable Caffeine Cocrystals: Implications for the Design of Flexible Organic Materials. *Angew. Chem. Int. Ed.* **2012**, *51*, 10319–10323. [CrossRef]
20. Gupta, P.; Karothu, D.P.; Ahmed, E.; Naumov, P.; Nath, N.K. Thermally Twistable, Photobendable, Elastically Deformable, and Self-Healable Soft Crystals. *Angew. Chem. Int. Ed.* **2018**, *57*, 8498–8502. [CrossRef]
21. Commins, P.; Desta, I.T.; Karothu, D.P.; Panda, M.K.; Naumov, P. Crystals on the move: Mechanical effects in dynamic solids. *Chem. Commun.* **2016**, *52*, 13941–13954. [CrossRef]
22. Kakkar, S.; Bhattacharya, B.; Reddy, C.M.; Ghosh, S. Tuning mechanical behaviour by controlling the structure of a series of theophylline co-crystals. *CrystEngComm* **2018**, *20*, 1101–1109. [CrossRef]
23. Colmenero, F. Mechanical properties of anhydrous oxalic acid and oxalic acid dihydrate. *Phys. Chem. Chem. Phys.* **2019**, *21*, 2673–2690. [CrossRef]
24. Mishra, M.K.; Mishra, K.; Narayan, A.; Reddy, C.M.; Vangala, V.R. Structural Basis for Mechanical Anisotropy in Polymorphs of a Caffeine–Glutaric Acid Cocrystal. *Cryst. Growth Des.* **2020**, *20*, 6306–6315. [CrossRef]
25. Bhattacharya, B.; Roy, D.; Dey, S.; Puthuvakkal, A.; Bhunia, S.; Mondal, S.; Chowdhury, R.; Bhattacharya, M.; Mandal, M.; Manoj, K.; et al. Mechanical-Bending-Induced Fluorescence Enhancement in Plastically Flexible Crystals of a GFP Chromophore Analogue. *Angew. Chem.* **2020**, *132*, 20050–20055. [CrossRef]
26. Colmenero, F. Negative linear compressibility in nanoporous metal–organic frameworks rationalized by the empty channel structural mechanism. *Phys. Chem. Chem. Phys.* **2021**, *23*, 8508–8524. [CrossRef]
27. Masunov, A.E.; Wiratmo, M.; Dyakov, A.A.; Matveychuk, Y.V.; Bartashevich, E.V. Virtual Tensile Test for Brittle, Plastic, and Elastic Polymorphs of 4-Bromophenyl 4-Bromobenzoate. *Cryst. Growth Des.* **2020**, *20*, 6093–6100. [CrossRef]
28. Thomas, S.P.; Shi, M.W.; Koutsantonis, G.A.; Jayatilaka, D.; Edwards, A.J.; Spackman, M.A. The Elusive Structural Origin of Plastic Bending in Dimethyl Sulfone Crystals with Quasi-Isotropic Crystal Packing. *Angew. Chem. Int. Ed.* **2017**, *56*, 8468–8472. [CrossRef]
29. Reddy, C.M.; Gundakaram, R.C.; Basavoju, S.; Kirchner, M.T.; Padmanabhan, K.A.; Desiraju, G.R. Structural basis for bending of organic crystals. *Chem. Commun.* **2005**, *1*, 3945. [CrossRef]
30. Reddy, C.M.; Padmanabhan, K.A.; Desiraju, G.R. Structure−Property Correlations in Bending and Brittle Organic Crystals. *Cryst. Growth Des.* **2006**, *6*, 2720–2731. [CrossRef]
31. Worthy, A.; Grosjean, A.; Pfrunder, M.C.; Xu, Y.; Yan, C.; Edwards, G.; Clegg, J.K.; McMurtrie, J.C. Atomic resolution of structural changes in elastic crystals of copper(II) acetylacetonate. *Nat. Chem.* **2017**, *10*, 65–69. [CrossRef] [PubMed]
32. Bag, P.P.; Chen, M.; Sun, C.C.; Reddy, C.M. Direct correlation among crystal structure, mechanical behaviour and tabletability in a trimorphic molecular compound. *CrystEngComm* **2012**, *14*, 3865. [CrossRef]
33. Colmenero, F. Organic acids under pressure: Elastic properties, negative mechanical phenomena and pressure induced phase transitions in the lactic, maleic, succinic and citric acids. *Mater. Adv.* **2020**, *1*, 1399–1426. [CrossRef]
34. Arkhipov, S.G.; Losev, E.A.; Nguyen, T.T.; Rychkov, D.A.; Boldyreva, E.V. A large anisotropic plasticity of L-leucinium hydrogen maleate preserved at cryogenic temperatures. *Acta Crystallogr. Sect. B Struct. Sci. Cryst. Eng. Mater.* **2019**, *75*, 143–151. [CrossRef]
35. Nguyen, T.T.; Arkhipov, S.G.; Rychkov, D.A. Simple crystallographic model for anomalous plasticity of L-Leucinium hydrogen maleate crystals. *Mater. Today Proc.* **2020**, *25*, 412–415. [CrossRef]
36. Barrio, M.; Maccaroni, E.; Rietveld, I.B.; Malpezzi, L.; Masciocchi, N.; Céolin, R.; Tamarit, J.-L. Pressure-temperature state diagram for the phase relationships between benfluorex hydrochloride forms I and II: A case of enantiotropic behavior. *J. Pharm. Sci.* **2012**, *101*, 1073–1078. [CrossRef]
37. Fabbiani, F.P.A.; Allan, D.R.; David, W.I.F.; Moggach, S.A.; Parsons, S.; Pulham, C.R. High-pressure recrystallisation—A route to new polymorphs and solvates. *CrystEngComm* **2004**, *6*, 504–511. [CrossRef]
38. Drebushchak, V.A.; McGregor, L.; Rychkov, D.A. Cooling rate "window" in the crystallization of metacetamol form II. *J. Therm. Anal. Calorim.* **2017**, *127*, 1807–1814. [CrossRef]

39. Fabbiani, F.P.A.; Pulham, C.R. High-pressure studies of pharmaceutical compounds and energetic materials. *Chem. Soc. Rev.* **2006**, *35*, 932. [CrossRef]
40. Moggach, S.A.; Marshall, W.G.; Parsons, S. High-pressure neutron diffraction study of L-serine-I and L-serine-II, and the structure of L-serine-III at 8.1 GPa. *Acta Crystallogr. Sect. B Struct. Sci.* **2006**, *62*, 815–825. [CrossRef]
41. Wood, P.A.; Francis, D.; Marshall, W.G.; Moggach, S.A.; Parsons, S.; Pidcock, E.; Rohl, A.L. A study of the high-pressure polymorphs of L-serine using ab initio structures and PIXEL calculations. *CrystEngComm* **2008**, *10*, 1154. [CrossRef]
42. Zakharov, B.A.; Kolesov, B.A.; Boldyreva, E.V. Effect of pressure on crystalline L- and DL-serine: Revisited by a combined single-crystal X-ray diffraction at a laboratory source and polarized Raman spectroscopy study. *Acta Crystallogr. Sect. B Struct. Sci.* **2012**, *68*, 275–286. [CrossRef]
43. Fisch, M.; Lanza, A.; Boldyreva, E.; Macchi, P.; Casati, N. Kinetic Control of High-Pressure Solid-State Phase Transitions: A Case Study on l-Serine. *J. Phys. Chem. C* **2015**, *119*, 18611–18617. [CrossRef]
44. Millar, D.I.A.; Oswald, I.D.H.; Francis, D.J.; Marshall, W.G.; Pulham, C.R.; Cumming, A.S. The crystal structure of β-RDX-an elusive form of an explosive revealed. *Chem. Commun.* **2009**, *5*, 562–564. [CrossRef]
45. Munday, L.B.; Chung, P.W.; Rice, B.M.; Solares, S.D. Simulations of High-Pressure Phases in RDX. *J. Phys. Chem. B* **2011**, *115*, 4378–4386. [CrossRef]
46. Hunter, S.; Sutinen, T.; Parker, S.F.; Morrison, C.A.; Williamson, D.M.; Thompson, S.; Gould, P.J.; Pulham, C.R. Experimental and DFT-D Studies of the Molecular Organic Energetic Material RDX. *J. Phys. Chem. C* **2013**, *117*, 8062–8071. [CrossRef]
47. Ma, Y.; Zhang, A.; Xue, X.; Jiang, D.; Zhu, Y.; Zhang, C. Crystal packing of impact-sensitive high-energy explosives. *Cryst. Growth Des.* **2014**, *14*, 6101–6114. [CrossRef]
48. Millar, D.I.A. *Energetic Materials at Extreme Conditions*; Springer: Berlin/Heidelberg, Germany, 2012; ISBN 9783642231315.
49. Gude, V.; Choubey, P.S.; Das, S.; Bhaktha, B.N.S.; Reddy, C.M.; Biradha, K. Elastic orange emissive single crystals of 1,3-diamino-2,4,5,6-tetrabromobenzene as flexible optical waveguides. *J. Mater. Chem. C* **2021**, *9*, 9465–9472. [CrossRef]
50. Das, S.; Mondal, A.; Reddy, C.M. Harnessing molecular rotations in plastic crystals: A holistic view for crystal engineering of adaptive soft materials. *Chem. Soc. Rev.* **2020**, *49*, 8878–8896. [CrossRef]
51. Matveychuk, Y.V.; Bartashevich, E.V.; Tsirelson, V.G. How the H-Bond Layout Determines Mechanical Properties of Crystalline Amino Acid Hydrogen Maleates. *Cryst. Growth Des.* **2018**, *18*, 3366–3375. [CrossRef]
52. Arkhipov, S.G.; Rychkov, D.A.; Pugachev, A.M.; Boldyreva, E.V. New hydrophobic L-amino acid salts: Maleates of L-leucine, L-isoleucine and L-norvaline. *Acta Crystallogr. Sect. C Struct. Chem.* **2015**, *71*, 584–592. [CrossRef] [PubMed]
53. Rychkov, D.A.; Arkhipov, S.G.; Boldyreva, E.V. Simple and efficient modifications of well known techniques for reliable growth of high-quality crystals of small bioorganic molecules. *J. Appl. Crystallogr.* **2014**, *47*, 1435–1442. [CrossRef]
54. Boehler, R. New diamond cell for single-crystal x-ray diffraction. *Rev. Sci. Instrum.* **2006**, *77*, 115103. [CrossRef]
55. Piermarini, G.J.; Block, S.; Barnett, J.D. Hydrostatic limits in liquids and solids to 100 kbar. *J. Appl. Phys.* **1973**, *44*, 5377–5382. [CrossRef]
56. Zakharov, B.A.; Achkasov, A.F. A compact device for loading diamond anvil cells with low-boiling pressure-transmitting media. *J. Appl. Crystallogr.* **2013**, *46*, 267–269. [CrossRef]
57. Piermarini, G.J.; Block, S.; Barnett, J.D.; Forman, R.A. Calibration of the pressure dependence of the R1 ruby fluorescence line to 195 kbar. *J. Appl. Phys.* **1975**, *46*, 2774–2780. [CrossRef]
58. Forman, R.A.; Piermarini, G.J.; Barnett, J.D.; Block, S. Pressure Measurement Made by the Utilization of Ruby Sharp-Line Luminescence. *Science* **1972**, *176*, 284–285. [CrossRef]
59. Frisch, M.J.; Trucks, G.W.; Schlegel, H.B.; Scuseria, G.E.; Robb, M.A.; Cheeseman, J.R.; Scalmani, G.; Barone, V.; Mennucci, B.; Petersson, G.A.; et al. GAUSSIAN 09. Revision D.01. Gaussian Inc.: Wallingford, CT, USA, 2016.
60. Mirone, P.; Chiorboli, P. Infrared and Raman spectra and vibrational assignment of maleic anhydride. *Spectrochim. Acta* **1962**, *18*, 1425–1432.
61. Shakhse Emampour, J.; Suh, J.S.; Moskovits, M. Raman Study of the Photochemistry of Maleic Acid Adsorbed on the Surface of Colloidal Silver. *Iran. J. Chem. Chem. Eng.* **1994**, *13*, 30–36.
62. Façanha Filho, P.F.; Freire, P.T.C.; Lima, K.C.V.; Mendes Filho, J.; Melo, F.E.A.; Pizani, P.S. Raman spectra of L-leucine crystals. *Brazilian J. Phys.* **2007**, *38*, 12.
63. Façanha Filho, P.F.; Freire, P.T.C.; Lima, K.C.V.; Mendes Filho, J.; Melo, F.E.A.; Pizani, P.S. High temperature raman spectra of L-leucine crystals. *Brazilian J. Phys.* **2008**, *38*, 131–137. [CrossRef]
64. Zhu, G.; Zhu, X.; Fan, Q.; Wan, X. Raman spectra of amino acids and their aqueous solutions. *Spectrochim. Acta Part A Mol. Biomol. Spectrosc.* **2011**, *78*, 1187–1195. [CrossRef]
65. Bellamy, L.J. *The Infrared Spectra of Complex Molecules*, 2nd ed.; Springer: Dordrecht, The Netherlands, 1980; Volume 2, ISBN 9789401165228.
66. Socrates, G. Infrared Characteristic Group Frequencies. Tables and Charts. *Anal. Chim. Acta* **1994**, *296*, 221–222.
67. Hemalatha, A.; Arulmani, S.; Sanjay, P.; Deepa, K.; Madhavan, J.; Senthil, S. Growth and Characterization of L-Leucenium Hydrogen Maleate Single Crystals for Nonlinear Optical Applications. *IOP Conf. Ser. Mater. Sci. Eng.* **2018**, *360*, 012044. [CrossRef]
68. Galkina, Y.A.; Vershinin, M.A.; Kolesov, B.A. Raman Spectra of Molecular Crystals with Strong Hydrogen Bonds N-H· · · N in the Temperature Range of 5–300 K. *J. Struct. Chem.* **2019**, *60*, 398–404. [CrossRef]
69. Kolesov, B.A. IR and Raman spectra of strong OHO hydrogen bonds. *J. Mol. Struct.* **2021**, *1233*, 130093. [CrossRef]

70. Kolesov, B. Hydrogen Bonds: Raman Spectroscopic Study. *Int. J. Mol. Sci.* **2021**, *22*, 5380. [CrossRef]
71. Zakharov, B.A.; Tumanov, N.A.; Boldyreva, E.V. β-Alanine under pressure: Towards understanding the nature of phase transitions. *CrystEngComm* **2015**, *17*, 2074–2079. [CrossRef]
72. Anis, B.; Börrnert, F.; Rümmeli, M.H.; Kuntscher, C.A. Role of the pressure transmitting medium on the pressure effects in DWCNTs. *Phys. Status Solidi* **2013**, *250*, 2616–2621. [CrossRef]
73. Gaydamaka, A.A.; Arkhipov, S.G.; Zakharov, B.A.; Seryotkin, Y.V.; Boldyreva, E.V. Effect of pressure on slit channels in guanine sodium salt hydrate: A link to nucleobase intermolecular interactions. *CrystEngComm* **2019**, *21*, 4484–4492. [CrossRef]
74. Yang, S.Y.; Butler, I.S. Pressure-tuning infrared and Raman microscopy study of the DNA bases: Adenine, guanine, cytosine, and thymine. *J. Biomol. Struct. Dyn.* **2013**, *31*, 1490–1496. [CrossRef]

MDPI
St. Alban-Anlage 66
4052 Basel
Switzerland
Tel. +41 61 683 77 34
Fax +41 61 302 89 18
www.mdpi.com

Crystals Editorial Office
E-mail: crystals@mdpi.com
www.mdpi.com/journal/crystals